Electronic Engineering Series

Editor: Professor W. A. Gambling
DSc, CEng, FIEE, FIERE
University of Southampton

Gunn-effect Electronics

Electronic Engineering Series

This series of monographs provides a coverage of advanced topics of a specialist nature, supplemented with more fundamental topics for the early years of under-graduate study.

Titles already published include:

Electronic Power Supplies, G. N. Patchett (*University of Bradford*);
Microwave Semiconductor Devices, H. V. Shurmer (*University of Warwick*);
Physical Electronics, J. Seymour (*Woolwich Polytechnic*).

Gunn-effect Electronics

B. G. BOSCH

Dipl.-Ing, PhD
Professor of Electronics
Ruhr-Universitat, Bochum
Federal German Republic

AND

R. W. H. ENGELMANN

Dipl.-Phys, Dr rer net
Hewlett-Packard Laboratories
Palo Alto, California
United States of America

A HALSTED PRESS BOOK

JOHN WILEY & SONS
New York

First published in Great Britain in 1975
by Pitman Publishing

Published in the U.S.A.
by Halsted Press, a Division
of John Wiley & Sons, Inc.,
New York.

Library of Congress Cataloging in Publication Data

Bosch, B G
 Gunn-effect electronics.

 "A Halsted Press book."
 Includes bibliographical references and indexes.
 1. Gunn effect. 2. Semiconductors.
3. Oscillators, Microwave. 4. Transistor circuits.
I. Engelmann, R. W. H., joint author. II. Title.
TK7872.G8B67 1975   621.3815   74–6974
ISBN 0–470–08970–9

Text set in 10/12 pt. Monotype Times New Roman, printed by photolithography
and bound in Great Britain at The Pitman Press, Bath

Preface

Along with the emergence of other new solid-state devices, Gunn-effect electronics has made an almost revolutionizing impact on the microwave scene. In recent years progress in the understanding, development and application of Gunn devices has been very rapid, but of late the rate of advance has begun to slow down somewhat. Certain specific fields in Gunn-effect electronics, like those of sinusoidal oscillators and broadband amplifiers have reached a fair degree of maturity while in the more recent branch of pulse generation and processing the basic principles have been established. For these reasons the time seemed to be opportune to collate the extensive amount of knowledge in the field into an organized form, thereby drawing upon our own personal experience and research work in Gunn-effect electronics of recent years.

Choosing a suitable, terse title for the book which has evolved was not entirely easy. Our treatment covers the theory and applications of what—in precise scientific terms—has come to be known as the 'transferred-electron effect', a description originally introduced in 1962 by C. Hilsum who, together with, but independently of, B. K. Ridley and T. B. Watkins, first proposed the basic operating principle. On the other hand, the term 'Gunn effect' is commonly applied nowadays to a variety of devices based on the electron transfer mechanism, although strictly it only refers to that particular mode, first experimentally observed by J. B. Gunn in 1963, in which a space-charge dipole domain travels across the semiconductor sample. Eventually we decided to adopt the more popular and impressive, though admittedly less accurate, title of 'Gunn-effect electronics' which is assumed to encompass all possible operational modes relying on the basic electron-transfer effect, with or without the occurrence of dipole domains.

The book covers the whole range of Gunn-effect, alias transferred-electron, devices but with differing emphasis given to the various aspects. A division into four main parts can be differentiated. The first two are of a primarily theoretical nature, dealing in chapter two with the relevant semiconductor physics and in chapters three to five with the description of the numerous operational modes exhibited by Gunn devices. The third part, comprising chapter six, deals with device technology, while the fourth covers, in chapters seven and eight, applications in the

v

field of microwave devices and systems and that of digital circuitry.

Device physics, here specifically meaning the electron-transfer mechanism, is treated only as extensively as is thought to be necessary for a proper understanding of the subsequent chapters. Particular emphasis is then given to the description of the almost bewildering array of operational modes found in transferred-electron devices, and an attempt has been made to classify and compare them as well as to identify their limitations. On the more practical side the problems of device technology with the restrictions they impose and, finally, the applications of various kinds have also received a relatively broad treatment. We hope that the organization of the material we have chosen will find approval.

The approach in the methodological presentation has been that of providing a reviewing monograph and general reference source (which extensively refers to the published literature by including a bibliography of more than 900 titles) but nevertheless with the firm intention of retaining the explanatory and introductory functions of a textbook. This aim obviously necessitated a compromise. The textbook character is exemplified by the preferences for explaining device behaviour in terms of physical arguments rather than by merely quoting the results of sophisticated numerical calculations obtained by computer. The reader will find that more special points are dealt with in supplementary small-print sections which can be passed over at first reading.

The book is intended for the student as well as for the practising engineer and scientist in electronics who wants to acquaint himself with the whole theme of Gunn-effect electronics or requires a reference source for it. Moreover, we hope that this book will also prove to be useful to those workers wishing to familiarize themselves only with specific parts of the subject. Readers are assumed to have already acquired an introductory understanding of solid-state device electronics.

During the course of preparing the script we have benefited from the critical comments and suggestions of several colleagues, primarily of R. Becker, W. Frey, P. Russer, H. Thim, and H. Yanai, who read preliminary versions of chapters. Furthermore we are particularly grateful to W. Heinle for carrying out some calculations and for his kind assistance at various other points. Some part of the script was completed while we were both engaged in research work on Gunn devices at the AEG-Telefunken Forschungsinstitut, Ulm/Donau, and our thanks are therefore due to the Management of this institution for providing the facilities. Furthermore we wish to express our gratitude to all those who helped in the production of the book by typing the manuscript or preparing the numerous drawings. Finally, we are heavily indebted to the Series Editor, Professor W. A. Gambling, for generously giving his advice and, last but not least, for his untiring patience when dealing with the script of two authors whose mother tongue is not English.

B. G. BOSCH
R. W. H. ENGELMANN

Acknowledgements

Acknowledgement is expressed to the bodies listed below for permission to reproduce copyright material:

Her Majesty's Stationery Office
American Institute of Physics
Assoziazione Elettrotecnica ed Elettronica
Electrochemical Society
The Institute of Electronics and Communication Engineers of Japan
The Institute of Electrical and Electronics Engineers
The Institute of Physics
The Institution of Electrical Engineers
American Institute of Metallurgical Engineers
The Physical Society of Japan
University of Tokyo
International Business Machines Corporation
ITT Europe
Microforms International Marketing Corporation
Mullard Ltd.
Nippon Electric Company Ltd.
Radio Corporation of America

Akademie Verlag, GmbH
Hayden Publishing Company, Inc.
Kluwer Technische Boeken B.V.
McGraw-Hill, Inc.
North-Holland Publishing Company
Schiele & Schön, GmbH
Scripta Publishing Corporation
Springer Verlag
VDE-Verlag

John Wiley & Sons, Inc.

Archiv für Elektronik und Übertragungstechnik
Microwave Journal

Contents

Contents

1

Introduction

1.1 ACTIVE MICROWAVE SEMICONDUCTOR JUNCTION DEVICES

Since the invention of the transistor in 1948, semiconductor electron devices have to a large extent replaced the vacuum electron tube for the generation and amplification of electromagnetic waves at frequencies below a few hundred megahertz. At higher frequencies, i.e. in the microwave part of the spectrum, the electron tube as an 'active device' in the form of the microwave triode, the klystron, the magnetron, the travelling-wave tube, and so on, was not so quickly ousted by its semiconductor competitor. This was due partly to technological difficulties and partly to the transit-time barrier which in semiconductor devices is felt at lower frequencies than in the vacuum tube since the attainable charge-carrier velocities are smaller as a result of the scattering processes in the crystal. Stimulated by the ever-increasing importance of the microwave region—with more recent applications such as space-craft com-munications or microwave traffic control—attempts have naturally been made to improve concepts and technology in order to enable semiconductor devices to operate at higher frequencies because of their inherent potentialities for smallness, cheapness and reliability. Remarkable success in this direction has been achieved over the last ten to fifteen years with the emergence and increasing sophistication of such devices as the microwave transistor, the tunnel diode, and the varactor diode.

The maximum oscillating frequency of *bipolar* transistors is now in the region of 10GHz. These transistors require small junction areas to obtain the required small device capacitances. This results in a high current density and leads to the typical comb-shaped emitter structure. In addition, the base width must be made extremely thin if the base transit time of the carriers is to remain sufficiently small. On the other hand, decreasing the base width causes an increase in base resistance which in conjunction with the collector-base capacitance forms a detrimental RC network, and thus a compromise must be made in the design for highest frequencies. The upper frequency limit of *unipolar* (field-effect) transistors has lately been raised by improved technology to even higher frequencies. Here, it is mainly the time constant associated with the charging of the gate electrode from the source, i.e. the magnitude

1

of the gate-channel capacitance and of the channel resistance, which determines the high-frequency performance.

Tunnel diodes exhibit a negative differential conductance region in their current-voltage characteristic, and are capable of operating up to frequencies in excess of 100 GHz because of the high speed of the underlying mechanism, i.e. the tunnelling of electrons through the potential barrier of a p-n junction. However since the tunnel effect, and hence the range of negative conductance, is restricted to voltages less than the band-gap potential of the particular semiconductor material (≈ 1 V), the attainable r.f. power levels are disappointingly low.

The varactor microwave generator, being a 'secondary' power source in that it requires a radio frequency input supply, is able to produce higher-frequency power by a parametric conversion process which relies on the non-linear relationship between the electric charge stored in a reverse-biased p-n junction and the voltage applied. Output frequencies of a few hundred GHz have been achieved, the maximum or cut-off frequency of an ideal varactor diode being inversely proportional to the dielectric relaxation time (see Section 3.2) of the semiconductor material. For efficient high-order frequency conversion a cascaded chain of individual varactor multipliers, which normally contain additional idler circuits, is necessary, resulting in a fairly complicated circuit. Making use of charge-storage effects, i.e. driving the varactor diode into its conducting state for part of each r.f. cycle, has been shown generally to improve the performance, and here the time required to remove the stored carriers during the reverse-bias phase mainly determines the high-frequency limit.

Conventional semiconductor devices operate by controlling the flow of charge carriers at a p-n junction and therefore, even with improved technology, are subjected to the fundamental transit-time limitations at high frequencies. Thus the possibility was considered, from the early 1950s onwards, of *using* the effects of a finite carrier transit time as had been done with microwave electron-beam devices of the klystron and travelling-wave tube type, rather than allowing it to remain as a limiting factor.

One proposal in this direction, dating from 1958, was the avalanche transit-time diode structure suggested by Read [1.1]. Here, an avalanche breakdown inside a small, reverse-biased p-n junction is controlled by an r.f. field, and the generated carriers are then made to drift through an adjacent homogeneously-doped semiconductor region in which the electric field is high enough to establish the saturated carrier drift velocity. To the 90° phase lag of the current with respect to the field, which is characteristic of an avalanche breakdown, another 90° lag is added by a proper choice of the drift-region length and hence an external negative r.f. conductance results. While an experimental verification of such a device was beyond the reach of the technology at that time, avalanche transit-time (Impatt) diodes, mostly however with a configuration somewhat modified from the original Read type, are now very successful in generating microwave power at frequencies up to a few hundred gigahertz.

2

1.2 NEGATIVE DIFFERENTIAL BULK CONDUCTIVITY IN SEMICONDUCTORS

Quite early on in the invasion of electronics by semiconductor devices, consideration was also given to the possibility of producing a bulk negative conductance in a
homogeneous piece of semiconductor (i.e. without p-n junctions) for the generation
and amplification of r.f. electromagnetic waves, particularly at high frequencies. It
seemed likely that by abandoning p-n junctions, with their restricted active volume
and their associated capacitances, it might be possible to achieve higher r.f. powers
and frequencies, in addition to the advantage of simpler manufacture.

A true negative conductance at the sample terminals can by no means be obtained
over the whole voltage or current amplitude range due to energy conservation
requirements. A negative differential value extending only over a restricted region
may be expected, since power must be delivered to the device in order to fix the
working point, and so the total (d.c.) resistance always remains positive. Problems
associated with negative differential conductances or resistances are well known in
electronics and date back to the 'humming' and 'musical' arc discharge of the final
years of the last century [1.2].

Let us consider a homogeneous and highly extrinsic n-type semiconductor
sample, i.e. a specimen for which the current is due solely to majority carriers. Then
the differential conductivity σ', defined by the derivative of drift current density
$J = env$ with respect to electric field E, can be expressed as*

$$\sigma' = \frac{\mathrm{d}J}{\mathrm{d}E} = ev\frac{\mathrm{d}n}{\mathrm{d}E} + en\frac{\mathrm{d}v}{\mathrm{d}E} \tag{1.1}$$

which has to be distinguished from the (always positive) total conductivity
$\sigma = J/E$. It will assume negative values either if the electron density n or if the mean
carrier drift velocity v falls sufficiently fast with an increase in E. Such a behaviour
over a certain field range will result in a strongly non-linear $J(E)$ characteristic as
indicated in Fig. 1.1, which depicts a voltage-controlled device of the dynatron type
[1.3], i.e. a characteristic being many-valued in voltage (electric field).

In 1958 Kroemer [1.4] put forward a proposal for a semiconductor amplifier in
which a negative differential conductivity (n.d.c.) would be obtained by utilizing
the fact that the effective mass of charge carriers depends on the shape of the
electronic band structure of the semiconductor and thus should become negative at
sufficiently high carrier energies (see Section 2.1). If the required energy could be
conveyed to the carriers by an electric field, a range with negative differential
mobility $\mu' = \mathrm{d}v/\mathrm{d}E < 0$ would be expected in such a device. An amplifier of this
kind ought to work at frequencies up to about 1 000 GHz since, to first order, only
the scattering relaxation time of the charge carriers (see Section 2.2) determines the

---

*Primed symbols are to denote differential quantities. We adopt the familiar method of assigning
a *positive* charge, viz. the elementary charge e, to the electron in order to avoid a predominance
of confusing minus signs in the equations.

upper frequency limit. Kroemer's idea has not yet been realized, and it appears unlikely that a 'negative-mass amplifier' can ever be achieved as other counter-acting effects will set in before the necessary high electron energies are reached. But his proposal has nevertheless induced further work on n.d.c. in bulk semiconductors.

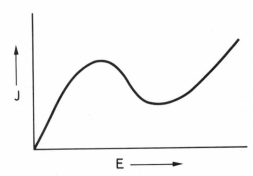

Fig. 1.1 Current-density/electric-field characteristic of a dynatron-type device. Current is carrier-current density J_c.

In particular a field-dependent 'electron capture mechanism' was postulated in 1961 by Ridley and Watkins [1.5] in which the free-electron density is reduced by trapping electrons at negatively-charged impurity centres where they are kept in non-conducting states. At low values of electric field the free electrons have, on the average, insufficient energy to overcome the potential barrier surrounding the impurity centres which thus do not affect the conduction mechanism. With increasing electric field the electrons would gain more and more energy and eventually could penetrate this barrier and therefore become trapped and immobile. At very high fields, however, the effect must vanish again since then the free electrons obtain such high energies that they remove previously trapped electrons from their locations by impact ionization. Thus $dn/dE < 0$ should result over a certain range of electron energies. This mechanism is related to the one put forward by Böer already in 1959 [1.6] in an attempt to explain field inhomogeneities observed by him in photo-conducting cadmium sulphide (CdS). Böer expected trapped holes to be freed by an increasing field and a subsequent recombination with conduction electrons, again causing $dn/dE < 0$.

Yet another mechanism for producing n.d.c. was proposed by Ridley and Watkins in 1961 [1.7] and also by Hilsum in 1962 [1.8]. It is based on the existence of a 'many-valley' conduction-band structure of such shape that there is, at a particular electron energy level, a fairly sudden increase in effective electron mass (see Section 2.1). If the transfer from the low-mass to the high-mass state is sufficiently rapid when increasing the applied electric field, the condition $dv/dE < 0$ should be possible beyond a critical field value. In particular the gallium and indium arsenides (GaAs, InAs) and antimonides (GaSb, InSb) were thought to possess the band structure necessary for exhibiting this 'transferred-electron effect' [1.8].

4

Theoretical work by Shockley in 1954 [1.9] had indicated that in a real n-type semiconductor sample with n.d.c. properties the boundary conditions at the cathode side (negative bias terminal) are such that no negative conductance can be observed externally in the static case or at low frequencies. This is so because electron injection from the cathode contact produces $dn/dE > 0$ and overcompensates any $dv/dE < 0$ or $dn/dE < 0$ resulting from a pure bulk effect. Only at very high frequencies is the charge readjustment from the cathode unable to follow and any inherent n.d.c. effect becomes noticeable.

In a semiconductor sample of finite dimensions exhibiting n.d.c. a second important effect will occur unless special precautions are taken. As Reik in 1961 [1.10] and especially Ridley in 1963 [1.11] pointed out, and as Böer *et al.* had already found experimentally in 1959 [1.13], the electric-field distribution in such a crystal will become non-uniform when the previously homogeneous field enters the n.d.c. range. In particular, a high-field region, or 'domain', may develop in an n-type sample at the cathode contact while the field in the remaining crystal falls to a subcritical value. This leads simultaneously to a reduced current density as a consequence of the $J(E)$ characteristic shown in Fig. 1.1 (see Sections 3.1 and 3.2). The high-field domain generally consists of a dipole space-charge layer as indicated by Poisson's equation [see eqn (3.2)], and moves in the direction of electron drift through the sample towards the anode where it is extinguished; then, in turn, a new domain can be formed at the cathode. Since domain formation and extinction is accompanied by a fall and rise in sample current, respectively, a cyclic current variation, or relaxation oscillation, results. An explanation of the instability of the homogeneous sample field in the n.d.c. region follows from thermodynamics because a break-up into two such 'phases' of different field constitutes minimum entrophy production (which here means minimum production of Joule heat), see [1.11] and [1.12].

The velocity of domain movement may vary considerably for the various n.d.c. mechanisms. Böer [1.13] observed domain velocities in CdS of the order of 10^{-1} cm s^{-1} or even lower, values which were also measured by Ridley and Pratt [1.14] when performing experiments with their trapped-electron mechanism. The domain velocity, not surprisingly, is so low because trapping and detrapping time constants are involved in these electron-capture processes. In the transferred-electron effect trapping generally does not play a role and the domains consist of free charge carriers only. They thus move nearly with the average carrier drift velocity. Since the electron velocities are of the order of 10^7 cm s^{-1} the frequency of the current oscillation due to domain generation and disappearance will reach the microwave region for sufficiently short, but still practicable, sample lengths.

Still another configuration of high-field domains, moving with the velocity of sound (approximately 10^5 cm s^{-1}), has been observed in GaAs [1.15] and other piezoelectric semiconducting materials; this is the result of n.d.c. associated with acoustic-wave amplification by supersonic carriers (acousto-electric domains), see for example [1.16].

1.3 GUNN'S DISCOVERY

After having experimentally studied the non-ohmic high-field behaviour of the elemental semiconductor germanium (Ge), J. B. Gunn, in the early 1960s, turned his attention to binary compound semiconductors like GaAs where he expected to find at most only a small difference in their high-field properties from that of Ge due to the polar character of the compounds [1.17].

In 1963, when investigating n-type GaAs samples several millimetres long with ohmic tin contacts attached to them (Fig. 1.2), Gunn found to his surprise that on reaching a threshold field value in the region of 2–4 kV cm^{-1}, random noise-like oscillations appeared in the sample current with amplitudes of the order of amperes.

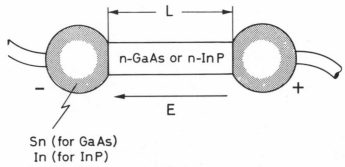

Fig. 1.2 Schematic outline of original Gunn-effect semiconductor sample with ball-shaped ohmic contacts.

He soon verified that this behaviour must be caused by a true bulk effect, since the contacts or surface properties were not important (at least to a first approximation). With shorter sample lengths, below 0·2 mm, the oscillations of some samples became coherent with fundamental oscillation frequencies of several gigahertz. The frequency was approximately inversely proportional to the sample length, thus pointing to a transit-time effect [1.18]. Figure 1.3 shows the measured current-voltage relationship of such a short GaAs sample. Specimens or indium phosphide (InP) behaved similarly. By an elegant experiment in which he measured the potential distribution along a GaAs sample 270 μm in length using a 15-μm wide capacitive probe, Gunn established that a marked electric-field inhomogeneity was moving from the cathode to the anode in a cyclic manner [1.19]. Thus an obviously new phenomenon was discovered which promised to be of great interest because of the powers and frequencies involved.

Retrospectively it seems astonishing that after the original discovery by Gunn, the underlying mechanism of the effect was the subject of intense speculation for almost a year. Eventually Kroemer [1.20] pointed out that the transferred-electron mechanism, as proposed by Ridley, Watkins and Hilsum, was the most probable explanation. This could soon be confirmed experimentally. In one of these experiments the GaAs crystal was subjected to high hydrostatic pressure [1.21], which according to theory alters the electronic band structure in such a way as to decrease the threshold field for the onset of the current oscillation. This was indeed observed. In an analogous experiment the threshold field was reduced by alloying phosphorus (P) into GaAs to form the mixed crystal $GaAs_xP_{1-x}$ [1.22].

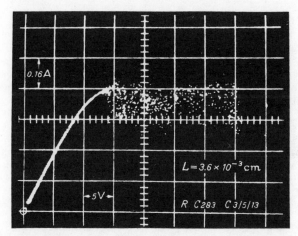

Fig. 1.3 Current (vertical) against voltage (horizontal) of an n-type GaAs device as measured on a sampling oscilloscope. After Gunn [1.19].

Hence a true negative-conductance semiconductor bulk effect, operating at microwave frequencies, had finally been found. The transit-time Gunn oscillations, however, were to be only the beginning of an evolution of a broad spectrum of possible operating modes and applications performed by devices based on the transferred-electron effect. This monograph is dedicated to these devices.

More or less comprehensive reviews of the field of transferred-electron or Gunn devices have previously been given in, for example, [1.23] to [1.38]. A Gunn-effect bibliography has been compiled in [1.39]; a reprint volume of pertinent publications is also available [1.40].

Frequently the term 'transferred-electron diode' or 'Gunn diode' is used. According to IEC Publ. 147-0 (and to most national standards) a 'diode' is defined to possess an asymmetrical current-voltage characteristic. Since, however, transferred-electron semiconductor samples exhibit in general an *anti*symmetrical characteristic like resistors (see Fig. 3.16), we prefer not to call them 'diode'. Rather we adopt the terms 'TE (transferred-electron) element' and 'TE device', or 'Gunn device' in cases where they operate in the true Gunn operational mode which is characterized by the appearance in the sample of space-charge dipole domains (see Chapter 3).

REFERENCES

1.1   Read, W. T., A proposed high-frequency, negative-resistance diode. *Bell Syst. Techn. J.* **37** (1958), p. 401.

Gunn-effect Electronics

1.2 Duddell, W., On rapid variations in the current through the direct-current arc. *Electrician* **46** (1900), pp. 269 and 310.

1.3 Hull, A. W., A vacuum tube possessing negative electric resistance. *Proc. Inst. Rad. Engrs* **6** (1918), p. 5.

1.4a Kroemer, H., Proposed negative-mass microwave amplifier. *Phys. Rev.* **109** (1958), p. 1856.

1.4b Kroemer, H., The physical principle of a negative-mass amplifier. *Proc. Inst. Rad. Engrs* **47** (1959), p. 397.

1.5 Ridley, B. K. and Watkins, T. B., Negative resistance and high electric field capture rates in semiconductors. *J. Phys. Chem. Solids* **22** (1961), p. 155.

1.6a Böer, K. W., Inhomogeneous field distribution in CdS single crystals in the high-field range (in German). *Z. Physik* **155** (1959), p. 184.

1.6b Böer, K. W., Layer-like inhomogeneities in homogeneous semiconductors in the range of negative differential resistivity. *Phys. Rev.* **139** (1965), p. A 1949.

1.7 Ridley, B. K. and Watkins, T. B., The possibility of negative-resistance effects in semiconductors. *Proc. Phys. Soc. (London)* **78** (1961), p. 293.

1.8 Hilsum, C., Transferred electron amplifiers and oscillators. *Proc. Inst. Rad. Engrs* **50** (1962), p. 185.

1.9 Shockley, W., Negative resistance arising from transit time in semiconductors. *Bell Syst. Techn. J.* **33** (1954), p. 799.

1.10 Reik, H. G., cited in [1.7].

1.11 Ridley, B. K., Specific negative resistance in solids. *Proc. Phys. Soc. (London)* **82** (1963), p. 954.

1.12 Kiess, H. and Stöckmann, F., A new mechanism for the saturation of photo currents (in German). *Phys. Stat. Sol.* **4** (1964), 117.

1.13a Böer, K. W., Hänsch, H. J. and Kümmel, U., Application of opto-electric effects for the analysis of the electric conduction mechanism in CdS single crystals (in German). *Z. Physik* **155** (1959), p. 17Q.

1.13b Böer, K. W., Field and current inhomogeneities in insulators and photo conductors at high bias fields (in German). *Festkörperprobleme* **1** (1962), p. 38, Vieweg, Braunschweig.

1.14a Ridley, B. K. and Pratt, R. G., A bulk differential negative resistance due to electron tunnelling through an impurity potential barrier. *Phys. Lett.* **4** (1963), p. 300.

1.14b Ridley, B. K. and Pratt, R. G., Hot electrons and negative resistance at 20° K in n-type germanium containing Au$^=$ centres. *J. Phys. Chem. Solids* **26** (1965), p. 21.

1.15 Hervouet, C., Lebailley, J., Leroux-Hugon, P. and Veilex, R., Current oscillations in GaAs under acoustic amplification conditions. *Solid-State Commun.* **3** (1965), p. 413.

1.16 White, D. L., Amplification of ultrasonic waves in piezoelectric semiconductors. *J. Appl. Phys.* **33** (1962), p. 2547.

1.17 Gunn, J. B., The Gunn effect. *Int. Sci. Technol.* (1965), Oct., p. 43.

1.18 Gunn, J. B., Microwave oscillations of current in III–V semiconductors. *Solid-State Commun.* **1** (1963), p. 88.

1.19 Gunn, J. B., Instabilities of current in III–V semiconductors. *IBM J. Res. Dev.* **8** (1964), p. 141.

1.20 Kroemer, H., Theory of the Gunn effect. *Proc. Inst. Electr. Electron. Engrs* **52** (1964), p. 1736.

1.21 Hutson, A. R., Jayaraman, A., Chynoweth, A. G., Coriell, A. S. and Feldman, W. L., Mechanism of the Gunn effect from a pressure experiment. *Phys. Rev. Lett.* **14** (1965), p. 639.

1.22 Allen, J. W., Shyam, M., Chen, Y. S. and Pearson, G. L., Microwave oscillations in Ga(As$_x$P$_{1-x}$) alloys. *Appl. Phys. Lett.* **7** (1965), p. 78.

1.23 Butcher, P. N. and Hilsum, C., Transferred electron oscillators. "Progress in Radio Science 1963–1966", Proc. XVth General Assembly of URSI, pt. II, Munich, Sept. 1966, p. 2290.

1.24 Bosch, B. G., Gunn-effect electronics (in German). *Telefunken-Röhre* **47** (1967), p. 13.

1.25   Bott, I. B. and Fawcett, W., The Gunn effect in gallium arsenide. In: Young, L. (Ed.), *Advances in Microwaves*, vol. 3, p. 223 (New York and London, Academic Press, 1968).

1.26   Uenohara, M., Bulk gallium arsenide devices. In: Watson, H. A. (Ed.), *Microwave Semiconductor Devices and their Circuit Applications*, p. 497 (New York, McGraw-Hill, 1969).

1.27   Hartnagel, H., *Semiconductor Plasma Instabilities*, including Gunn-effect and avalanche oscillations (London, Heinemann, 1969).

1.28   Sze, S. M., *Physics of Semiconductor Devices* (New York, J. Wiley and Sons, 1969), Part V: Bulk-effect devices, p. 731.

1.29   Kataoka, S. and Tateno, H., *Bulk Effect Semiconductors* (in Japanese; Tokyo, Nikkan Kogyo Shinbunsha, 1969).

1.30   Ikoma, T., Sugeta, T., Torizuka, H. and Yanai, H., Characteristics of the transferred electron devices. *J. Fac. Engng, Tokyo Univ.* (*B*) **30** (1970), p. 347.

1.31   Carroll, J. E., *Hot Electron Microwave Generators* (Maidenhead, Edward Arnold, 1970).

1.32   Copeland, J. A. and Knight, S., Applications utilizing bulk negative resistance. In: Willardson, R. K. and Beer, A. C. (Eds), *Semiconductors and Semimetals*, vol. 7, pt. A, p. 3 (New York and London, Academic Press, 1971).

1.33   Heime, K., The Gunn effect (in German). *Der Fernmelde-Ingenieur* **25** (1971), No. 11, p. 1 (pt. I); and No. 12, p. 1 (pt. II).

1.34   Shur, M. S., *Gunn Effect* (in Russian; Leningrad, Energiya, 1971).

1.35a   Sterzer, F., Transferred electron (Gunn) amplifiers and oscillators for microwave applications. *Proc. IEEE* **59** (1971), p. 1155.

1.35b   Narayan, Y. S. and Sterzer, F., Transferred electron amplifiers and oscillators. *IEEE Trans. Microwave Theory and Techniques* **MTT—18** (1970), p. 773.

1.36   Unger, H. G. and Harth, W., *High-frequency Semiconductor Electronics* (in German; Stuttgart, Hirzel, 1972).

1.37   Bulman, P. J., Hobson, G. S. and Taylor, B. C., *Transferred Electron Devices*. (London and New York, Academic Press, 1972).

1.38   Kroemer, H., Gunn-effect bulk instabilities. In: Hershberger, W. D. (Ed.), *Topics in Solid State and Quantum Electronics*, p. 20 (New York, John Wiley and Sons, 1972).

1.39a   Gaylord, T. K., Shah, P. L. and Rabson, T. A., Gunn effect bibliography. *IEEE Trans. Electron Dev.* **ED–15** (1968), p. 777.

1.39b   Gaylord, T. K., Shah, P. L. and Rabson, T. A., Gunn effect bibliography supplement. *IEEE Trans. Electron Dev.* **ED–16** (1969), p. 490.

1.40   Eastman, L. F. (Ed.), *Gallium Arsenide Microwave Bulk and Transit-time Devices* (Dedham, Mass., Artech House Inc., 1972).

2
The Transferred-Electron Effect (Generalized Gunn Effect)

In this chapter we discuss the electron transfer mechanism leading to negative differential conductivity (n.d.c.) in a homogeneous, bulk semiconductor material. This mechanism was originally proposed by Ridley and Watkins [1.7] as well as by Hilsum [1.8] and was subsequently, after the experimental verification by Gunn [1.18], improved on and refined by numerous other workers.

As briefly mentioned in Section 1.2, a semiconducting crystal which possesses a suitably-shaped band structure, with a many-valley conduction band of a particular kind, is required for the transferred-electron (TE) effect. To present knowledge, the compound semiconductors GaAs, InP, CdTe, ZnSe, and the ternary alloys GaAs$_x$P$_{1-x}$, In$_x$Ga$_{1-x}$Sb, Ga$_x$Al$_{1-x}$As and InAs$_x$P$_{1-x}$, i.e. semiconductors made up from elements of either the III and V, or of the II and VI, groups of the periodic system, exhibit the proper band structure under normal environmental conditions (see Section 6.1.1). Others take on the necessary band configuration when their crystal lattice is deformed by the influence of high pressure which alters the relative position of the band valleys (Section 6.1.1). For applications of the TE effect in the form of devices of various kinds, GaAs presently still constitutes the optimum choice (Section 6.1.1) and, thus, is almost exclusively used. For this reason, only GaAs is considered in this chapter and also, unless stated otherwise, in the remainder of this book.

For an explanation or more detailed description of the general terms from semiconductor physics which appear in the following sections the reader is referred to textbooks such as [2.1, 2.2 and 6.29]. A comprehensive review particularly of high-field phenomena in semiconductors is given in [2.3].

2.1 SEMICONDUCTOR BAND STRUCTURE

In classical mechanics a free-space electron is a particle and possesses the kinetic energy $W = p^2/2m_0$, p being its momentum and m_0 its mass, as shown in Fig. 2.1 by the (thin-lined) parabola. In the quantum-mechanical approach the free electron is associated with a matter wave of wavelength $\lambda = h/p$, where $h = 6.625 \times 10^{-34}$ Js denotes Planck's constant. Introducing the wave propagation number $k = 2\pi/\lambda$, the kinetic energy of the electron wave in free space is then $W = \hbar^2 k^2/2m_0$ (thin-lined parabola in Fig. 2.1), where $\hbar = h/2\pi$.

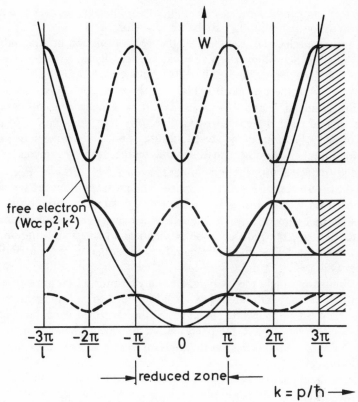

Fig. 2.1 Energy W against momentum p, or wave number $k = p/\hbar$, for free-space electron (thin parabola) and for electron moving in one-dimensional periodic crystal potential (thick curves).

Dealing here with conduction processes in solids, the electrons under considera-tion are by no means free but are acted upon by the surrounding crystal lattice which consists of arrays of interacting atoms with a resulting periodic potential. The existence of this coupled periodic structure has the effect, as schematically shown in Fig. 2.1 by the full-line segments, of splitting up the previously continuous energy curve. Thus allowed energy bands (hatched regions) are separated by inaccessible intervals known as the forbidden bands (band gaps) which appear at k values being an integer of π divided by the basic period l of the structure. This basic period has a certain fixed relation to the lattice constant a_0 depending on the exact arrangement of the atoms in the particular type of crystal structure, with $l \equiv a_0$ in the one-dimensional lattice, for instance. The reader familiar with electromagnetic-wave propagation will notice that a similar behaviour is found in the case of a coupled system consisting of an infinite number of identical resonators having a number of differing eigen-values (resonant frequencies), or to the case of the homogeneous, but periodically disturbed, transmission line, see [2.4]. These

11

structures form a filter whose dispersion diagram (frequency against wave-propagation number) shows alternating pass bands and stop bands.

The quantum-mechanical treatment of the electron motion in a periodic lattice potential (based on Schrödinger's wave equation) actually leads to electronic matter waves and hence to energy eigen-values $W(k)$ that are *periodic* in the wave-propagation number k with the period $2\pi/l$ (Fig. 2.1, thick curves). This means that the propagation numbers k and $k' = k + i2\pi/l$ ($i = \pm$ integer) lead to physically undistinguishable results. Hence it is possible to restrict the k values to the solid segments of Fig. 2.1 as done in the preceding paragraph. This defines a 'free-electron' propagation number. Another possible and mathematically simpler restriction leads to the 'reduced' propagation number defined by $-\pi/l < k < \pi/l$. This is shown in the centre part of Fig. 2.1 ('reduced zone') where the dashed segments can be obtained from the solid ones by an appropriate lateral translation procedure. Always in the one-dimensional case, but also for many three-dimensional, more complicated lattices, the reduced zone is identical to the first Brillouin zone, a Brillouin zone being defined by the Bragg reflection properties of the periodic lattice for plane waves.

In semiconductors, only the two highest energy bands which are partially filled with electrons at temperatures above the absolute zero, are of concern for the electronic transport mechanism. The upper of these, termed the conduction band, contains mobile electrons and the band below, the valence band, contains mobile holes (empty electronic states); both kinds of charge carrier contribute to the conduction process.

Because of the higher complexity of the actually three-dimensional, anisotropic semiconductor crystals, the allowed bands do not have, in general, the relatively simple shape as shown in Fig. 2.1. Rather, the edge (extremum) of a band may not be positioned at the centre of the reduced zone or lie exactly above the edge of the adjoining band. Also, a certain band may be composed of more than one branch. Furthermore, the electron-wave propagation, and consequently the band structure, may differ for various crystal directions so that one has to introduce a wave propagation *vector* $\vec{k}$, where $|\vec{k}| = 2\pi/\lambda$. Finally, there may be additional (satellite) minima of a particular band within the reduced zone, a case found, for instance, in the conduction band of GaAs.

The special form of the conduction band C and the valence band V of GaAs is shown in Fig. 2.2 for the [100] and [$\bar{1}$00] crystal directions*. This knowledge of the

---

*The orientation of a *plane* in a crystal with a cubic lattice structure is specified, in terms of the lattice constant, by a set of three numbers known as Miller indices which are enclosed in parentheses, e.g. (100). A family of equivalent planes is represented by using waved parentheses, e.g. {100}. *Directions* in a cubic crystal system are identified by the Miller indices, shown with rectangular brackets, of the plane normal to the direction, e.g. [100]. For denoting a family of equivalent directions the indices of a member of this family are given in pointed brackets, e.g. ⟨100⟩ for the family consisting of [100], [010], [001], [$\bar{1}$00], [0$\bar{1}$0] and [00$\bar{1}$] where $\bar{1} \equiv -1$. The position of *points* in a crystal is similarly specified in terms of the lattice constant also using three numbers which are, generally, enclosed in parentheses cf. Fig. 6.1.

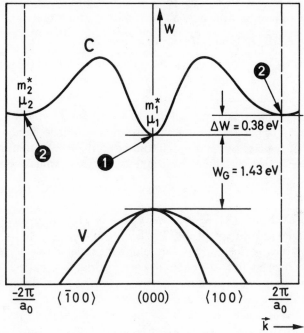

Fig. 2.2 Conduction-band (C) and valence-band (V) structures for GaAs in the ⟨100⟩ directions.

band structure relies on experimentally obtained information (e.g. [2.5] and [2.6]) as well as on the results of calculations [2.7] employing, for example, the pseudo-potential [2.8a] and the '$k \cdot p$' methods [2.8b]. The semiconductor GaAs constitutes a direct-transition material since the conduction-band and valence-band edges, separated by a band gap of approximately $W_G = 1 \cdot 43$ eV at 300 K [2.6], lie at the same $\vec{k}$ value, namely at the centre of the reduced zone, i.e. at the point (000), ('direct' band gap). The valence band consists of two branches which join, i.e. are degenerate, at (000). Of importance for the electron-transfer mechanism is the fact that the conduction band, in addition to the central minimum (valley) at (000), shows satellite valleys in the ⟨100⟩ directions. These seem to be positioned at the edge of the reduced zone [2.8], by an amount of about $\Delta W = 0 \cdot 38$ eV above the bottom of the central valley [2.5b]. Because of the six existing ⟨100⟩ directions there are three full (composed of six half) satellite valleys.

In semiconductor physics it is customary to assign to the mobile charge carriers an effective mass m^* which differs from the gravitational free-carrier mass m_0, the latter being $m_0 = 9 \cdot 108 \times 10^{-28}$ g for the electron. This effective mass is introduced to allow for the influence of the crystal-lattiee forces on the carriers. Using the concept of the effective mass m^*, the dynamical behaviour of the entire carrier

13

ensemble in response to an external force can, then, simply be described in terms of Newton's law of motion. It is possible to show that, for a proper choice of the co-ordinate axes, $m^\star$ depends on $W(\vec{k})$ through the relation

$$\frac{1}{m_i^\star} = \frac{1}{\hbar^2} \frac{\partial^2 W}{\partial k_i^2} \tag{2.1}$$

with i indicating a principal crystal direction; i.e. $m_i^\star$ is inversely proportional to the band curvature. Hence for the special case of a parabolic band structure $m_i^\star$ is independent of the electron energy. According to the band structure (see Figs 2.1 and 2.2) the effective mass may even take on negative values. Since, as already mentioned, $W(\vec{k})$ is likely to differ for the various crystal directions, $1/m^\star$ in general is a tensor.

For mobile electrons in GaAs the (conduction-band) central-valley effective mass has been determined by experiment [2.5c and d] to be nearly $m_1^\star = 0.066\, m_0$ (sub-scripts 1 and 2 refer to the central valley and to the satellite valley, respectively). Up to moderately high energies the constant-energy surfaces in k space as obtained for the somewhat *non-parabolic* central valley are *spheric* so that the effective mass can there be regarded as a slightly energy-dependent scalar [2.5–2.8]. In the *non-spheric* $\langle 100 \rangle$ satellite valleys [2.8] the effective mass is a tensor. It is, however, sufficient for most purposes to use a mean value, which is called combined-density-of-states mass, $m_2^{\star(N)} \approx 0.85\, m_0$ [2.5b]. This mass would be obtained if all the equivalent $\langle 100 \rangle$ satellite valleys are replaced by a single spheric valley possessing the same density of states $N^\star$. The density of states of a valley denotes the number of allowed energy states per unit volume and energy interval available to the conduction electrons and is proportional to $[m^{\star(N)}]^{3/2}$. Thus, because $m_1^{\star(N)} = m_1^\star$, the density-of-states ratio between the satellite and central valleys is approximately $N_2^\star/N_1^\star = 46$.

The proportionality factor between average electron drift velocity v and the applied electric field E is termed the (total) electron drift mobility $\mu = v/E$. The mobility shows a field dependence since the relationship between v and E is, in general, non-linear (see Section 2.3). With a mean free travelling time between two randomizing collisions of τ_p (momentum relaxation time) and since the acceleration is $eE/m^\star$ one then has

$$\mu = e\tau_p/m^\star. \tag{2.2}$$

The low-field mobilities for GaAs at a lattice temperature $T_L = 300$ K have been calculated to be $\mu_1 = 11\,000\text{–}12\,000$ cm² (Vs)$^{-1}$ [2.9–2.11] and $\mu_2 = 150\text{–}175$ cm² (Vs)$^{-1}$ [2.11–2.13]. This implies that τ_p is markedly smaller in the satellite valleys. The mobility values obtained by measurement are, not surprisingly, somewhat lower with, e.g., $\mu_1 = 7\,500\text{–}9\,300$ cm² (Vs)$^{-1}$ [2.14] (for material with electron concentrations in the $10^{14}\text{–}10^{15}$ cm^{-3} range) and $\mu_2 = 110$ [2.15] or 150 cm² (Vs)$^{-1}$ [2.16].

14

2.2 SCATTERING PROCESSES AND ELECTRON DISTRIBUTION

To obtain information on the conduction properties of a semiconductor as a function of an applied electric field, and to determine whether a negative differential conductivity region can be expected, the various scattering processes to which the mobile charge carriers are subjected in the crystal lattice have to be investigated. In addition, the influence of scattering on the carrier-density distribution with respect to energy must be studied.

The average rate at which an electron gains energy from the electric field is

$$\langle dW/dt \rangle_E = eEv = e\mu E^2$$

and this must be compensated by a corresponding loss rate $-\langle dW/dt \rangle_{sc}$ due to the scattering processes, i.e. $\langle dW/dt \rangle_E + \langle dW/dt \rangle_{sc} = 0$. Without such a relaxation of electron energy $W = (m^\star/2)u^2$ (u being the instantaneous electron velocity) by scattering, the electrons would gain, even at relatively low electric fields, so much energy that a steady-state electron distribution could no longer be maintained and a catastrophic breakdown condition would result.

In a real semiconductor, assuming a reasonably perfect and pure one, interactions of the charge carriers with the thermally-excited lattice vibrations play the most important role (see, e.g. [2.2] and [2.17]). Through the coupling of the harmonic oscillators formed by the vibrating lattice atoms, elastic waves propagate inside the crystal. According to quantum mechanics such a system can attain only discrete energy values $(i + 1/2)\,hf$, with i being an integer and f the vibration frequency. The energy quantum of the elastic wave is termed a phonon. During the interaction between a conduction electron and the lattice, one phonon, in general, is either emitted (if the energy of the lattice wave increases) or absorbed (if it decreases). Although phonons are, thus, generated or annihilated during the interaction process it is customary to speak of the *collision* of an electron with a phonon, the latter behaving like a neutral particle. For such a collision the laws of energy, momentum, and charge-carrier number, conservation have to be fulfilled.

A quantitative picture of the phonon scattering process can be obtained from the phonon dispersion curves such as those shown in Fig. 2.3 for GaAs at room temperature in the $\langle 100 \rangle$ directions [2.18]. The curves, which were obtained by neutron spectroscopy, give the dependence of frequency f on the (reduced) wave number $k\phi$ for the longitudinal (L) and the transverse (T) elastic waves. A lower-frequency 'acoustical' branch (A) is obtained when adjacent lattice atoms—in GaAs a Ga atom and its neighbouring As atom—oscillate nearly *in* phase, and a higher-frequency 'optical' branch (O) when they oscillate nearly in *anti*phase. Multiplying the ordinate of the dispersion diagram in Fig. 2.3 by h and the abscissa by $\hbar$ yields a representation of phonon energy $W\phi = hf$ against phonon momentum $p\phi = \hbar k\phi$.

There are basically two types of mechanism for the interaction between electrons and phonons: due to the resulting strain, lattice vibrations cause a deformation of the crystal band structure, leading to a deformation potential which actually induces scattering of the electrons. Additionally, the vibrations cause a displacement of the ionic charges giving rise to a Coulomb potential. Deformation-potential scattering involves primarily acoustical phonons since they produce the strongest strain in the crystal. To the contrary, for Coulomb-type scattering optical phonons are mainly involved. In compound semiconductors (with their partially polar chemical bond) the optical mode with its antiphase oscillations of the adjacent, somewhat differently charged, atoms leads to particularly pronounced electric dipole centres. Electron scattering at this latter type of inhomogeneity is called polar scattering whereas the somewhat weaker Coulomb-type interaction involving acoustical phonons is termed piezoelectric scattering.

Phonon scattering processes may involve energy and momentum exchanges such that the electron still remains inside a certain conduction-band valley ('intravalley scattering'), or these changes might be sufficient that in many-valley semiconductors a transfer of electrons into a different valley is possible ('intervalley scattering'). If this transfer occurs between valleys of the same type (e.g. the $\langle 100 \rangle$ satellite valleys in GaAs) one speaks of 'equivalent' intervalley scattering, otherwise it is termed 'non-equivalent'.

Due to the high degree of semiconductor perfection and purity required for TE devices (Section 6.1) we should expect to be able to neglect scattering at crystal-structure imperfections and at ionized impurity centres cf. [2.19]. However in a real crystal these effects are not always negligible (see Section 6.1.3). Electron–electron scattering is not significant as long as the electron concentration remains below about 10^{17} cm^{-3} [2.20], a condition which again is generally satisfied in TE devices (Section 6.1.1).

Gaining quantitative results on the influence of an applied electric field, and of the scattering mechanisms, on the electron transport properties requires the solution of a position-independent Boltzmann equation which, for the steady state, can be written symbolically as

$$\left(\frac{\partial F_\nu(\vec{k})}{\partial t} \right)_{\vec{E}} + \sum_i \left(\frac{\partial F_\nu(\vec{k})}{\partial t} \right)_i = 0 \qquad (2.3)$$

with $\nu = 1, 2, \ldots$ referring to the various valleys and i to the different scattering processes involved. The term

$$\left(\frac{\partial F_\nu(k)}{\partial t} \right)_{\vec{E}} = - \frac{e\vec{E}}{\hbar} \vec{\nabla}_k F_\nu(\vec{k})$$

denotes the rate of change of the electron 'distribution function' $F_\nu(\vec{k})$ due to an electric field $\vec{E}$ in the formalism of the Nabla operator $\vec{\nabla}_k = (\partial/\partial k_x, \partial/\partial k_y, \partial/\partial k_z)$. The function $F_\nu(\vec{k})$ described how the electron ensemble is distributed in momentum $\hbar\vec{k}$ and hence, for a given direction in $\vec{k}$-space, in kinetic energy $W(\vec{k})$. The second term in eqn (2.3) is the rate of change of $F_v(\vec{k})$ produced by the scattering processes. The distribution functions for different valleys v are coupled due to the intervalley scattering terms contained in $\sum_i$. A knowledge of this distribution is required for computing the average properties of the ensemble such as average energy $\langle W \rangle$ or drift momentum $\langle p \rangle$ as a function of the electric field.

Equation (2.3) states that in the steady-state case the sum of all time-dependent variations of the electron distribution $F_v(\vec{k})$ in a certain conduction-band valley v, as caused by the field $\vec{E}$ and by the various scattering processes i, must compensate each other. Written out in full, one obtains a rather complicated integro-differential equation (see, e.g., [2.3]) which has not yet been solved in its general form. Approximate solutions have, however, been obtained either under certain simplifying assumptions concerning the distribution function or for restricted electric-field ranges.

Solutions of the Boltzmann equation for GaAs in the electric-field range where the TE effect occurs, have been obtained by two types of approximations. The first of these [2.20] takes into account polar and acoustical deformation-potential *intra*valley scattering and deformation-potential *inter*valley scattering. It is assumed that at sufficiently high fields many of the electrons in the central and satellite valley attain such high energies $W \gg W\phi$ as to justify considering all the phonon scattering processes as being quasi-elastic. The distribution function can then be expressed by two new functions depending only on energy W. In this case it is possible to simplify the scattering terms of the Boltzmann equation by expanding the distribution-function expression for the *scattered* electrons into a Taylor series up to second order in phonon energy $W\phi$. Thereby, the Boltzmann equation is converted into a pair of coupled differential equations. This method has the disadvantages that a quasi-elastic approach is of doubtful validity at low fields and also that only two terms are retained in the spherical-harmonics expansion of the distribution function. The calculation has subsequently been extended [2.21 and 2.22] through considering the non-parabolicity of the central conduction-band valley and has been modified [2.23 and 2.24] by assuming the electron distribution in the satellite valleys to be Maxwellian at the lattice temperature for electric fields up to a few kV cm^{-1}. This latter assumption seems to be justified since in the satellite valleys no appreciable heating of the electrons takes place even at such high fields [2.28] (see also Fig. 2.5).

In a different approach [2.12, 2.25–2.27] the distribution function for each valley was expressed by a Maxwellian distribution which is considered to be displaced in momentum space, due to the average electron (drift) momentum, and to which an 'electron temperature' greater than the crystal-lattice temperature can be assigned. The same electron-phonon scattering processes as in the first above-mentioned approach [2.20] were taken into account in Boltzmann's equation. The postulation of this displaced Maxwellian distribution implicitly requires, in addition, strong

17

electron–electron scattering to establish thermal equilibrium *within* the electron gas. Although this assumption is rather unrealistic since, as stated above [2.20], inter-carrier collisions are hardly expected to be effective at the electron concentrations usually encountered in TE devices, the shape of the drift-velocity/field characteristic obtained by these computations shows a rather good agreement with most of the experimental results (Section 2.3). An extended transport calculation in the displaced-Maxwellian approximation, taking into account the non-parabolicity of the central valley, was carried out in [2.29] and [2.30].

Furthermore two successful numerical methods of determining the electron distribution function at high fields have been employed which do not rely on approximations and *a-priori* assumptions about its structure as mentioned above. One is a Monte-Carlo calculation [2.28 and 2.31] in which the motion of a single electron in $\vec{k}$-space is simulated and repetitively investigated, thus treating the case as a statistical problem. The other approach is an iterative technique [2.32]. It exploits the stability of the steady state, since after a long enough time the distribution function will tend to the steady-state form irrespective of its initial value.

After these general remarks we now describe more particularly the processes encountered in a hypothetical extrinsic n-type GaAs crystal with mathematically perfect doping homogeneity and of unlimited dimensions, when subjected to an electric field. In a real GaAs sample of finite size, complications will arise which are dealt with in subsequent chapters.

Without an applied electric field and at an assumed (fixed) lattice temperature $T_{\mathrm{L}} = 300$ K, most of the mobile electrons rest at the bottom of the central conduction-band valley 1 (Fig. 2.2) as it constitutes the lowest possible energy state. There, the energy and momentum exchange between the electrons and the lattice is balanced out by equal phonon absorption and emission. When the field is applied the electron gas takes up energy from it, leading to an occupation of higher energy states in the central valley. Then energy is transferred to the lattice, mainly by polar scattering with longitudinal optical phonons for which the energy quantum is about $5 \cdot 5 \times 10^{-21}$ J $\approx 0 \cdot 035$ eV according to Fig. 2.3. Since GaAs is a polar semiconductor this type of scattering is the most effective one at not too high fields for symmetry and energy reasons.

With a further increase in field, and thereby an increase in mean electron energy, the energy relaxation by polar scattering becomes less effective. This is seen in Fig. 2.4 from the calculated curve for the average energy relaxation $-\langle \mathrm{d}W/\mathrm{d}t \rangle_{\mathrm{sc, po}}$ in the central valley since it exhibits a saturation effect. As a result, the average electron kinetic energy $\langle W \rangle$ rises very rapidly above the value corresponding to thermal equilibrium. The higher-energy electrons thus formed are said to be 'hot' and an effective electron temperature T_{e} ($>$ lattice temperature T_{L}) is assigned to them. In analogy to the thermal-equilibrium case, $\langle W(T_{\mathrm{L}}) \rangle = 3k_{\mathrm{B}}T_{\mathrm{L}}/2$, for a Maxwellian distribution and a parabolic valley one writes

$$T_{\mathrm{e}} = 2\langle W \rangle / 3k_{\mathrm{B}}. \tag{2.4}$$

18

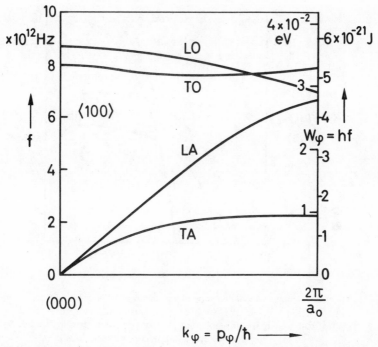

Fig. 2.3 Measured phonon dispersion diagram for GaAs in the $\langle 100 \rangle$ directions, showing phonon frequency f (energy W_φ) against wave number k_φ (momentum p_φ) after Waugh and Dolling, [2.18]. LO longitudinal optical branch, TO transverse optical branch, LA longitudinal acoustical branch, TA transverse acoustical branch.

(Here $k_B = 1 \cdot 38 \times 10^{-23}$ J K^{-1} is Boltzmann's constant.)

The calculated dependence of the electron temperature on electric field in GaAs is demonstrated in Fig. 2.5. It shows a marked increase for the central-valley electrons at fields of a few kV cm^{-1} where the energy loss rate levels off. The curve of $-\langle dp/dt \rangle_{sc, i}$ for polar scattering in Fig. 2.4 shows that in the central valley the average momentum relaxation also becomes less effective beyond fields of a few kV cm^{-1}. The loss rate $\langle dp/dt \rangle_{sc}$ causes a randomization of the electron velocity distribution in momentum space and thus a finite electron mobility. In fact, the momentum relaxation time τ_p introduced in eqn (2.2) follows directly from $-\langle dp/dt \rangle_{sc} = \langle p \rangle / \tau_p$, where $\langle p \rangle$ is the electron drift momentum.

Before the relatively rapid increase in electron energy (temperature) and momentum can lead to a catastrophic situation, a reasonable fraction of the electrons in the central valley will have attained energies $\gtrsim 0 \cdot 38$ eV which enables them to transfer to the high-mass $\langle 100 \rangle$ satellite valleys (see Fig. 2.2). This non-equivalent intervalley scattering is accompanied by a pronounced new loss in energy and momentum (Fig. 2.4), and thus by a reduction in the temperature rise of the central-valley electrons (Fig. 2.5). The intervalley scattering process goes along with the emission

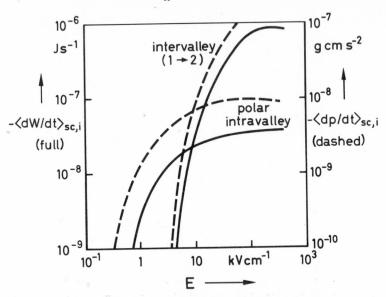

Fig. 2.4 Calculated dependence on electric field E of the average electron energy loss rate $-\langle dW/dt \rangle_{sc,i}$, and of the average momentum loss rate $-\langle dp/dt \rangle_{sc,i}$ in the central valley of GaAs for polar intravalley and for intervalley scattering into the $\langle 100 \rangle$ valleys. Adapted from Heinle [2.30].

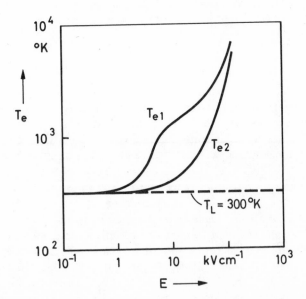

Fig. 2.5 Calculated dependence on electric field E of the electron temperatures T_{e1} and T_{e2} in the central valley and in the $\langle 100 \rangle$ satellite valleys, respectively, in GaAs. Lattice temperature $T_L = 300$ K. Adapted from Heinle [2.30].

or absorption of an optical or acoustical 'intervalley phonon' ($W\phi = 0 \cdot 029$ eV) which must possess such momentum and energy values that the momentum and energy balance conditions

$$\hbar \vec{k}_2 = \hbar \vec{k}_1 \mp \hbar \vec{k} \phi, \; _{1 \rightarrow 2} \qquad \begin{cases} - \text{ emission} \\ + \text{ absorption} \end{cases} \qquad (2.5)$$

$$W_2 = W_1 \mp W\phi$$

are fulfilled. Here $\hbar \vec{k}_\phi, \; _{1 \rightarrow 2}$ denotes the momentum of the intervalley phonon. Since valley 1 is positioned at the centre and the satellite valleys 2 at the edges of the reduced zone, the intervalley phonons involved lie at the extreme right of the phonon dispersion diagram (Fig. 2.3).

The probability of a transfer back into the low-mass central valley is rather low because of the high relative density of states in the satellite valleys ($N_2^\star \approx 46 \, N_1^\star$; see Section 2.1). For a sufficiently high electric field the satellite valleys are therefore, in the steady-state case, much more populated than the central valley. In the satellite valleys the electron mobility is limited mainly by equivalent intervalley scattering (between the various $\langle 100 \rangle$ valleys, involving both optical and acoustical phonons) and by polar intravalley scattering.

The effectiveness of the various scattering mechanisms for the momentum relaxation process can also be judged by an inspection of their relevant relaxation times. Solving the Boltzmann eqn (2.3) is sometimes, over certain parameter ranges, facilitated if the (right-hand) collision terms can be expressed by a scattering relaxation time τ_{sc}:

$$\left(\frac{\partial F(\vec{k})}{\partial t} \right)_i = - \frac{F(\vec{k}) - F_0(\vec{k})}{\tau_{sc,i}}. \qquad (2.6)$$

The quantity $\tau_{sc, i}$ is a measure of how rapidly after a disturbance the distribution function is restored by the scattering processes i to its equilibrium value $F_0(\vec{k})$. Accordingly, the scattering is the more effective the smaller $\tau_{sc, i}$ is. The scattering times calculated for the central-valley electrons in GaAs are shown in Fig. 2.6 as a function of the instantaneous electron energy for $T_L = 300$ K. As we have already seen in Fig. 2.4, polar intravalley scattering dominates at low energies (fields), with the intervalley scattering setting in at about $0 \cdot 38$ eV, i.e. at the energy value at which the electron transfer to the $\langle 100 \rangle$ satellite valleys becomes possible. Acoustic-phonon intravalley scattering remains negligible throughout. The *total* scattering relaxation time τ_{sc} leads to the momentum relaxation time τ_p of eqn (2.2) by averaging it over the electron distribution in momentum space: $\tau_p = \langle \tau_{sc} \rangle$, where $\tau_{sc}^{-1} = \sum_i \tau_{sc,i}^{-1}$.

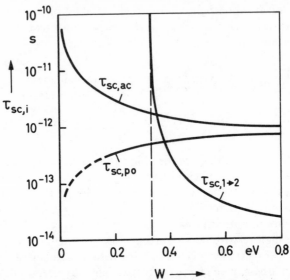

Fig. 2.6 Calculated dependence on electron energy W of the scattering relaxation times for acoustical ($\tau_{sc,\ ac}$) and for polar ($\tau_{sc,\ po}$) intravalley scattering in the central valley, as well as for intervalley scattering from the central to the $\langle 100 \rangle$ satellite valleys ($\tau_{sc,\ 1 \rightarrow 2}$; intervalley-scattering deformation potential, Ξ, assumed to be 5×10^8 eV cm^{-1}) in GaAs. After Conwell and Vassell [2.22].

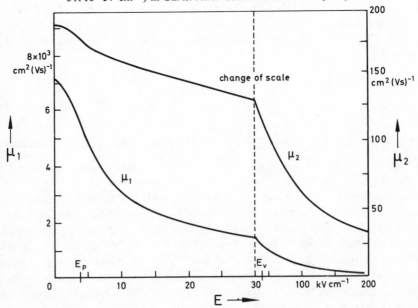

Fig. 2.7 Calculated central- and satellite-valley mobilities, μ_1 and μ_2, of electrons in GaAs as a function of electric field E at $T_L = 300$ K, after Heinle [2.30].

2.3 THE DRIFT-VELOCITY/FIELD CHARACTERISTIC

As we have seen in the preceding section, the electron transfer between the conduction-band valleys leads to a fairly rapid increase in the electron concentration
(n_2) of the high-mass satellite valleys, with a corresponding reduction of the low-mass
central-valley concentration (n_1). Figure 2.8 shows the field dependence of the

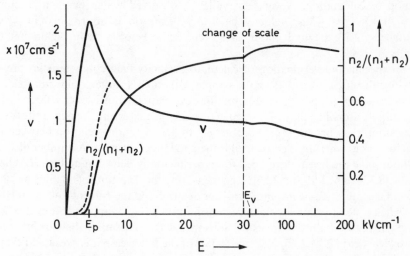

Fig. 2.8 Calculated average electron drift velocity v, and relative electron population
$n_2/(n_1 + n_2)$ in the $\langle 100 \rangle$ satellite valleys, as a function of electric field E for n-type
GaAs (full curve), after Heinle [2.30]. Relative electron population obtained by Hall
measurements (dashed curve), after Acket [2.50 and 2.51]. $T_L = 300$ K.

relative population $n_2/(n_1 + n_2) = n_2/n$ as calculated by the displaced-Maxwellian
approach including the effect of the central-valley non-parabolicity (full curve [2.30]).
Also shown (dashed curves [2.30]) is an experimental result for the relative electron
population obtained by Hall measurements up to 7 kV cm^{-1} using microwave
heating [2.50 and 2.51]. (In the experiment the assumption was made that the
electron drift mobility in the central valley equals their Hall mobility, which may be
invalid at higher fields and thus partially account for the discrepancy with the
theoretical result. See also [2.54]). The rapid growth of n_2, in conjunction with the
fact that the electron mobilities in the valleys 2 and 1, respectively, are such that
$\mu_2 \ll \mu_1$ (Section 2.1), then leads to the crucial result that beyond a certain field
value the average electron drift velocity

$$v = \mu E = \left(\frac{n_1}{n} \mu_1 + \frac{n_2}{n} \mu_2 \right) E \tag{2.7}$$

falls with increasing field, i.e. negative differential conductivity (n.d.c.) is obtained.
The field dependence of the two mobilities μ_1 and μ_2 again as obtained in the

displaced-Maxwellian approximation [2.30], is shown in Fig. 2.7. Thus the velocity-field characteristic of the steady-state case can be plotted according to eqn (2.7), see Fig. 2.8. The curve exhibits a pronounced negative slope above a peak field $E_p = 3.95$ kV cm^{-1}, with a maximum negative differential mobility of $\mu' = -2\,000$ cm^2 (Vs)$^{-1}$. The mean value of the low-field mobility is $\mu_1 \approx 6\,300$ cm^2 (Vs)$^{-1}$ and thus is lower than the one quoted in Section 2.1 due to the different method of calculation. A very flat valley is observed which has a minimum at $E_v = 39$ kV cm^{-1}. The velocity peak-to-valley ratio is approximately $v_p/v_v = 2.08/0.98 \approx 2.13$. Beyond about 50 kV cm^{-1} the velocity drops again but only slightly.

An experimental verification of calculated velocity-field characteristics presents certain difficulties. In a real GaAs sample with such a large equilibrium electron concentration n_0 (doping*) as required for device applications space-charge instabilities (dipole domains) may occur which obscure the picture by the non-uniform field distribution they produce; cf. Sections 3.1 to 3.3. A unique relation between drift velocity (or current) and electric field is then no longer possible. A number of special methods have thus been developed for experimentally determining the $v(E)$ characteristic [2.33–2.44, 2.55 and 2.56]: e.g., measuring the drift time of an injected carrier pulse in a sample of low doping [2.36]; or determining the current-voltage relationship when a steady-state dipole domain is in transit ([2.55] and [2.56]; evaluation is based on the domain dynamics of Section 3.4.4); or heating the carriers with a microwave field [2.38–2.43, 2.57 and 2.58]. In the latter case it is presumed that the field variations are so fast that the time is too short for instabilities to build up. The curves obtained in the various experiments, however, are not in complete agreement and neither are the curves computed from the different theoretical approaches. This is shown in Fig. 2.9 by three measured $v(E)$ characteristics (curves 1 to 3) together with some of the computed ones (curve 4: modified quasi-elastic approximation; curve 5: Monte-Carlo calculation; curve 6: displaced-Maxwellian calculation, same as in Fig. 2.8. The last two show the best agreement with the experimental curves).

Details of the characteristic are still open to debate (see also [2.13] and [2.45–2.47]). In particular the exact position of the peak and valley fields do not yet appear to be definitely known. Furthermore, it has still to be ascertained whether the drift velocity at high fields, up to values at which impact ionization effects set in, remains nearly constant, increases, or even drops again as suggested by the computation leading to Fig. 2.8. Experimental evidence indicates that at fields of at least $E \approx 130$ kV cm^{-1} the drift velocity has not yet risen again to its value at $E \approx E_p/2$ [2.48]. To facilitate then a really exact determination of the characteristic, a better knowledge of the fundamental parameters involved, particularly of the scattering deformation-potential constants, must be awaited. Also, the effect of the

---

*We assume that in our extrinsic semiconductor the donor sites are shallow with energies only just below the conduction band. In this case the donors are ionized completely at room temperature, and the equilibrium electron concentration n_0 can be considered to equal the doping concentration N_d.

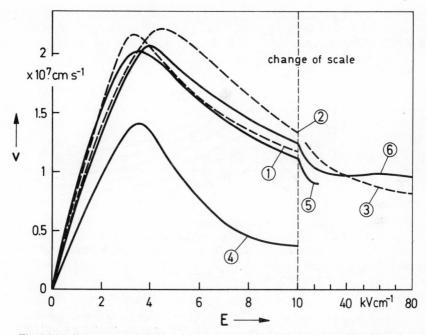

Fig. 2.9 Various measured (dashed curves) and calculated (full curves) drift-velocity v against electric-field E characteristics for n-type GaAs at $T_L = 300$ K. Curve 1: after Ruch and Kino [2.36a]; 2: after Acket *et al.* [2.40]; 3: after Bastida *et al.* [2.55]; 4: after Heinle [2.24] (for intervalley scattering: $\Xi = 5 \times 10^8$ eV cm^{-1}); 5: after Fawcett *et al.* [2.31] ($\Xi = 1 \times 10^9$ eV cm^{-1}); 6: after Heinle [2.30] ($\Xi = 5 \cdot 3 \times 10^8$ eV cm^{-1}).

$\langle 111 \rangle$ satellite valleys, (cf. small-print note at end of this chapter) which exist in GaAs at the edge of the reduced zone and probably lie at a slighly higher energy than the $\langle 100 \rangle$ valleys [2.49 and 2.22], has so far not been incorporated. Finally, an exact calculation will have to take account of the fact that the constant-energy surface of the $\langle 100 \rangle$ valleys are ellipsoidal as mentioned in Section 2.1.

For device application the *temperature* dependence of the velocity-field characteristic is of special interest (see Sections 5.2.2 and 6.3). As an example of a theoretical evaluation Fig. 2.10 shows the results obtained from the Monte-Carlo calculation [2.59] which served as the basis for curve 5 of Fig. 2.9 at 300 K. One observes a general drop of the drift velocity with increasing temperature. Whereas the velocity-peak field E_p increases slightly from $3 \cdot 1$ kV cm^{-1} at 77 K to $3 \cdot 7$ kV cm^{-1} at 500 K, the maximum negative differential mobility decreases from 4 200 to 1 000 cm^2 (Vs)$^{-1}$ over the same temperature range.

As a consequence of the scattering processes the electrons are subjected to diffusion which may play an important role when considering the growth of space charge in an

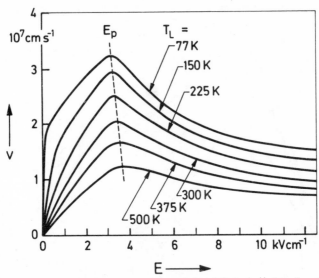

Fig. 2.10 Calculated temperature dependence of the $v(E)$ characteristic for electrons in GaAs. After Ruch and Fawcett [2.59].

n.d.c. medium (see Chapters 4 and 5). A diffusion constant D of the electrons can be defined that is directly related to their mobility μ according to the Einstein equation

$$D = \frac{k_B T_{ev}}{e}\, \mu_v. \tag{2.8}$$

Here, $v = 1,2$ refers to the central and satellite valleys. Similar to the average drift velocity v of eqn (2.7) an average diffusion constant follows from

$$D = \frac{n_1}{n} D_1 + \frac{n_2}{n} D_2. \tag{2.9}$$

The field dependence of D at 300 K is shown in Fig. 2.10: curve 1 was determined experimentally [2.36b], whereas curve 2 is based on a Monte-Carlo calculation [2.53] and curve 3 on the displaced-Maxwellian calculation already made use of above. Curve 3 is simply obtained from Figs 2.5, 2.7 and 2.8 by using the relationships (2.8) and (2.9). As can be seen, the discrepancy between experiment and theory is much more pronounced than in the case of the $v(E)$ characteristic. [Partly due to neglect of 'transfer diffusion' ('straggling diffusion', see Sec. 5.3.1) caused by randomness of electron transfer between the different valleys; see Hammer, C. and Vinter, B. *Electron. Lett.* **9** (1973), p. 9.]

To summarize the processes described in this chapter, a semiconductor material in general must fulfil the following requirements in order to be suitable for the electron-transfer mechanism and, hence, to show an n.d.c. region:

1. The conduction band must have a many-valley structure with the curvature of a valley 1 exceeding that of a valley 2 ($m_2^* \gg m_1^*$) where
2. the less-curved valley 2 must be positioned higher in energy by a few kT_L

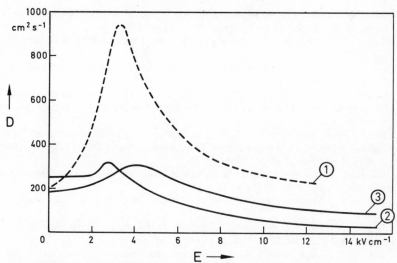

Fig. 2.11 Measured (dashed curve) and calculated (full curves) average electron diffusion coefficient D as a function of electric field E in GaAs. Curve 1: after Ruch and Kino [2.36b]; 2: after Fawcett and Rees [2.53]; 3: after Heinle [2.30].

than valley 1 in order to be essentially depopulated from electrons in the field-free case ($\Delta W \gtrsim 4k_B T_L$).

3. Additionally, to prevent impact ionization across the band gap setting in before the electron transfer commences, the ionization energy W_i must exceed the separation of the valley minima ($\Delta W < W_i$; $W_i \approx 1{\cdot}25 W_G$ for GaAs, [2.52]).

A possible, more complex electron-transfer mechanism involving an additional set of satellite valleys, is briefly described at the end of Section 5.2.1 ('three-level' mechanism).

REFERENCES

2.1   Spenke, E., *Electronic Semiconductors* (New York, McGraw-Hill, 1958). 2nd German Edn (Berlin/Heidelberg/New York, Springer, 1965).

2.2   McKelvey, J. P., *Solid State and Semiconductor Physics* (New York and Tokyo, Harper and Row/Weatherhill, 1966).

2.3   Conwell, E. M., *High Field Transport in Semiconductors* (New York and London, Academic Press, 1967).

2.4   Brillouin, L., *Wave Propagation in Periodic Structures*. Electric filters and crystal lattices (New York, McGraw-Hill, Dover Publ., 1953).

2.5a   Ehrenreich, H., Band structure and electron transport of GaAs. *Phys. Rev.* **120** (1960), p. 1951.

2.5b   Pitt, G. D. and Lees, J., Electrical properties of the GaAs X_{1C} minima at low electric fields from high-pressure experiment. *Phys. Rev.* **B2** (1970), p. 4144. Also: The X_{1C} conduction band minimum in high-purity epitaxial n-type GaAs. *Solid-State Commun.* **8** (1970), p. 491.

2.5c   Vrehen, Q. H. F., Cyclotron resonance and Hall experiments with high-purity epitaxial GaAs. *J. Phys. Chem. Solids* **29** (1968), p. 129.

2.5d   Chamberlain, J. M. and Stradling, R. A., Interband magneto-optical absorption in gallium arsenide. *Solid-State Commun.* **7** (1969), p. 1275.

2.6   Sturge, M. D., Optical absorption of gallium arsenide between 0·6 and 2·75 eV. *Phys. Rev.* **127** (1967), p. 768.

2.7   Gray, A. M., Evaluation of electronic energy band structures of GaAs and GaP. *Phys. Stat. Sol.* **37** (1970), p. 11.

2.8a   Cohen, M. L. and Bergstresser, T. K., Band structures and pseudopotential form factors for fourteen semiconductors of the diamond and zincblende structures. *Phys. Rev.* **141** (1966), p. 789.

2.8b   Pollak, F. H., Higginbotham, C. W. and Cardona, M., Band structures of GaAs, GaP, InP and AlSb: The k·p method. *Proc. Int. Conference Phys. Semiconductors, Kyoto* 1966; *J. Phys. Soc. Japan* **21** (1966), Suppl. p. 20.

2.9   Weisberg, L. R., Woolston, J. R. and Glicksman, M., Electron mobilities in gallium arsenide. *J. Appl. Phys.* **29** (1958), p. 1514.

2.10   Hilsum, C., Polar scattering in III–V compounds. *Proc. Phys. Soc. (London)* **76** (1960), p. 414.

2.11   Heinle, W., Modified variational calculation of the zero-field mobility of polar semi-conductors. *Z. Physik* **217** (1968), p. 150.

2.12   Butcher, P. N. and Fawcett, W., Calculation of the velocity-field characteristic for gallium arsenide. *Phys. Lett.* **21** (1966), p. 489.

2.13   Conwell, E. M., Field dependence of mobility in the (100) conduction band minima of GaAs. *Phys. Lett.* **21** (1966), p. 368.

2.14   Kang, C. S. and Greene, P. E., Preparation and properties of high-purity epitaxial GaAs grown from Ga solution. *Appl. Phys. Lett.* **11** (1967), p. 171.

2.15   Hutson, A. R., Jayaraman, A. and Coriell, A. S., Effects of high pressure, uniaxial stress, and temperature on the electrical resistivity of n-GaAs. *Phys. Rev.* **155** (1967), p. 786.

2.16   Lees, J., Wasse, M. P. and King, G., Interband carrier transfer in GaAs at high pressures. *Solid-State Commun.* **5** (1967), p. 521.

2.17   Ziman, J. M., *Electrons and Phonons* (Oxford, Clarendon Press, 1960).

2.18   Waugh, J. L. T. and Dolling, G., Crystal dynamics of gallium arsenide. *Phys. Rev.* **132** (1963), p. 2410.

2.19   Heinle, W., Influence of ionized impurity scattering on the Gunn-effect characteristic in GaAs in the displaced Maxwellian approximation. *Phys. Lett.* **29A** (1969), p. 131.

2.20   Conwell, E. M. and Vassell, M. O., High-field distribution function in GaAs. *IEEE Trans. Electron Dev.* **ED–13** (1966), p. 22.

2.21   Conwell, E. M. and Vassell, M. O., Effect of nonparabolicity on drift velocity in GaAs. *Phys. Lett.* **25A** (1967), p. 302.

2.22   Conwell, E. M. and Vassell, M. O., High-field transport in n-type GaAs. *Phys. Rev.* **166** (1968), p. 797.

2.23   Heinle, W., Calculation of the Gunn threshold in GaAs. *Phys. Lett.* **27A** (1968), p. 628.

2.24   Heinle, W., Calculation of hot electron distribution functions and of the threshold field in the Gunn effect. *Z. Physik* **222** (1969), p. 301.

2.25   Butcher, P. N. and Fawcett, W., Intervalley transfer of hot electrons in gallium arsenide. *Phys. Lett.* **27** (1965), p. 216.

2.26   Butcher, P. N. and Fawcett, W., The intervalley transfer mechanism of negative resistivity in bulk semiconductors. *Proc. Phys. Soc. (London)* **86** (1965), p. 1205.

2.27   Butcher, P. N., The Gunn effect. *Rep. Prog. Phys.* **30** (1967), pt. 1, p. 97.

2.28   Boardman, A. D., Fawcett, W. and Rees, H. D., Monte Carlo calculation of the velocity-field relationship for gallium arsenide. *Solid-State Commun.* **6** (1968), p. 305.

2.29   Heinle, W., Influence of nonparabolicity on the Gunn-effect characteristic in GaAs in the displaced Maxwellian approximation. *Phys. Lett.* **27A** (1968), p. 629.

2.30 Heinle, W., Displaced Maxwellian calculation of transport in n-type GaAs. *Phys. Rev.* **178** (1969), p. 1319.

2.31 Fawcett, W., Boardman, A. D. and Swain, S., Monte Carlo determination of electron transport properties in gallium arsenide. *J. Phys. Chem. Solids* **31** (1970), p. 1963.

2.32a Rees, H. D., Calculation of steady state distribution functions by exploiting stability. *Phys. Lett.* **26A** (1968), p. 416.

2.32b Rees, H. D., Calculation of distribution functions by exploiting the stability of the steady state. *J. Phys. Chem. Solids* **30** (1969), p. 643.

2.33 Gunn, J. B. and Elliott, B. J., Measurement of the negative differential mobility of electrons in GaAs. *Phys. Lett.* **22** (1966), p. 369.

2.34 Chang, D. M. and Moll, J. L., Direct observation of the drift velocity as a function of the electric field in gallium arsenide. *Appl. Phys. Lett.* **9** (1966), p. 283.

2.35 Thim, H. W., Potential distribution and field dependence of electron velocity in bulk GaAs measured with a point-contact probe. *Electron. Lett.* **2** (1966), p. 403.

2.36a Ruch, J. G. and Kino, G. S., Measurement of the velocity-field characteristic of gallium arsenide. *Appl. Phys. Lett.* **10** (1967), p. 40.

2.36b Ruch, J. G. and Kino, G. S., Transport properties of GaAs. *Phys. Rev.* **174** (1968), p. 921.

2.37 Fay, B. and Kino, G. S., A new measurement of the velocity-field characteristic of GaAs. *Appl. Phys. Lett.* **15** (1969), p. 337.

2.38 Acket, G. A., Determination of the negative differential mobility of n-type gallium arsenide using 8 mm microwaves. *Phys. Lett.* **27A** (1967), p. 200.

2.39 Acket, G. A. and De Groot, J., Measurements of the current-field strength characteristic of n-type gallium arsenide using various high-power microwave techniques. *IEEE Trans. Electron Dev.* **ED–14** (1967), p. 505.

2.40 Acket, G. A., 'T Lam, H. and Heinle, W., The low temperature velocity-field characteristics n-type GaAs. *Jap. J. Appl. Phys.* **7** (1968), p. 1084.

2.41a Braslau, N., Velocity-field characteristic of gallium arsenide from measurement of the conductivity in a microwave field. *Phys. Lett.* **24A** (1967), p. 531.

2.41b Braslau, N. and Hauge, P. S., Microwave measurement of the velocity-field characteristic of GaAs. *IEEE Trans. Electron Dev.* **ED–17** (1970), p. 616.

2.42 Hamaguchi, C., Kono, T. and Inuishi, Y., Microwave measurement of differential negative conductivity due to intervalley transfer of hot electrons in n-type GaAs. *Phys. Lett.* **24A** (1967), p. 500.

2.43 Cohen, L. D., A microwave evaluation of velocity-field characteristic in different regions of individual epitaxial gallium-arsenide layers. *Proc. IEEE* **57** (1969), p. 1299.

2.44 Yamashita, A., Space-charge-limited currents and the velocity-field characteristic in n-type GaAs. *Jap. J. Appl. Phys.* **7** (1968), p. 1084.

2.45 Hilsum, C., Field dependence of mobility in the (100) conduction band minima of GaAs. *Phys. Lett.* **20** (1966), p. 136.

2.46 Hilsum, C., Comments on the field dependence of mobility in the (100) conduction band minima of GaAs. *Phys. Lett.* **21** (1966), p. 605.

2.47 Conwell, E. M. and Vassell, M. O., Variation of drift velocity with field in GaAs. *Appl. Phys. Lett.* **9** (1966), p. 411.

2.48 Gunn, J. B., On the shape of travelling domains in gallium arsenide. *IEEE Trans. Electron Dev.* **ED–14** (1967), p. 720.

2.49 Hilsum, C., Band structures, effective charge and scattering mechanisms in III–V compounds. In: Hulin, M. (Ed.), *Physics of Semiconductors*, p. 1127, (New York, Academic Press, 1964).

2.50 Acket, G. A., Determination of the Hall mobility of hot electrons in gallium arsenide using 8 mm microwaves. *Phys. Lett.* **25A** (1967), p. 374.

2.51   Acket, G. A., Microwave Hall mobility of hot electrons in gallium arsenide. *Philips Res. Repts* **22** (1967), p. 541.

2.52   Chynoweth, A. G., Charge multiplication phenomena. In: Willardson, P. K. and Beer, A. C. (Eds), *Semiconductors and Semimetals*, vol. 4, p. 263, (New York & London, Academic Press, 1968).

2.53   Fawcett, W. and Rees, H. D., Calculation of the hot electron diffusion rate for GaAs. *Phys. Lett.* **29A** (1969), p. 578.

2.54   Kar, R. K. and Mukherjee, M. N., Plasma frequency and intervalley population transfer in n-GaAs. *Phys. Lett.* **30A** (1969), p. 355.

2.55   Bastida, E. M., Fabri, G., Svelto, V. and Vaghi, F., Indirect electron drift velocity versus electric field measurement in GaAs. *Appl. Phys. Lett.* **18** (1971), p. 28.

2.56a Pokorny, J. and Jelinek, F., Measurement of the velocity/field characteristic of GaAs sample in dipole-mode operation. *Electron. Lett.* **7** (1971), p. 528.

2.56b Pokorny, J. and Jelinek, F., Experimental nonsaturating velocity-field characteristic of GaAs. *Proc. IEEE* **60** (1972), p. 457.

2.57   Glover, G. H., Error in microwave measurements of the velocity-field characteristic of n-type GaAs due to energy relaxation effects. *Appl. Phys. Lett.* **18** (1971), p. 290.

2.58   Glover, G. H., Microwave measurement of the velocity-electric-field characteristic of inhomogeneous GaAs with conducting substrate. *J. Appl. Phys.* **42** (1971), p. 4025.

2.59   Ruch, J. G. and Fawcett, W., Temperature dependence of the transport properties of gallium arsenide determined by a Monte Carlo method. *J. Appl. Phys.* **41** (1970), p. 3843.

3

Space-Charge Dipole Domains (Gunn Effect in Strict Sense)

3.1 INTRODUCTION

In the general case, a finite-sized sample of a semiconducting material possessing a velocity-field characteristic like that shown in Fig. 2.8 is thermodynamically unstable when biased in the negatively-sloped part of its characteristic [1.10, 1.11 and 3.1]. (Stability criteria are investigated in Sections 4.4.1.2 and 4.4.2.2.) Such a system makes an attempt to establish a steady state characterized by minimum entropy production, which here corresponds to the generation of a minimum of Joule heat. This energetically more favourable state is reached if, instead of having a homogeneous field distribution over the sample length, the field splits up into a high-field region and a low-field region. An example of such a situation is schematically indicated in Fig. 3.1 for an n-type semiconducter. In the high-field region ('domain') of this case the field tends to a value which, in general, far exceeds the peak field E_p of the $v(E)$ characteristic (Fig. 2.8), whereas the outside field drops below it, such that, on reaching the steady state, a reduced current density is obtained. For a device with constant terminal voltage $V = V_0$ this naturally means a reduced power dissipation. A possible steady-state situation is qualitatively illustrated in Fig. 3.2b, by the high and low fields $E_{2\infty} \gg E_p$ and $E_{1\infty} < E_p$, respectively, and a reduced drift velocity $v(E_{1\infty}) \equiv v_{1\infty}$ which is equal to the domain drift velocity v_D (detailed treatment in Section 3.4.4.2).

The fields $E_{1\infty}$ and $E_{2\infty}$ denote steady-state values, i.e. they will be reached at the time $t \to \infty$. Furthermore, in a real semiconductor the outside field approaches $E_{1\infty}$ in space only asymptotically, owing to diffusions of the mobile charge carriers.

Poisson's equation demands that at the interface of the regions with differing field strengths there must exist space-charge layers which constitute a drain or a source for field lines (see Figs 3.1b and 3.7a: space-charge dipole domain). The formation of a domain by field break-up is thus only possible if a sufficient departure from electrical charge neutrality can occur in the sample. Since the space-charge layers are made up of mobile electrons which, in GaAs, drift with a velocity of the order of 10^7 cm s^{-1} (see Fig. 2.8), a domain can develop fully during the carrier transit time $\tau = L/v$

31

(with L denoting the sample length) only if the time needed for the charge rearrangement is $|\tau_R| \leqslant \tau$ (see Section 3.2).

When domain formation and the associated departure from field homogeneity become severe, the negative conductance of the static $v(E)$ characteristic, which we wish to use, may not be obtainable externally at the terminals of the device. However, as we will see, the formation, movement and extinction of dipole domains is a repetitive process giving rise to current relaxation oscillations which have proved to be very useful by themselves (Gunn dipole oscillations; [1.18, 1.19]). Thus, in the application of transferred-electron devices two principal cases have to be distinguished: either (a) the development of any substantial space charge and field inhomogeneity is prevented by special means so that negative conductivity is indeed directly available for amplifier or oscillator purposes—the case covered in Chapter 4; or (b) use is made intentionally of the space-charge instabilities with their oscillatory behaviour. The latter is the subject of this chapter which starts with a discussion on departure from electrical neutrality and the mechanisms of space-charge layer nucleation.

3.2 DEPARTURE FROM ELECTRICAL NEUTRALITY

We consider a uniformly-doped n-type semiconductor of positive conductivity σ at thermal equilibrium. Owing to a small local perturbation of the majority-carrier concentration from its constant equilibrium value n_0, the neutrality provided by the counter-balancing opposite charges of the uncompensated ionized impurity atoms shall have been offset. Starting from the one-dimensional continuity equation for the charge carriers

$$\frac{\partial \rho}{\partial t} + \frac{\partial J_c}{\partial z} = 0 \qquad (3.1a)$$

we ask for the response of the locally-created space-charge density $\rho = e(n - n_0)$. Differentiating the expression for the carrier current density

$$J_c = J - D \frac{\partial \rho}{\partial z} = \sigma E - D \frac{\partial \rho}{\partial z}$$

with respect to z and inserting Poisson's equation

$$\frac{\partial E}{\partial z} = \frac{\rho}{\varepsilon} \qquad (3.2a)$$

yields

$$\frac{\partial J_c}{\partial z} = \frac{\sigma}{\varepsilon} \rho + D \frac{\partial^2 \rho}{\partial z^2}$$

in the limit of small electric fields $E \to 0$. Making use of this result, the continuity eqn (3.1a) becomes

$$\frac{\partial \rho}{\partial t} + D \frac{\partial^2 \rho}{\partial z^2} + \frac{\sigma}{\varepsilon} \rho = 0 \qquad (3.1b)$$

with ε being the permittivity. Restricting ourselves first to the *temporal* response, eqn (3.1b) has the solution

$$\rho(t) = \rho(t = 0) \cdot \exp\left(-t/\tau_R\right)$$

where the 'dielectric relaxation time'

$$\tau_R = \varepsilon/\sigma = \varepsilon/en\mu \approx \varepsilon/en_0\mu \tag{3.3a}$$

represents the *time constant* for the decay of the space charge to neutrality. For practical semiconductors τ_R is of the order of 10^{-12}s, i.e. it appears that a departure from electrical neutrality is only possible over extremely short time intervals.

Turning now to the *spatial* response, we find as the solution of eqn (3.1b)

$$\rho(z) = \rho(z = 0) \cdot \exp(-z/L_{Db})$$

where

$$L_{Db} = (D\tau_R)^{1/2} = \left\{\frac{k_B T_e \varepsilon}{e^2 N_d}\right\}^{1/2}. \tag{3.3b}$$

The parameter L_{Db}, which involves the diffusion coefficient D, is called Debye length and is seen to determine the *distance* over which a small unbalanced charge decays. The second expression for L_{Db} in eqn (3.3b) results since for the uncompensated ionized donor density we can assume $N_d = n_0$ and use is made of Einstein's eqn (2.8). For GaAs with a low-field conductivity of 1 $(\Omega \text{ cm})^{-1}$ one has $L_{Db} \approx 0.15\ \mu\text{m}$.

If the semiconductor exhibits a *negative* (differential) conductivity σ', any charge imbalance will *grow* with a time constant equal to $|\tau_R|$ instead of decay. For example, in TE GaAs biased in the negative-mobility region of the $v(E)$ characteristic space-charge growth takes place with the time constant $|\tau_R| \approx 1.3$ ps if the typical parameters $\mu' = -2\,000\ \text{cm}^2\ (\text{Vs})^{-1}$, $n_0 = 5 \times 10^{15}\ \text{cm}^{-3}$ and a relative permittivity of $\varepsilon/\varepsilon_0 = 12.5$ are assumed. The modulus of τ_R is a measure of charge growth during the initial 'small-signal' phase (cf. Section 4.3). The build-up of the charge eventually leads to the 'large-signal' steady domain state, characterized by minimum entropy production.

3.3 NUCLEATION AND INITIAL GROWTH OF SPACE-CHARGE LAYERS

The only moving domains that have, during experiments, so far been definitely identified in a homogeneous negative-mobility semiconductor sample with $\tau/|\tau_R|$ sufficiently large for mature domains to develop, and with an apparently linear low-resistivity (ohmic) contact at each end, are of the aforementioned space-charge dipole type (Fig. 3.1). This is not necessarily to be expected as the following discussion will show.

3.3.1 Idealized Sample

We first consider a sample as schematically shown in Fig. 3.2a where, however, we

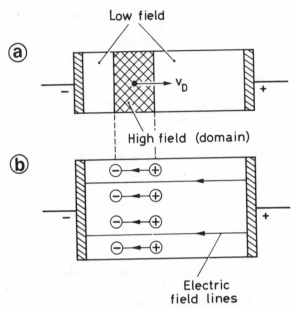

Fig. 3.1 Schematic representation of an n-type semiconductor sample with domain formation, where the domain is of the high-field type (space-charge dipole domain) which drifts with the velocity v_D: (a) showing the two field regions, (b) indicating electric field lines (note: the electric field E as defined in the text points into the opposite direction because of the assumed positive charge of the electron).

neglect any possible influence which the boundary parallel to the electric field might exert on the conduction process (one-dimensional problem). The sample is supposed to consist of an extrinsic n-type semiconductor with negative-mobility properties and with a band gap sufficiently large that minority carriers are of no concern. Metal contacts are attached to the semiconductor body at both ends. Without an externally applied voltage, the electron concentration n well within the semiconductor must be equal to $N_d = n_0$ (neutrality condition) as indicated in the carrier profile of Fig. 3.2a (full curve). Near the metal-semiconductor boundaries, however, the concentration is determined rather by the requirement of thermal equilibrium with the metal. Depending on the relative values of the work functions* for the semi-conductor and the metal, either a static electron accumulation layer or a depletion layer is formed in the semiconductor border zone (besides the rather improbable case of the work functions having just such values that $n(z)$ remains constant at n_0). In the former case, which is assumed to exist here, we have the desired ohmic contact, whereas in the latter case a non-linear Schottky-barrier contact results.

---

*The work function is a characteristic property of the material describing the potential barrier that an electron must overcome to be emitted.

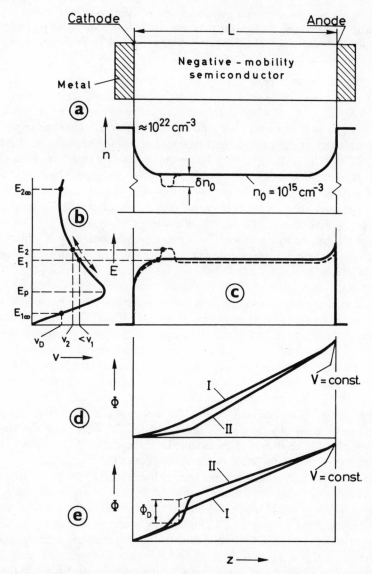

Fig. 3.2 Illustration of domain formation in an n-type TE semiconductor with ideal ohmic contacts. (a) Schematic representation of sample and spatial distribution of electron concentration with assumed notch in net-doping level (δn_0) near cathode, (b) velocity-field characteristic, (c) spatial field distribution $E(z)$, (d) spatial potential distribution $\Phi(z)$ for the case without net-doping notch indicating the formation of an accumulation-layer domain, (e) formation of space-charge dipole domain if net-doping notch is present.

35

Usually, with GaAs an ohmic contact has to be obtained technologically from a Schottky barrier by an alloying process which increases the net doping concentration near the contact and hence changes the semiconductor work function in the appropriate way. Achieving ohmic contacts on the relatively pure GaAs required for TE devices is, however, not a straight-forward task (see Section 6.2.2).

Firstly, we consider a perfectly homogeneous doping throughout the semi-conductor (full profile curve in Fig. 3.2a) and apply a fixed space-averaged field $E_h = \text{const} = E_{h0} = V_0/L$ which lies in the negative-mobility range of the $v(E)$ characteristic [E_h = homogeneous component of the field $E(z)$ in the sample]. The accumulation layer at the cathode end which is associated with the field increase sketched in Fig. 3.2c starts to grow, because of the negative sign of τ_R [eqn (3.3a)], and moves towards the anode since it consists of mobile electrons. The corresponding spatial potential distributions $\Phi(z)$ for two subsequent instants of time are shown in Fig. 3.2d. The nucleation and movement of such a pure accumulation domain was first analysed by Kroemer [3.2] and by McCumber and Chynoweth [3.4]. However, as already mentioned, only domains of the *dipole* type, i.e. consisting of an accumulation layer *plus* a depletion layer (deficiency of electrons as compared to neutral distribution), have been definitely identified*.

To explain the formation of *dipole* domains at an ohmic cathode contact of our hypothetical one-dimensional sample, a secondary, *depletion*-type nucleation centre must be introduced somewhat removed from the primary centre, i.e. from the contact accumulation layer. In Fig. 3.2a this secondary nucleation centre is indicated by the zone of reduced net doping (doping notch), by an amount σn_0 (not to scale).

The process of *dipole* nucleation will now be illustrated with the help of Figs 3.2b, c and e. Here it will again be assumed that the sample is biased in the negative-mobility region of the $v(E)$ characteristic (Fig. 3.2b) by a constant-voltage source $V_B = V_0$. Before any charge rearrangement begins, the electric-field distribution along the sample will be as shown by the dashed curve of Fig. 3.2c, which includes an initial zone of higher field E_2 according to the assumed $n_0(z)$ dependence. The corresponding spatial potential distribution $\Phi(z)$ is indicated by curve I in Fig. 3.2e.

Since the field inhomogeneity (i.e. the initial charge dipole) lies in the negatively-sloped part of the characteristic, it must grow. At the field maximum E_2 the electron drift velocity is lower than outside the inhomogeneity where the field has approxi-mately the value $E_1 (v_1 > v_2$ in Fig. 3.2b). This difference in velocity has the result that at the cathode side of the field maximum further negative space charge accumu-lates, whereas at the anode side positive space charge (electron depletion) grows. This formation and increase of space charge in the two zones adjacent to the high-field disturbance requires an increased field difference, i.e. a growth of E_2, which has to be associated with a decrease in E_1 (since $V = \text{const}$). Hence the velocity difference increases further and thus a space-charge dipole is rapidly formed. Since the domain

---

*Experimental observation of pure accumulation-mode operation has been claimed for bulk n-type Ge samples at 77 K and biased along $\langle 100 \rangle$ crystal directions [3.72]. For TE behaviour of Ge under special environmental conditions see Section 6.1.1.

consists of mobile electrons, it will move from the nucleation centre and drift towards the anode as it grows. Curve II in Fig. 3.2e shows the spatial potential $\Phi(z)$ when the dipole domain has already grown somewhat and begun to travel away.

In computer simulations of the electrical behaviour of n.d.c. semiconductors with uniform doping and ideal ohmic contacts, a doping notch like that in Fig. 3.2a must be included if dipole-layer instabilities are to develop (cf., e.g. [3.4]). Dipole-layer domains are also predicted by computer calculations if, instead of including the notch, the realistic assumption of random spatial doping fluctuations along the semiconductor is made [3.2].

3.3.2 Practical Samples

In practice, dipole-domain nucleation directly at the cathode is generally observed if a metallic cathode is used. It has been proposed [3.3a] that the alloying process* necessary for forming metallic contacts could result in a damaged layer, introducing across the sample area a non-uniformity of the kind shown in Fig. 3.2a, which would have a particularly pronounced nucleating effect. To be effective for the nucleation process, a non-uniformity must spatially extend over at least a few Debye lengths.

On the other hand, it is found that dipole domains are nucleated even if a non-uniformity (doping deficiency) of only *limited* transverse extension exists or if such a localized non-uniformity near the cathode merely results from boundary effects or from the unavoidable damaged region at the transverse edges of the actual, three-dimensional sample. The domain-formation process in this case operates as follows [3.67].

Consider a sample which has a doping notch σn_0 of limited lateral extension and localized at the sample edge near the cathode as shown in Fig. 3.3. After applying

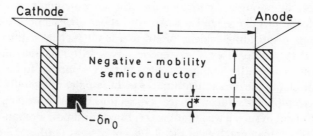

Fig. 3.3 N-type TE semiconductor sample with ohmic contacts and laterally confined net-doping notch near cathode.

a sufficiently high bias field E_{ho} a domain will start to develop from this point since the required combination of electron accumulation (at ohmic contact) and depletion (at the doping notch) are present. A dipole domain initially limited in its lateral extension causes a deformation of the current paths in its immediate surrounding. This is schematically depicted in Fig. 3.4 which shows the resulting squeeze on a

*See Section 6.2.2.

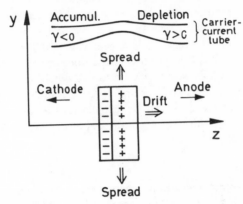

Fig. 3.4 Schematic illustration of lateral dipole-domain spreading as a result of the squeezing of carrier-current tubes. After Yanai *et al.* [3.67].

representative carrier-current tube nearby. If $A^\star(z)$ denotes the cross-sectional area of the tube and we define a parameter

$$\gamma = \frac{1}{A^\star} \frac{\mathrm{d}A^\star}{\mathrm{d}z}$$

the deformation of the tube is such that $\gamma < 0$ towards the cathode and $\gamma > 0$ towards the anode (Fig. 3.4). In regions where $\gamma \neq 0$ the electric field has to change and, hence, space charge appears in spite of a constant net doping n_0. Carriers deplete in the region of $\gamma > 0$ and accumulate where $\gamma < 0$. Consequently a dipole domain has formed in the squeezed carrier tube. In this manner the domain nucleation spreads laterally along equipotentials until the semiconductor surface is reached. Computer calculations [3.67–3.69] and experimental evidence [3.68, 3.70] indicate a transverse domain spreading velocity of 10^8 to a few 10^9 cm s^{-1}, thus suggesting a propagation at the speed of electromagnetic waves. After having spread laterally across the entire sample, the domain grows and travels parallel to the current path in a manner similar to the one-dimensional case. In samples of laterally non-uniform geometry (e.g. co-planar contact device, cf. Section 6.2.1) or doping, the current paths might be bent in such a way that the condition for dipole-domain nucleation is fulfilled even if a doping notch as in Fig. 3.3 is not present [3.67 and 3.68].

In early investigations of space-charge growth from an initial charge fluctuation of finite dimensions [3.56 and 6.7b] the lateral propagation of the fluctuation had been neglected. In this case it turns out that the laterally finite nucleation dipole domain will generally grow much *slower* than suggested by the one-dimensional calculation in which the charge layer necessarily extends to infinity perpendicular to the bias field. This reduction in growth results from the fact that the differential conductivity of TE semiconductors is always *positive* for the (small-signal) electric-field components ΔE_y of the nucleation dipole *perpendicular* to the bias field ($\mu_z' = \mathrm{d}v/\mathrm{d}E$, whereas $\mu_y' = v/E$; [3.57 and 3.58]).

Neglecting the transverse spread of domain nucleation is equivalent to considering only a small segment of the sample of transverse extension $d^\star$ ($d^\star/L \ll 1$, Fig. 3.3) separated from the rest of the semiconductor body. The encountered problem is, then, directly related to that found in the case of the 'transversely thin sample' ($d/L \ll 1$; Section 4.4.2) in which transverse components of the electric field lines originating from the space-charge play an important role. Such samples show a reduction of space-charge growth which may lead to an inhibition of mature dipole domains (Section 4.4.2.2). When, on the other hand, in a practical sample with dimensions $d \approx L$ the domain nucleation has spread transversely right across, the transverse (fringing) domain field can be neglected compared to the longitudinal components, and further domain growth together with the propagation towards the anode does not differ appreciably from the one-dimensional case.

On the other hand, in a simplified investigation of the influence on domain growth of transverse doping gradients, as they are often experienced in practical TE material (Sections 6.1.2 and 6.1.3), it was found that such inhomogeneities tend to speed up the transient process [3.73]. In this case the rate of electric-field change is increased as a result of transverse fields that exist in the domain region.

In samples with highly-doped contact regions characterized by the cathode boundary being removed from the metal contact to the interface between the highly-doped semiconductor region and the main negative-mobility section of lower doping (e.g. n-type epitaxial layer on n$^+$ substrate; see Sections 6.1.2 and 6.2.1, Figs 6.6c and d), space-charge domains often nucleate away from the cathode boundary, somewhere inside the bulk [3.3b]. This seeming inability to nucleate domains at the cathode itself occurs even though the material has, of course, statistical doping fluctuation.

It is interesting to note that, on the other hand, by a rapid switch-on of the bias voltage, multiple domains may initially be formed along the sample over its entire length. In this case, in addition to the cathode boundary minor nucleating centres, formed by the doping fluctuations or possibly by other irregularities, become effective. However, the nucleating effect of the major nucleation centre, i.e. commonly the cathode, dominates generally from the second domain-transit cycle onward. An operational mode utilizing the effect of multiple-dipole formation is described in Section 3.5.3.

In conclusion, there seems to be a large degree of variance in the domain nucleation mechanism operating in actual TE samples. For dipole-mode devices it would certainly be desirable to have a nucleation mechanism relying on a pronounced non-uniformity, which acts similarly as the 'notch' of Fig. 3.2a immediately following the cathode. On the other hand, for the dipole-free operation discussed in Chapter 4, an ohmic contact of ideal behaviour on a semiconductor of high homogeneity is favourable [3.9]. In neither case should there be uncontrolled macroscopic doping irregularities or other crystal inhomogeneities along the sample length because they might act as undesired nucleation centres for domains.

The obvious uncertainty about the exact nature of the nucleation mechanism stems largely from the fact that the properties of the 'ohmic' contacts obtained in practice are not yet known sufficiently. As already mentioned, there is then a feeling [3.3a, 3.5a and 3.6] that more often than not the empirically-designed ohmic contacts applied to GaAs (Section 6.2.2) are by no means 'well-behaved', i.e. they do not show low-resistance properties [3.3, 3.5 and 3.6]. Rather it is suggested that between the contact metal and the negative-mobility semiconductor material a thin high-resistivity layer is often formed during the contact processing. This layer may have a linear or even a non-linear current-voltage characteristic. A high-resistivity layer is sometimes also found in the form of a narrow lightly-doped zone at the interface between epitaxially-grown TE GaAs and a highly-doped substrate (Fig. 6.5). For samples exhibiting such a high-resistivity layer at the cathode interface, a rather convincing model of the domain nucleation process [3.5a] will be described below. The model relies on the control of the process by the 'imperfect' boundary region for which several possible actual embodiments have been put forward [3.5a] including a very shallow Schottky or hetero-junction barrier and an embedded p-layer.

3.3.3 Sample with Cathode Control Region

Figure 3.5 illustrates in a highly simplified manner the formation of a dipole-layer domain in a negative-mobility semiconductor if a *linear* high-resistivity control region exists [3.5a]. For convenience only the steady-state case is considered, i.e. the displacement current is neglected, and space-charge layers are supposed to have only a minor extension along the carrier-drift direction.

A schematic representation of the sample configuration is given in Fig. 3.5a, where the length L_c of the control section is exaggerated as compared to the actual other dimensions. In practice, L_c might be expected to be of the order of 1 μm, cf. [3.5a]. The control and the main sections are thought to have drift-current-density/electric-field characteristics as indicated by Fig. 3.5b.

We now assume that a bias field $E_{h0} = V_0/L$, with L being the overall sample length, is applied and raised until the current density is J_1, corresponding to the peak current of the main characteristic (Fig. 3.5b). The field and potential distributions for this case I are shown in Figs 3.5c and d, respectively. According to the particular slopes of the two characteristics, the field in the control region (E_c) exceeds that in the main region (E_1) and thus an electron depletion layer must exist at the interface of the two regions. There the field transition will be a gradual one extending at least over a few Debye lengths, i.e. we have a field tail at above-peak values ($E > E_p$) reaching into the negative-mobility semiconductor body over a certain distance (Fig. 3.5c).

As we have discussed above, any space-charge (here the electron depletion) in a part of a semiconductor where the field is in the negative-conductivity range, tends to grow (and generally to move if the space charge is due to mobile carriers). This means in our case that a potential distribution like the one labelled I_a in Fig. 3.5d develops. Since constant-voltage operation is again assumed, the split-up in field which necessarily takes place when the depletion layer starts to grow and move, is of the form indicated by the arrows in Fig. 3.5b, i.e. the field E_1 at the anode side of the layer must drop. This, in turn, causes the current density and hence the control field E_c, to decrease.

It is now of importance to know how far the current can fall. We aspire to have a

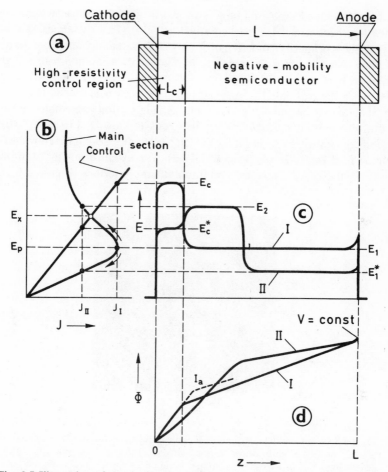

Fig. 3.5 Illustration of dipole-domain formation in an n-type TE semiconductor with high-resistivity control region at cathode. (a) Schematic representation of sample, (b) drift-current-density/electric-field characteristics of control and main regions, (c) spatial field distribution $E(z)$ for two consecutive time instants I and II, (d) corresponding spatial potential distribution $\Phi(z)$.

drop *below* the current value existing at the cross-over point of the two characteristics, denoted in Fig. 3.5b by the field E_x. In this case, as for instance depicted by current density J_{II}, the field in the control region (E_c^*) is *lower* than that in the adjacent part of the main semiconductor body (E_2), thus forming an accumulation layer at the interface. This means that the high-field domain, associated with a space-charge dipole, has developed; a result which we were looking for (curves II in Figs 3.5c and d). This dipole domain will subsequently grow and travel towards the anode contact.

Whether or not the cross-over point will be passed, depends on the location of the 'dynamic' branch of the current-voltage characteristic which is valid if, and as long as, a fully-developed dipole domain is in transit, i.e. if the steady state has been reached (see Section 3.4.4.2, particularly Fig. 3.15). This dynamic branch of the characteristic is primarily a function of doping level and sample length. If the dynamic branch lies *above* the cross-over point, no dipole formation is possible in the way investigated here.

Even when the bias field is somewhat *below* the peak field value ($0.8E_p \lesssim E_{ho} \lesssim E_p$, [3.7] there may be a depletion layer in the main section immediately adjacent to the control region at fields $E > E_p$ (analogous to case I_a; [3.7 and 3.3b]). This depletion domain will be *static** if the field at the anode side of it is below the domain sustaining field ($\approx E_p/2$; see Section 3.4.4.2). The extension of the static domain depends on the exact boundary conditions at the interface. Figure 3.6 shows a

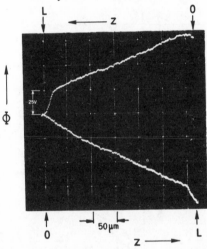

Fig. 3.6 Spatial potential distribution $\Phi(z)$ measured in an n-type GaAs sample with bias $E_{ho} = V_0/L$ slightly below the peak field E_p, indicating existence of 'cathode drop' region. Left arrows show position of cathode for upper curve. Lower curve was obtained by reversal of bias polarity. After Gunn [3.7].

particularly pronounced example of such a 'cathode drop' as obtained by a measurement of the spatial potential distribution along a GaAs sample. When the bias polarity was reversed in this case, the potential drop appeared at the opposite contact which then served as the cathode.

3.4 DIPOLE-DOMAIN DYNAMICS

3.4.1 General Remarks

The dynamic behaviour of TE semiconductors has been theoretically treated by numerous authors in various phenomenological or semi-phenomenological ways.

---

*For an investigation of domain trapping in general, on the basis of minimum Joule heat production, see [3.8].

These analyses can roughly be divided into two main groups. In one of these groups the complete space-charge dynamics including steady-state and transient behaviour are covered, with the assumption of particular boundary conditions, which automatically leads to the nucleation of space-charge layers and eventually to the large-signal instabilities observed by Gunn. This approach [3.2, 3.4, 3.10 and 3.11] yields detailed and convincing results. But from a didactical point of view it suffers from the fact that even the evaluation of the basic equations (partial differential equations with boundary conditions) has to be carried out with the aid of a computer. Thus only a limited insight into the physical processes can be obtained.

In the other approach certain initial assumptions and simplifications are introduced. These make it possible to treat the dipole-domain dynamics by largely elementary analytical methods [1.11, 3.12–3.14, 3.16–3.21], thus allowing a physical discussion of the basic equations. Even the transient case can be included [3.19–3.23], i.e. domain formation and decay, though only the solution of a system of *ordinary* differential equations is required. The simplifying assumptions involve using a phenomenological $v(E)$ characteristic and a defined initial microscopic nucleation space-charge dipole at the cathode contact. Furthermore diffusion must be neglected. The treatment of the *steady-state* case can be refined by taking into account a field-*independent* diffusion coefficient [3.14 and 3.65]. Inclusion of the realistic field-*dependent* diffusion, however, affords a numerical solution [3.15]. Finally, the time dependence of the $v(E)$ characteristic at high frequencies cannot be taken into account [see Section 5.2.5 for a discussion of the influence of the finite electron relaxation times on the $v(E)$ characteristic]. Owing to such simplifications, obviously no exact quantitative agreement with experiment can be expected, but the semi-quantitative results obtained are sufficiently accurate for many purposes. We now consider dipole-domain growth, steady-state propagation, and extinction, for a voltage-controlled n-type GaAs element, following an analysis which involves the above-listed simplifications including the neglect of diffusion.

3.4.2 Basic Equations
The fundamental expression for treating the dipole dynamics is derived here for the case where the doping is uniform throughout the sample [3.19–3.21]. The $v(E)$ relationship is assumed to be as shown in Fig. 2.8 and is approximated by the analytical expression (Fig. 3.9; [3.11])*

$$v = v(E) \equiv [\mu_1 E + v_v (E/E_a)^4] [1 + (E/E_a)^4]^{-1} \qquad (3.4)$$

with the parameters $\mu_1 = 8\,000$ cm^2 (Vs)$^{-1}$, $v_v = 8\cdot5 \times 10^6$ cm s^{-1} as an asymptotic value for $E \to \infty$, and $E_a = 4$ kV cm^{-1}, yielding a peak velocity of $v_p = 2\cdot08 \times$

---

*In the original analysis [3.21 and 3.26] a less realistic approximation was assumed because the shape of the $v(E)$ curve was then not so well known. W. Heinle has kindly performed a re-calculation using eqn (3.4) instead, and has supplied the authors with the new results which are presented in Figs 3.10–3.12, 3.17, 3.18, and 4.4.

10^7 cm s^{-1} at $E_p = 3.47$ kV cm^{-1} or $v_p/v_v = 2.45$. (For choice of sign see footnote in Section 1.2, p. 3.)

Restricting ourselves to the one-dimensional case, the carrier current density J_c is described by the continuity eqn (3.1), i.e. in our case with diffusion neglected by

$$\frac{\partial}{\partial t}(en) + \frac{\partial}{\partial z}(env) = 0 \tag{3.1c}$$

since $J_c = J$. As the third relation Poisson's eqn (3.2), i.e.

$$\frac{\partial E}{\partial z} = \frac{e}{\varepsilon}(n - n_0) \tag{3.2b}$$

has to be used. Combining eqns (3.1c) and (3.2b), and carrying out a first integration with respect to z, immediately yields

$$\varepsilon \frac{\partial E}{\partial t} + e\left\{n_0 + \frac{e}{\varepsilon}\frac{\partial E}{\partial z}\right\}v = J_t(t) \tag{3.5}$$

with

$$J = e\left\{n_0 + \frac{\varepsilon}{e}\frac{\partial E}{\partial z}\right\}v$$

being the drift current density. The integration function $J_t(t)$ of eqn (3.5), the 'basic equation' of our problem, obviously describes the total current density. If the behaviour of the sample is determined by the terminal voltage V—the case to be considered here—it is useful to convert eqn (3.5) into a more convenient form. A further integration, over the total sample length L, leads to

$$J_t L = \varepsilon L \frac{dE_h}{dt} + e \int_0^L \left\{n_0 + \frac{\varepsilon}{e}\frac{\partial E}{\partial z}\right\}v \, dz \tag{3.6}$$

with the mean value of the field (homogeneous component):

$$E_h = E_h(t) \equiv \frac{1}{L}\int_0^L E(z,t)\,dz = \frac{V(t)}{L}. \tag{3.7}$$

Equation (3.6) permits us to eliminate $J_t(t)$ from eqn (3.5) so that a non-linear integro-differential equation is obtained for the electric field:

$$\frac{\partial E}{\partial t} + \left(\frac{en_0}{\varepsilon} + \frac{\partial E}{\partial z}\right)v = \frac{dE_h}{dt} + \frac{en_0}{\varepsilon L}\int_0^L v\,dz + \begin{cases} \dfrac{1}{L}\int_0^L \dfrac{\partial E}{\partial z}v\,dz & (3.8a) \\[2ex] \dfrac{1}{L}\int_{E(z=0)}^{E(z=L)} v\,dE & (3.8b) \end{cases}$$

3.4.3 Form of Stable Domains

Closed solutions of the basic eqn (3.5), or (3.8), can be obtained only for special cases, for instance for domains having grown to the steady state, i.e. to a distribution $E(z^\star) = E(z - v_D t)$ which moves across the sample with a constant velocity v_D and exhibits a constant field $E = E_{1\infty}$ (asymptotic value) outside of the field inhomogeneity. A constant current, determined by $E_{1\infty}$, is then flowing through the external circuit.

Making use of the relations

$$\frac{\partial E}{\partial t} = -v_D \frac{dE}{dz^\star}, \quad \frac{\partial E}{\partial z} = \frac{dE}{dz^\star}, \tag{3.9}$$

eqn (3.5) then becomes

$$-\varepsilon(v_D - v)\frac{dE}{dz^\star} = J_t - en_0 v. \tag{3.10}$$

In the region of constant field $E_{1\infty}$ outside of the domain, one has

$$J_t = J = en_0 v_{1\infty} \text{ where } v_{1\infty} \equiv v(E_{1\infty}). \tag{3.11}$$

Combining eqns (3.10) and (3.11) yields a differential equation for the electric field distribution

$$\varepsilon(v_D - v)\frac{dE}{dz^\star} = -en_0(v_{1\infty} - v). \tag{3.12}$$

Steady-state domains require $v_D = v_{1\infty}$. This requirement seems to be plausible since otherwise a continuing growth and decay of space charge must take place, a process not compatible with the steady state [3.21a] (see also Section 3.4.4.1). With $v_D = v_{1\infty}$, eqn (3.12) becomes

$$(v_{1\infty} - v)\left(\frac{dE}{dz^\star} + \frac{en_0}{\varepsilon}\right) = 0. \tag{3.13}$$

This expression has the solution

$$v = v_{1\infty} \text{ or } \frac{dE}{dz^\star} = -\frac{en_0}{\varepsilon} \tag{3.14, 3.15}$$

from which one can construct the simple field distributions shown in Fig. 3.7a. One has either a triangular domain with a field discontinuity and a linear field gradient according to eqn (3.15) or a trapezoidal domain with an additional constant-field region as determined by eqn (3.14). A zone fully depleted of electrons, i.e. with a charge density equalling the ionized donor charge density, is associated with the linear field gradient, whereas the electrons collect in the accumulation layer of infinitely small extension. (The complete removal of electrons from the depletion zone and the infinitely high electron density in the accumulation layer result from the neglect of diffusion. For the sake of clarity, the accumulation layers in Fig. 3.7a have been drawn to show the actually expected *finite* extension and electron density.)

$E_{2\infty}$ is the maximum field strength of the domain. For the particular $v(E)$ relation assumed here [eqn (3.4)], where v has no minimum but drops monotonically beyond the peak field towards an asymptotic value, no *trapezoidal* domain shape is, however, possible in the steady state and thus we will be concerned only with triangular domains in the following (cf. [2.48]).

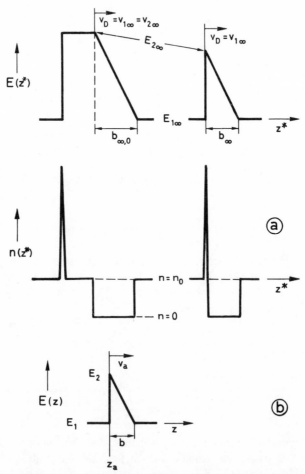

Fig. 3.7 Principal forms of (a) steady-state domain field configuration $E(z^*)$ with associated carrier distribution $n(z^*)$ and (b) of field configuration $E(z)$ for domain during formation.

3.4.4 Dynamics of Triangular Domains

To investigate the dynamics of space-charge dipole domains, we first carry out a formal mathematical analysis and, subsequently, discuss the steady-state as well as the transient cases by interpreting the derived formulae. It is assumed in the following that an arbitrarily small initial dipole domain exists and, furthermore, that the non-steady-state domain has qualitatively the same triangular shape as derived for the steady-state one (see Fig. 3.7b). Additionally the linear field gradient is assumed to remain unchanged during the domain extinction process, i.e. when the domain reaches the anode (see Fig. 3.8).

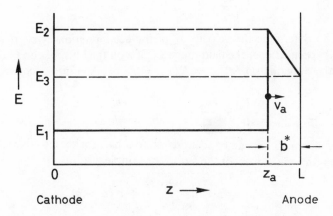

Fig. 3.8 Definition of the electric-field parameter E_3 of a domain reaching the anode.

It would appear that the initial field inhomogeneity (non-steady-state domain) is characterized by the low field E_1 being smaller at the cathode side than at the anode side because of the inherent accumulation layer detaching from the metal contact. However, in an investigation of such a case [3.21b] it was shown that the fields outside of the domain attain nearly equal values before an appreciable current drop has occurred.

3.4.4.1 *General Mathematical Formulation*

(a) *Domain Growth.* Specifying the basic current density eqn (3.5) over the field range E_1 (see Fig. 3.7b) and using

$$\frac{\partial E_1}{\partial t} = \frac{\mathrm{d}E_1}{\mathrm{d}t} \quad \text{and} \quad \frac{\partial E_1}{\partial z} = 0$$

one obtains for the total current density

$$J_\mathrm{t}(t) = \varepsilon \frac{\mathrm{d}E_1}{\mathrm{d}t} + en_0 v_1 \tag{3.16}$$

with $v_1 \equiv v(E_1)$. Similarly, at the maximum domain field $E_2 = E_2(z_\mathrm{a}, t)$ the expression

$$J_\mathrm{t}(t) = \varepsilon \frac{\mathrm{d}E_2}{\mathrm{d}t} + en_0 v_\mathrm{a} \tag{3.17}$$

with $v_a = dz_a/dt$ is found by using the total derivative

$$\frac{dE_2}{dt} = \frac{\partial E_2}{\partial t} + v_a \frac{\partial E_2}{\partial z_a}$$

and the relation

$$\frac{\partial E_2}{\partial z_a} \equiv \left(\frac{\partial E}{\partial z}\right)_{z=z_a} = -\frac{en_0}{\varepsilon}.$$

In order to determine the velocity v_a of the accumulation layer, the basic field eqn (3.8b) is specified over the field range E_1: if z on the l.h.s. of eqn (3.8) is a point outside of the domain and $E_{(z=0)} = E_{(z=L)} = E_1$ is taken into account, eqn (3.8b) reduces to

$$\frac{dE_1}{dt} + \frac{en_0}{\varepsilon} v_1 = \frac{dE_h}{dt} + \frac{en_0}{\varepsilon L} \int_0^L v \, dz. \qquad (3.18)$$

The integral in eqn (3.18) can be separated into a part for the positive space-charge region of length b and a part for the rest of the sample, viz.

$$\int_0^L v \, dz = \int_{E_2}^{E_1} \frac{v \, dE}{\partial E/\partial z} + v_1(L - b)$$

where now the integration variable has been changed from z to E. Introducing, according to eqns (3.9) and (3.15), for the domain range

$$\frac{\partial E}{\partial z} = -en_0/\varepsilon$$

whence for the domain width

$$b = (\varepsilon/en_0) (E_2 - E_1) \qquad (3.19)$$

one has

$$\int_0^L v \, dz = v_1 L + \frac{\varepsilon}{en_0} f(E_1, E_2) \qquad (3.20)$$

with

$$f(E_1, E_2) \equiv \int_{E_1}^{E_2} v \, dE - v_1(E_2 - E_1) = \int_{E_1}^{E_2} (v - v_1) \, dE \qquad (3.21)$$

Hence, eqn (3.18) becomes

$$\frac{dE_1}{dt} = \frac{dE_h}{dt} + \frac{1}{L} f(E_1, E_2) \qquad (3.22)$$

From eqn (3.7) one obtains the voltage balance condition (Kirchhoff's law) for the device terminals

$$E_h = E_1 + \Phi_D/L \qquad (3.23)$$

where Φ_D denotes the domain excess potential (see Fig. 3.2e) which is given by

$$\Phi_D = \Phi_D(E_1, n_0) \equiv \int_0^L (E - E_1) \, dz = \frac{\varepsilon}{2en_0} (E_2 - E_1)^2. \qquad (3.24)$$

Differentiating eqn (3.23) and inserting it into eqn (3.22) yields the growth rate of the domain potential

$$\frac{d\Phi_D}{dt} = -f(E_1, E_2) \qquad (3.25)$$

or with eqns (3.24), (3.16) and (3.17), by eliminating Φ_D and J_t, the required accumulation-layer velocity

$$v_a = v_1 + \frac{f(E_1, E_2)}{E_2 - E_1} = \int_{E_1}^{E_2} \frac{v \, dE}{E_2 - E_1}. \qquad (3.26)$$

This expression can also be obtained by a different derivation [3.33]. It has been shown [3.12a] that eqn (3.26) applies quite generally to travelling accumulation layers with arbitrary boundaries.

For a given terminal voltage $V(t) = E_h(t)L$, the domain dynamics, i.e. $J_t(t)$, $E_1(t)$, $E_2(t)$ and $\Phi_D(t)$, are completely described by the current-density eqns (3.16) and (3.17) in conjunction with eqn (3.26) for v_a, and by the terminal condition (3.23) in conjunction with eqn (3.24) or (3.25) for Φ_D, where $f(E_1, E_2)$ is defined in eqn (3.21).

(b) *Steady-State Domain.* As the steady state (subscript ∞) is characterized by vanishing time derivatives, we obtain from eqn (3.16)

$$J_{t\infty} = J_\infty = en_0 v_{1\infty} \qquad (3.27)$$

and further

$$f(E_{1\infty}, E_{2\infty}) = 0 \qquad (3.28a)$$

because of eqn (3.22). With the definition (3.21) one obtains

$$\int_{E_{1\infty}}^{E_{2\infty}} (v - v_{1\infty}) \, dE = 0. \qquad (3.28b)$$

From eqns (3.28a) and (3.26) it follows that

$$v_{a\infty} \equiv v_D = v_{1\infty}. \qquad (3.29)$$

Equation (3.28) constitutes 'Butcher's equal-areas rule' which will be discussed in Section 3.4.4.2.

(c) *Domain Extinction.* Figure 3.8 schematically represents the triangular domain during the extinction phase at the anode. In this case the two current-density eqns (3.16) and (3.17) together with eqn (3.26) for v_a remain valid. From the basic eqn (3.5) one obtains for the field E_3 at the anode an additional current-density relation

$$J_t(t) = \varepsilon \frac{dE_3}{dt} \tag{3.30}$$

by taking into account

$$\frac{\partial E_3}{\partial t} = \frac{dE_3}{dt} \quad \text{and} \quad \frac{\partial E_3}{\partial z} = -\frac{en_0}{\varepsilon}.$$

In the terminal condition (3.23) Φ_D has to be replaced by

$$\Phi_D = \int_0^L (E - E_1) \, dz = \frac{\varepsilon}{2en_0} (E_2 - E_3)^2 + \frac{\varepsilon}{en_0} (E_2 - E_3)(E_3 - E_1) \tag{3.31}$$

and eqn (3.25) changes to

$$\frac{d\Phi_D}{dt} = -[f(E_1, E_2) + v_1(E_3 - E_1)]. \tag{3.32}$$

This latter expression results if the l.h.s. of eqn (3.8b) is again specified for the field range E_1 as in the case of domain growth. In doing so the following relations apply (cf. Fig. 3.8):

$$E_{(z=0)} = E_1, \quad E_{(z=L)} = E_3 \quad \text{and}$$

$$\int_0^L v \, dz = \int_{E_2}^{E_3} \frac{v \, dE}{\partial E / \partial z} + v_1 (L - b^\star)$$

with

$$\partial E / \partial z = -en_0 / \varepsilon$$

over the domain width

$$b^\star = \varepsilon(E_2 - E_3) / en_0.$$

Hence, for the complete description of the domain extinction process one has the current-density eqns (3.16), (3.17), and (3.30) together with eqn (3.26) for v_a, and the terminal condition (3.23) with eqn (3.31) or (3.32) for Φ_D, where $f(E_1, E_2)$ is defined in eqn (3.21).

3.4.4.2 *Steady-State Domain Propagation*
The most important relation governing the steady-state behaviour is Butcher's equal-areas rule, [eqn (3.28)], which can geometrically be interpreted as shown in Fig. 3.9. To every given value of $E_{1\infty}$ lying below the peak field E_p there belongs a domain field $E_{2\infty}$ on the dashed curve in such a way that the hatched areas are equal.

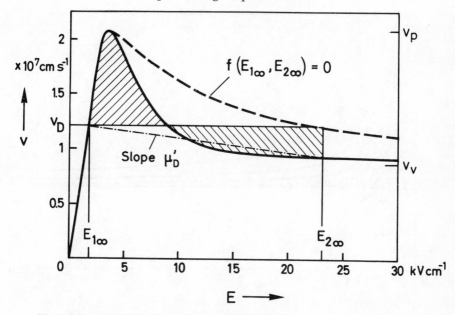

Fig. 3.9 Interpretation of Butcher's equal-areas rule in the v-against-E plane. The slope of the curve $f(E_{1\infty}, E_{2\infty}) = 0$ equals μ'_D which is geometrically constructed according to eqn (3.37).

The equal-areas rule, [eqn (3.28b)], together with the terminal condition (3.23), completely define the domain parameters $E_{1\infty}$, $E_{2\infty}$ and b_∞/L. Evaluations of these parameters as a function of the average field E_h are presented in Figs 3.10–3.12 for several values of the net-doping times length product $n_0 L$. For the higher $n_0 L$ values an upper limit of E_h is given when the maximum domain field $E_{2\infty}$ reaches the dielectric breakdown value* (of the order of 200 kV cm^{-1}). The curves indicate that steady-state domains can exist even if E_h is *below* the peak value of the $v(E)$ characteristic, provided E_h exceeds a *lower* critical value E_S (domain sustaining field; cf. [3.40]). If the terminal voltage drops below $V_S = E_S L$ the excess potential $\Phi_{D\infty}$ required by a steady-state domain can no longer be provided (cf. Fig. 3.13). To initiate a domain, however, an *upper* threshold field value $E_T \approx E_p$ (see below) has to be reached or exceeded in any case, i.e. the region of n.d.c. must have been entered (triggered mode of operation, [3.30–3.32]).

It is convenient to represent the equal-areas rule, [eqn (3.28b)], in terms of the domain potential $\Phi_{D\infty}(E_{1\infty})$ according to eqn (3.24) (domain characteristic); it can

---

*Dielectric breakdown, as it may occur in the high-field domain particularly of heavily-doped TE devices, is associated with electron-hole pair production. Spontaneous radiation of quantum energy by recombination of electron-hole pairs created in this way was observed in n-type GaAs [3.32, 3.74, 3.75 and 6.36], InP [3.77], and CdTe samples [6.30] near the band-gap wavelength (infra-red). If the pair production is large enough, the spontaneous emission can become stimulated emission (laser action) [3.76, 3.77].

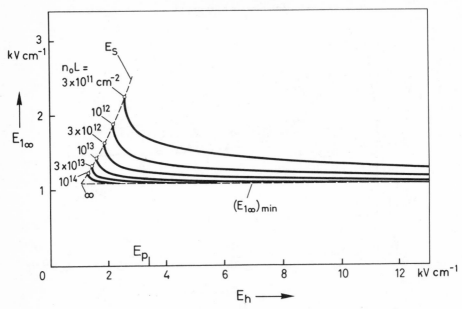

Fig. 3.10 Asymptotic low field $E_{1\infty}$ against space-averaged field E_h in the steady state for various values of the n_0L product. After Heinle [3.34].

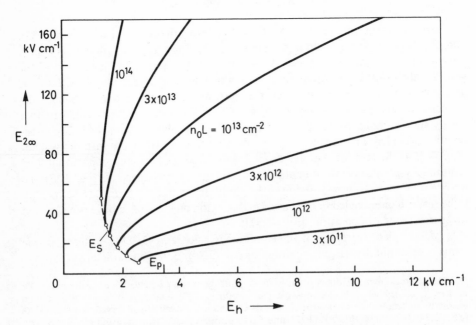

Fig. 3.11 Maximum domain field $E_{2\infty}$ against space-averaged field E_h in the steady state for various values of the n_0L product. After Heinle [3.34].

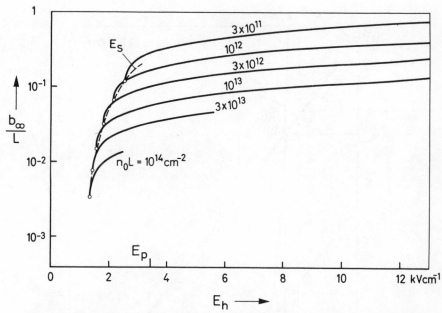

Fig. 3.12 Normalized domain width b_∞/L against space-averaged field E_h in the steady state for various values of the n_0L product. Curves are shown only in the range $E_{2\infty} \leq 200$ kV cm^{-1}. After Heinle [3.34].

then be plotted together with the 'device-line' eqn (3.23) in a diagram of Φ_D against E_1 to obtain as the intersection the working point W, i.e. the steady-state values $E_{1\infty}$ and $\Phi_{D\infty}$ for a particular set of sample parameters and bias voltage V_0 [3.7, 3.24 and 3.25]. In Fig. 3.13 this is carried out for the case of $n_0 = 3 \times 10^{14}$ cm^{-3}, $L = 20$ μm, and $V_0 = 10$ V (fully-drawn device line). The second (dashed) device line determines an operating condition ($V_0 = 5.3$ V) in which no self-initiated domain formation occurs but triggering of domains is possible. Of the two working points I and II only I, however, is stable. Measured values of domain potential $\Phi_{D\infty}$ [3.7, 3.29 and 3.78] agree fairly well with the calculated ones [3.14, 3.25], at least for a not too small $\Phi_{D\infty}$.

Figure 3.14 [3.27] shows the experimentally determined profile of steady-state triangular high-field domains [2.48, 3.29 and 3.55] during their travel. In contrast to the idealized configuration of Fig. 3.7, the accumulation layer (left field gradient) has a finite width and the gradient is not constant, mainly because of diffusion effects. For the same reason the corners are not sharp. The depletion layer is characterized by the almost linear field gradient (right) which approaches the ideal case of Fig. 3.7. The domain with the higher maximum field resulted when the bias voltage was increased. An increase in bias is mainly absorbed by the domain since $E_{1\infty}$ stays almost constant (cf. Fig. 3.13): the field gradient associated with the accumulation zone becomes steeper, and the depletion zone widens while its field gradient hardly

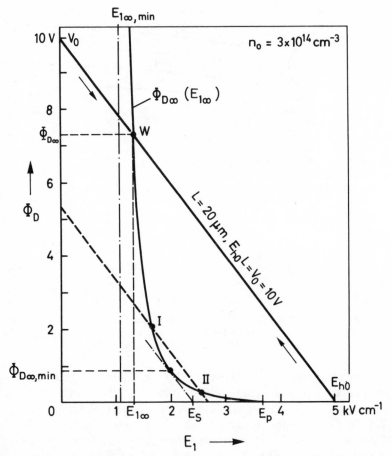

Fig. 3.13 Particular example of domain characteristic representing domain potential
$\Phi_{D\infty}$ against low field $E_{1\infty}$, together with 'device lines' Φ_D against E_1, for above-
threshold operation (full line) and for triggered operation (dashed line).

alters. The oscillograms shown in Fig. 3.14 have been obtained by placing a capacitive probe close to the sample surface, measuring the time dependence of the picked-
up potential $\Phi(t)$ when a domain was passing by, and forming the time derivative
$d\Phi/dt$ by an appropriate network. This latter quantity represents the required
spatial, $z^\star$-dependent profile of the domain because of $z^\star = z - v_D t$.

The knowledge of the steady-state domain parameters allows us to determine the
so-called '*time-dependent*' $I(V)$ or $v(E_h)$ characteristic [3.42, 3.48], which is useful for
instance in analysing the properties of the domain modes of operation (see Sections
3.5.2 and 3.5.3). This characteristic consists of a static and a steady-state or dynamic
branch as shown in Figs 3.15a (calculated) and 3.15b (measured with sampling
oscilloscope). The static part corresponds to the quasi-ohmic range of the general

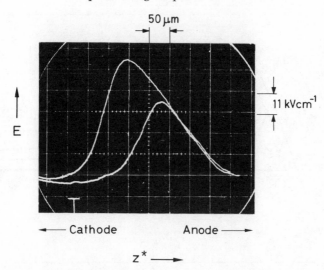

50 μm

11 kVcm⁻¹

E

◄── Cathode   Anode ──►

z* ──►

Fig. 3.14 Measured profiles of steady-state triangular high-field domains. The larger domain was obtained by an increase in bias field E_{h0}. After Gunn [3.27].

static $v(E)$ characteristic, [eqn (3.2) or Fig. 2.8], up to the threshold field E_T for domain formation and is valid if no dipole domain exists. Stability considerations (Section 4.4.1.2) show that E_T somewhat exceeds the peak field E_p. The exact position of the final point on the static branch depends on $n_0 L$; in Fig. 3.15a the threshold E_T is given for several $n_0 L$ values. Real GaAs TE devices sometimes show an apparent increase in threshold field which is not related to this $n_0 L$ dependence but rather may be due to high-resistivity interface layers and the like as discussed in Section 3.3.

In the range $E_p \leq E_h \leq E_T$ the drift current density J cannot be calculated from the drift velocity v according to $J = e n_0 v$ which would lead to a static differential negative resistance. Carrier injection occurring from the cathode leads to an average $\bar{n} > n_0$ and $J \gtrsim J_p$ in this range [1.9]. Since this process establishes a non-uniform field distribution, the static $v(E_h)$ branches of Fig. 3.15a should be considered only as approximate for $E_h > E_p$.

When E_T has been reached, a domain is formed and the drift velocity (current) drops to the dynamic branch. This branch applies if, and as long as, a steady-state domain is in transit. If the bias falls below the sustaining field E_S while a domain propagates, the domain is quenched and the drift velocity (current) reverts to the static branch. In Fig. 3.15b the traces connecting the static and dynamic branch, obtained during the transition phase, have been dashed for the sake of clarity (slope determined by load resistance). The dynamic branch can be constructed graphically for a certain $n_0 L$ product by plotting the appropriate $E_{1\infty}(E_h)$ curve of Fig. 3.10 in the form $v_D = v(E_{1\infty})$ against E_h using the $v(E)$ characteristic [eqn (3.4)]. Thus a series of dynamic branches is obtained (Fig. 3.15a).

The part of the dynamic branch lying at field values below E_T can be measured

55

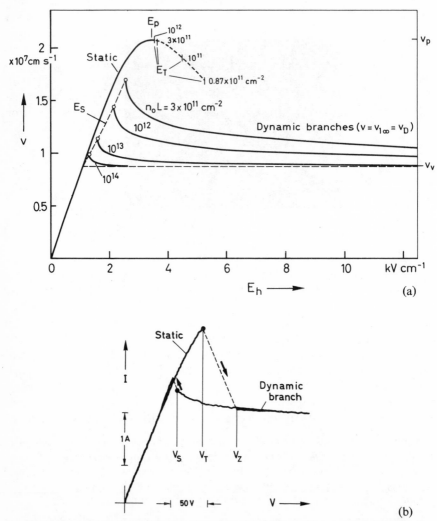

Fig. 3.15 Static and dynamic branches of 'time-dependent' drift-current/voltage velocity/average-field) characteristic of an n-type GaAs device: (a) calculated for different n_0L products (after Heinle [3.34]); (b) measured for $n_0 \approx 4 \times 10^{14}\,\text{cm}^{-3}$, $L \approx 350\,\mu\text{m}$ ($n_0L \approx 1\cdot4 \times 10^{13}\,\text{cm}^{-3}$), $A \approx 500\,\mu\text{m} \times 500\,\mu\text{m}$. (Courtesy of R. Becker, AEG-Telefunken Research Institute).

(Fig. 3.15b) by providing a bias in the range $E_S < E_{h0} < E_T$ and adding to it an overdrive field pip for initiating the dipole domain, e.g. [3.29, 3.64]. Without applying this special overdrive pip to the bias, plots like that shown in Fig. 3.16 are rather obtained when measuring on a sampling oscilloscope the $I(V)$ characteristic of a GaAs TE device which is being traversed by steady-state domains. (For $V > V_T$

one has $I = I_\infty = en_0 v_{1\infty} A$, with A being the cross-sectional area of the sample.) The particular $I(V)$ characteristic of Fig. 3.16 was measured in a low-impedance load circuit for a pulsed GaAs sample possessing two ohmic contacts of nearly equal cross-section (like the one shown in Fig. 6.6a). With such a contact configuration this typical antisymmetric characteristic is obtained (see, however, Section 6.4, particularly Figs 6.22a and 6.24). Beginning at the origin we have first the quasi-ohmic part, for both bias directions, up to the threshold value E_T at which the current drops fairly abruptly to its minimum value I_∞. [Displaying I_∞ (V) requires a coherent sampling process. For an oscillogram with incoherent sampling see Fig. 1.3.] A further increase in bias voltage eventually leads again to a rise in current, due to ionization effects (see Section 6.4).

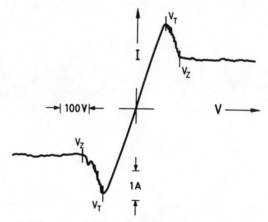

Fig. 3.16 Measured drift-current/voltage characteristic of an n-type GaAs device for both bias polarities (sample parameters similar as in Fig. 3.13b): $V < V_T$ static branch, $V > V_Z$ (load-dependent) dynamic branch. (Courtesy of R. Becker, AEG-Telefunken Research Institute).

3.4.4.3 *Transient Behaviour*

The dielectric relaxation time τ_R which has been introduced in Section 3.2, eqn (3.3a), as a measure for space-charge decay or growth, follows directly from the domain-growth eqn (3.25) by carrying out an expansion with respect to a small deviation $E_2 - E_1$ from a homogeneous field distribution E_h (linear approximation) after Φ_D has been expressed by eqn (3.24) and $f(E_1, E_2)$ by eqn (3.21). For the growth rate of the domain field amplitude $E_2 - E_1$, which is proportional to the charge per unit area $\varepsilon(E_2 - E_1)$ stored in the accumulation or depletion layer, one obtains

$$\frac{d(E_2 - E_1)}{dt} \approx -\frac{en_0}{\varepsilon} \mu_h'(E_2 - E_1) \tag{3.33}$$

where $\mu_h' \equiv \mu'(E_h) = (dv/dE)_{E_h}$ is the differential mobility at $E = E_h$. According to eqn (3.33) a space-charge dipole of small amplitude will decay in the ohmic part of the characteristic ($\mu' > 0$), and grow in the negatively-sloped part with a time constant equal to the absolute value of $\tau_R = \varepsilon/en_0\mu'_h$ at the working point E_h. In the typical example quoted in Section 3.2 we obtained $|\tau_R| \approx 1\cdot3$ ps.

In the large-signal range the growth behaviour of the domain becomes more complex. Figure 3.17 illustrates the complete dipole-domain formation process with the help of the $v(E)$ characteristic. The sample is assumed to be biased in the negatively-sloped part of the characteristic at a field $E_h = E_{h0} = 4$ kV cm^{-1}. To

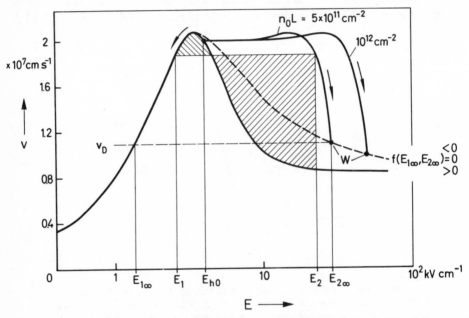

Fig. 3.17 Illustration of domain in v-against-E plane for two values of the n_0L product ('unequal-areas rule'). After Heinle [3.34].

every value of $E_1(< E_{h0})$, which lies on the characteristic, the corresponding value of E_2 can be determined with the aid of the terminal condition, [eqns (3.23) and (3.24)], so that a locus curve is obtained as shown in Fig. 3.17 for two different n_0L products. Such a curve corresponds to the device line of Fig. 3.13. The function $f(E_1, E_2)$ defined by eqn (3.21) geometrically denotes the difference (negative!) between the two hatched areas of Fig. 3.17 (note the semi-logarithmic plot when comparing the areas!). It determines the growth rate of the domain according to eqn (3.25) which consequently has been termed 'unequal-areas rule' [3.22, 3.23]. From eqn (3.22), with $dE_h/dt = 0$, follows that $E_1(t)$ monotonously decreases; hence $E_2(t)$ has to increase on the locus curve of Fig. 3.17 (arrow! cf. Fig. 3.13). The modulus $|f(E_1, E_2)|$ first increases with time and then passes through a maximum. The points where the curves for E_2 reach the dashed curve $f(E_{1\infty}, E_{2\infty}) = 0$

correspond to the steady-state condition. It is reached asymptotically since there $(dE_{1,2}/dt) \to 0$.

Dipole-domain build-up from an arbitrarily-chosen small initial domain, its movement across the sample, and eventually the extinction is illustrated in Fig. 3.18, using the equations we have derived in Section 3.4.4.1. The homogeneous outside field E_1, the maximum domain field E_2, and the corresponding total current density J_t, are shown as functions of the position z_a of the accumulation layer for two $n_0 L$

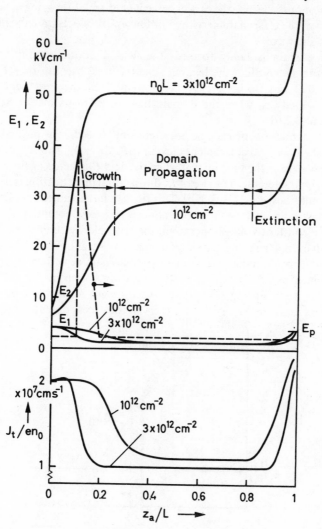

Fig. 3.18 Domain fields E_1, E_2 and normalized total current density J_t as a function of the position z_a of the accumulation layer for two $n_0 L$ values; bias field $E_{h0} = 4\ \mathrm{kV\ cm^{-1}}$. After Heinle [3.34].

59

products and the same bias field E_{ho}. The dashed field distribution indicates how to construct the domain shape during formation and propagation. The equations of Section 3.4.4.1 are invariant against the transformation $t \to \vartheta t$; $L \to \vartheta L$; $n_0 \to n_0/\vartheta$. Therefore all processes take place ϑ-times faster if the sample length is reduced and the net doping increased by the factor ϑ keeping the $n_0 L$ product constant.

The current falls appreciably only when domain build-up has already well advanced (cf. Section 8.2.1). Because of the displacement current, the total current decreases almost monotonously despite the carrier current (cf. Fig. 3.17) having to pass through the maximum of the $v(E)$ characteristic. For low $n_0 L$ products the current waveform approaches a sinewave.

While the domain is being formed, the field discontinuity must necessarily propagate more slowly than the leading edge $[v_1 > v_{\text{a}}$ because of $f(E_1, E_2) < 0$; eqns (3.21) and (3.26)]. Both velocities become equal and then represent the domain propagation velocity v_{D} when the domain has grown to its mature state $[f(E_{1\infty}, E_{2\infty}) = 0$; eqn (3.29)].

During the extinction phase, the total current exceeds the peak drift current $e n_0 v_{\text{p}}$ due to the flowing displacement current $(dE_1/dt > 0)$. The curves in Fig. 3.18 have been drawn up to the time instant when the field E_1 reaches again the peak field E_{p} of the $v(E)$ characteristic. The rise of E_2 during domain extinction is also found by computer simulations, e.g. [3.4, 3.39b]. This rise can be explained as follows [3.39b]. When the domain passes into the anode contact its excess potential Φ_{D} diminishes. Since we have constant-voltage operation, the low-field E_1 behind the domain and hence the drift current rise. In an attempt to maintain equilibrium, a charging current is supplied to the rest domain; consequently the domain peak field E_2 grows during the dissolution phase. This process is schematically shown in Fig. 3.19 where the time interval t_1 to t_2 during extinction is considered. As a result of the domain movement by a distance Δz within this interval, the excess potential required drops

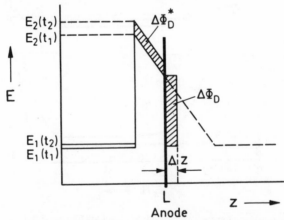

Fig. 3.19 Extinction component $\Delta\Phi_{\text{D}}$ and growth component $\Delta\Phi_{\text{D}}^*$ of domain excess potential during domain dissolution at anode contact. Adapted from Robrock [3.39b].

by an amount $\Delta\Phi_D$. On the other hand, owing to the increase in current (field E_1) created thereby, the potential of the rest domain grows by $\Delta\Phi^\star$. In detail this extinction process is rather involved.

When the increasing low field E_1 enters the negatively-sloped part of the characteristic ($E_1 \geq E_T$), a new domain is formed at the cathode, and the process of domain build-up, propagation, and extinction is repeated. In this way transit-time current oscillations are produced which have a fundamental frequency '

$$f_\tau = 1/\tau = \bar{v}_a/L \approx v_D/L \qquad (3.34)$$

if the domain width $b \ll L$. For GaAs samples of a length $L = 10$–100 μm, for example, the frequencies are 10 to 1 GHz since $v_D \approx 10^7$ cm s^{-1}.

Figure 3.20 shows the measured total-current waveform of a GaAs sample of $n_0 L = 6 \times 10^{12}$ cm^{-2} oscillating in this dipole-domain transit-time mode (Gunn

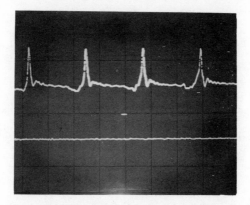

Fig. 3.20 Total-current waveform of n-type GaAs device ($n_0 \approx 3 \times 10^{14}$ cm^{-1}, $L \approx 220$ μm) oscillating in the dipole-domain transit-time mode. Horizontal scale: 1 ns per div. (Courtesy of R. Becker, AEG-Telefunken Research Institute.)

mode). The lower trace marks the base line. Fluctuations in current occur during the passage of the domain (current minimum), as in Fig. 3.20, if the domain drifts through regions of the sample which have (more or less pronounced) changes in cross-sectional area or doping level (cf. Section 8.2.3).

3.4.5 Equivalent Circuit

An approximate closed-form expression for the small-signal admittance of a TE semiconductor element while being traversed by a steady-state triangular domain, can be derived from the formulae obtained in Section 3.4.4.1 [3.35]. This is accomplished by investigating the response of the sample with regard to a small-signal perturbation superimposed on the bias field E_{h0}, i.e.

$$E_h = E_{h0} + \Delta E_h(t). \qquad (3.35a)$$

The domain fields and the total-current density, thus, deviate from the steady-state case like

$$E_i = E_{i\infty} + \Delta E_i, \qquad i = 1, 2$$

$$J_t = J_{t\infty} + \Delta J_t. \tag{3.35b}$$

The perturbations are assumed to be periodic in time so that

$$\Delta E_i = \hat{E}_i \exp(j\omega t), \qquad i = h\ 1, 2$$

$$\Delta J_t = \hat{J}_t \exp(j\omega t).$$

One then obtains for the small-signal current density specified in the *domain free* region (field E_1), from eqn (3.16), (3.27) and (3.35b), the expression

$$\hat{J}_t \approx j\omega\varepsilon\hat{E}_1 + en_0\mu'_{1\infty}\hat{E}_1 \tag{3.36}$$

since to first order

$$v_1 \equiv v(E_1) = v_{1\infty} + \Delta v_1 \approx v_{1\infty} + \mu'_{1\infty}\Delta E_1$$

with

$$\mu'_{1\infty} = \left(\frac{dv_1}{dE_1}\right)_\infty.$$

Similarly the small-signal current density specified for the *domain top* at E_2 is derived from eqn (3.17). By inserting eqn (3.35) and

$$v_a = v_{1\infty} + \Delta v_a$$

[see eqn (3.29)] into eqn (3.26) and using (3.28b) we obtain to first order

$$\Delta v_a \approx \frac{v_{2\infty} - v_{1\infty}}{E_{2\infty} - E_{1\infty}} \Delta E_2 = \mu'_D\Delta E_2 \tag{3.37}$$

and hence

$$v_a \approx v_{1\infty} + \mu'_D\Delta E_2.$$

Equation (3.17) thus yields

$$\hat{J}_t \approx j\omega\varepsilon\hat{E}_2 + en_0\mu'_D\hat{E}_2. \tag{3.38}$$

Furthermore we require the expression for the small-signal terminal voltage which is obtained as

$$\hat{V} \equiv \hat{E}_hL = \underbrace{(L - b_\infty)\hat{E}_1}_{\text{across}} + \underbrace{b_\infty\hat{E}_2}_{} = \hat{V}_r + \hat{V}_D \tag{3.39}$$

across
rest of sample   domain

by employing the terminal condition (3.23) in its small-signal form with the relevant relation

$$\hat{\Phi}_D \approx \frac{\varepsilon}{en_0}(E_{2\infty} - E_{1\infty})\,(\hat{E}_2 - \hat{E}_1) = b_\infty(\hat{E}_2 - \hat{E}_1)$$

for the small-signal domain excess potential which follows to first order from eqn (3.23).

The small-signal admittance of the region outside of the domain (Y_r) is then readily obtained as the current $\hat{J}_t A$, defined in eqn (3.36), divided by the voltage $\hat{V}_r$ from eqn (3.39); similarly the admittance of the domain region (Y_D) is found by making use of $\hat{J}_t A$ from eqn (3.38) and of $\hat{V}_D$ from eqn (3.39), viz.:

$$\left. \begin{aligned} Y_r &= \frac{\hat{J}_t A}{\hat{V}_r} = G_r + j\omega C_r \\[2mm] &\approx \frac{en_0\mu'_{1\infty}A}{L - b_\infty} + j\omega \frac{\varepsilon A}{L - b_\infty} \end{aligned} \right\} \quad (3.40a)$$

and

$$\left. \begin{aligned} Y_D &= \frac{\hat{J}_t A}{\hat{V}_D} = G_D + j\omega C_D \\[2mm] &\approx \frac{en_0\mu'_1 A}{b_\infty} + j\omega \frac{\varepsilon A}{b_\infty}. \end{aligned} \right\} \quad (3.40b)$$

The equivalent circuit, thus, consists of the series connection of two RC terms as shown in Fig. 3.21. The first term represents the differential conductance G_r of the part of the sample outside of the domain, which is shunted by its static capacitance C_r. The second term consists of the differential domain conductance G_D in parallel with the domain capacitance C_D which is due to the space-charge accumulation. One can easily verify that μ'_D as introduced in eqn (3.37) is the slope to the function $v_{1\infty}(E_{2\infty})$ as defined by the equal-areas rule [eqn (3.28), see Fig. 3.9] and hence G_D is always *negative*. This property can be used to obtain reflection-type amplification from a TE element oscillating in the dipole-domain mode (cf. Section 7.3). The negative conductance should be available up to frequencies which are a multiple of the domain transit-time frequency [3.35]; (see also Section 5.1.4.1). C_D, on the other hand, which is a function of the domain excess potential, is responsible for parametric effects (Section 5.1.4.2). The differential mobility given by the slope of the dymanic branches of the $v(E_h)$ characteristic (Fig. 3.15a) follows directly from this equivalent circuit at zero frequency [$\omega = 0$ in eqn (3.40)].

A small-signal equivalent circuit for the sample while being traversed by a domain, similar to that of Fig. 3.21, can also be derived using mainly plausibility arguments [3.36, 3.37]. Again by plausibility consideration [3.38, 3.39, cf. also 3.79] the equiva-

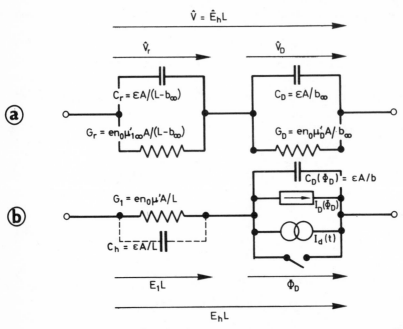

Fig. 3.21 Equivalent circuit of TE element: (a) small-signal case during transit of steady-state domain, and (b) extended circuit required for large-signal case.

lent circuit has been extended to include domain formation and annihilation, i.e. to cover the time-depending large-signal case (Fig. 3.21b). G_D is replaced by a parallel circuit of a non-linear element $I_D(\Phi_D)$, a current generator $I_a(t)$, and a switch. $I_D(\Phi_D)$ describes the drift current carried by the domain during formation and propagation and is a decreasing function of the domain excess potential Φ_D only. It is well approximated [3.66] by the current $I_\infty = J_\infty A$ carried in a steady-state domain at an excess potential Φ_D. $J_\infty(\Phi_D)$ follows from Butcher's equal-areas rule [eqn (3.28a)] in the representation of Fig. 3.13 with the abscissa $E_{1\infty}$ converted to current density using eqns (3.27) and (3.24). In the saturation region at high Φ_D values I_∞ is approximately a d.c. current generator. The current generator $I_a(t)$ constitutes the discharging current occurring when the domain is extinguished at the anode and depends on time explicitly. It is a current pulse generator since this current is present only when the domain reaches the anode. Its magnitude is proportional to the charge on the domain. The switch, finally, shorts out the domain part of the equivalent circuit when the domain is absent. It must open when the voltage across G_r reaches the threshold value V_T required for domain formation and closes again when the domain disappears at the anode. Since the domain part of this large-signal equivalent circuit consumes only the domain *excess* potential Φ_D, the low-field conductance G_1 (field range E_1) of the *entire* sample has to be placed in series. For most cases of interest the geometric capacitance C_h of the sample is negligible. Similar conclusions were obtained from more rigorous numerical calculations of the small-signal admittance of a dipole domain oscillator [3.81].

If, as usual, the semiconductor chip is put into an encapsulation (see Section 6.2.1), the equivalent circuit of the device has to be supplemented by the corresponding additional circuit elements due to the package, cf. [3.41], i.e. mainly by a series inducted (connecting lead) and parallel capacitance (case capacitance); see p. 262.

3.5 RESONANT-CIRCUIT OSCILLATOR OPERATION

3.5.1 General Remarks

As we have seen during the discussion of the space-charge dipole dynamics a sample of a semiconductor which exhibits the TE effect can break into large-signal instabilities manifested by current relaxation oscillations. According to eqn (3.34) these (transit-time) oscillations may have fundamental frequencies well within the microwave region. By connecting a resistive load circuit ($R_L > 0$) to the sample—this only quantitatively changing the behaviour as compared to the constant-voltage case ($R_L = 0$) considered so far—the oscillatory power could be put to use. However, the use of a *resonant* load circuit will have advantages, for example, with regard to efficiency (Section 5.2), frequency stability and noise (Sections 5.3 and 7.1.2.5), and also makes possible a tuning of the frequency (Sections 3.5.2, 3.5.3 and 7.1.2.2). Thus like other types of oscillator, transferred-electron oscillators are generally operated in this way.

Results of computer simulations [3.10, 3.54] of the behaviour of a GaAs element connected to a parallel RCL circuit as in Fig. 3.22, will be presented in Section 5.2.4.2.

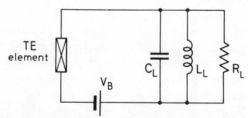

Fig. 3.22 Principal circuit of a biased TE device connected to a parallel resonant circuit containing a resistive component R_L.

Unfortunately, these calculations do not provide a particular clear understanding of the underlying effects when the sample is subjected to an a.c. voltage superimposed on the d.c. bias. For this reason the basic mechanism of the two principal modes of resonant-circuit operation will be analysed here in a mainly descriptive manner, showing that they rely on the control of the threshold field values E_T and E_S by the a.c. field, i.e. on the control of the nucleation and extinction processes. Special properties of these modes will be considered in Section 5.2.4.1, making use of the results of simplified-model calculations [3.46, 3.48].

The complications which arise when a GaAs sample, which is able to form mature dipole domains, is connected to an external circuit, were discussed in detail by Gunn [3.42]. He differentiates between (1) a fundamental transit-time mode, (2) a harmonic transit-time mode, (3) an under-voltage resonant mode, (4) an over-voltage resonant mode and finally (5) a mixed mode. The first two modes are obtained if the resonant circuit is tuned close to the transit-time frequency of the TE element or, since the current may be very non-sinusoidal (Fig. 3.20), to one of its higher harmonics, and if the overall bias field never drops below the threshold E_T. In the under-voltage resonant mode [3.43, 3.38] the domain is annihilated before it has reached the anode when the terminal voltage $E_h L$ swings below the sustaining voltage $E_S L$. This mode, which now is generally termed 'quenched dipole-domain mode', has proved to be of great practical interest and is described below in Section 3.5.3. Domain extinction during transit can also be achieved if the field E_h swings sufficiently high [mode (4)] to cause avalanche multiplication within the domain. Then the average current density will eventually exceed the peak value J_p, leading to the nucleation of a further domain while the first one is still present. The two domains compete with each other and, as can be shown by thermodynamical reasoning, the new one takes over. However, this mode is rather unstable and noisy [3.42]. The second operational mode of considerable practical interest is the mixed mode (5) which is now called 'delayed dipole-domain mode', sometimes also 'inhibited domain mode'. This mode [3.44, 3.45], dealt with in the following section, is characterized by the terminal voltage $E_h L$ being below $E_T L$ when the dipole domain arrives at the anode. Mode (1) can be regarded as a limiting case of the delayed-dipole mode, since the voltage amplitude becomes too small to interfere with the domain nucleation process.

When a TE device is operated with a resonant circuit in a dipole mode, the negative differential domain conductance G_D is externally observable only at frequencies below approximately f/Q, where f is the frequency of the circuit-controlled domain oscillation and Q the quality factor of the circuit. At higher frequencies, the amplitude of the a.c. voltage cannot respond to external voltage variations. An exception to this may occur in the delayed-dipole mode for particular operating parameters [3.63] where an external negative conductance may be obtained (see Section 5.1.4.1).

3.5.2 Delayed Dipole-Domain Mode

Figure 3.23, upper left, shows the 'time-dependent' $v(E_h)$ characteristic of a GaAs TE element calculated for $n_0 L = 10^{12}$ cm^{-3} (see Fig. 3.15a). As indicated in Fig. 3.23, we now assume a terminal voltage $E_h L$ consisting of a d.c. component $E_{h0} L$ ($= E_T L$ for simplicity) on which an a.c. component $\tilde{E}_h L$ is superimposed with a frequency equal to the resonant frequency of the overall circuit in Fig. 3.22. The a.c. circuit is assumed to support a pure sinusoidal voltage waveform, $\tilde{E}_h = \hat{E}_h \sin \omega t$, only. Additionally, the Q factor of the circuit is taken to have such a value that the terminal voltage does not swing below the domain sustaining value $V_S = E_S L$. For simplification it is further assumed, here and in the subsequent treatment of quenched-domain operation, that the domain formation and extinction times are negligibly short compared to the transit time τ of the domain and to the oscillation period T. If, additionally, the geometrical sample capacitance C_h is neglected the total current in the sample is solely given by the drift current of the 'time-dependent' $I(V)$ characteristic.

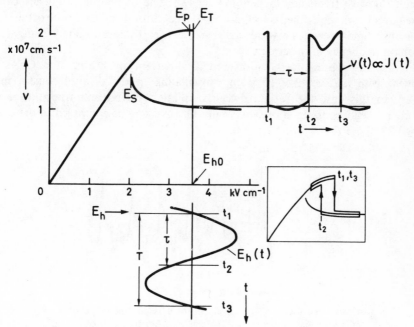

Fig. 3.23 Illustration of delayed dipole-domain operation using the static and the dynamic branches of the $v(E_h)$ characteristic ($n_0L = 10^{12}$ cm^{-2}).

Assuming, now, the relation

$$T/2 \leqq \tau \leqq T \tag{3.41}$$

between domain transit time and the oscillation period of the resonant circuit, we will have the following process. At the time instant t_1 when the threshold is reached, a domain is formed and the current drops to the corresponding value on the dynamic branch. For $E_T > E_p$ carrier injection from the cathode actually results in $\bar{n} > n_0$ (see Section 4.4.1.3). This effect has been neglected here, i.e. it is assumed $\bar{n} = n_0$, yielding $I(E_h) \propto v(E_h)$. Owing to the assumption of negligible domain formation time the transition from the static to the dynamic branch (and vice versa) takes place instantaneously. Later, when the domain has reached the anode after the time interval τ at $t = t_2$, the nucleation and formation of a second domain at the cathode is not possible since now the terminal voltage E_hL is below E_TL. Only after the period T has expired, a domain can form again. The inset in Fig. 3.23 illustrates which parts of the characteristic are traversed during one cycle.

67

Because of the delay in domain formation the sample current oscillates with a fundamental component of frequency $1/T$, and not $1/\tau$. As Fig. 3.23 shows, a (drift) current pulse train with a duty factor approaching 50% for $T \to 2\tau$ can be obtained. The TE element may be considered to be locked to the resonant frequency of the circuit ('passive frequency locking', cf. [3.47]). The oscillation frequency can be tuned over the range defined by relation (3.41), i.e. theoretically over one octave from the transit-time frequency towards lower frequencies, by appropriately changing the resonant frequency of the circuit.

Figure 3.24 represents the measured total-current waveform of a GaAs TE element with $n_0 L \approx 1{\cdot}2 \times 10^{12}$ cm^{-2} oscillating in the delayed-domain mode [3.49]. In this case the ratio between oscillation period and transit time was $T/\tau \approx 1{\cdot}5$, leading to the more spiky current waveform as compared to the one in Fig. 3.23 where $T/\tau \approx 1{\cdot}8$.

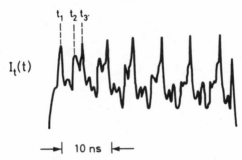

Fig. 3.24 Measured total-current waveform for delayed dipole operation of n-type GaAs device ($n_0 L \approx 1{\cdot}2 \times 10^{13}$ cm^{-2}). After Ikoma *et al.* [3.49].

3.5.3 Quenched Dipole-Domain Mode

The frequency can also be varied in another mode, involving dipole quenching. This mode [3.43] relies on the fact that the travelling dipole domain is annihilated if the terminal voltage $E_h L$ across the sample drops below the domain sustaining value $E_s L$ (Fig. 3.25). The electric energy stored in the domain is then dissipated by the ohmic sample resistance $R_1 = 1/G_1$ (see Fig. 3.21b). No steady-state domain can exist, according to the statements in Section 3.4.4.2, below the field E_s, but the suppression of an existing *non*-steady-state domain may require E_h to drop still further [3.34]. For most practical purposes it is, however, sufficient to consider E_s as the field value at which complete domain suppression occurs.

The mechanism of the quenched-domain mode is illustrated in Fig. 3.25 for which the assumptions are the same as for Fig. 3.23, with the exception of the resonant circuit now being less heavily loaded so that the a.c. space-averaged field $\tilde{E}_h$ swings *below* E_s. Then, at the instant t_1, a domain is nucleated and formed. This causes the (drift) current to drop to the value defined by the dynamic characteristic. The dynamic branch is subsequently passed (see inset) until at $t = t_2$ the domain is quenched before it has reached the anode ($\tau > T$ in Fig. 3.25). After the quenching of the domain, the drift velocity (current) is determined by the static branch until, at $t = t_3$, a new domain will be formed and the cycle repeats. Thus, in this mode too, the oscillation has a fundamental frequency $1/T$ which corresponds to the resonant frequency of the overall circuit.

Figure 3.26 shows the measured *total*-current waveform and the associated terminal voltage for quenched-dipole operation of the same GaAs TE element for which the oscillograms in Fig. 3.24 have been obtained [3.49]. The left (lower) current peak in the *drift*-current waveform shown in Fig. 3.25 has practically disappeared. This is probably the result of the finite time actually required for the quenching process.

Frequency tuning, by altering the resonant-circuit frequency, in the quenched mode with sinusoidal a.c. voltage swing is theoretically possible down to $f \approx 2/3\tau$. The lowest oscillating frequency attainable depends on the magnitude of the bias field E_{h0} and on the established amplitude $\hat{E}_h$ of the a.c. field. For frequencies above $1/\tau$ the length of the TE element traversed by the domain decreases. The upper frequency limit is determined by the discharging (quenching) time for the domain. This time, in turn, is governed by the (positive) dielectric relaxation time of the material outside of the domain since it equals the RC time constant of the circuit formed by the domain capacitance C_D and the positive resistance $1/G_1$ of the sample in series with the domain (see equivalent circuit Fig. 3.21b and Section 5.1.2).

It should be possible to raise the upper frequency limit of the quenched-domain mode by initiating multiple dipole domains instead of only a single domain [3.50, 3.51 and 3.11]. Multiple domains are obtained if the passage of the bias field from below to above-threshold values occurs sufficiently fast [3.2, 3.3b and 3.50]. In this case domains are nucleated not only near the cathode but also at the random doping fluctuations along the sample length. Such multiple-domain formation can be obtained during the first oscillation cycle in samples operated with resistive or sufficiently low-Q load circuits if the bias voltage is rapidly turned on ($\mathrm{d}V/\mathrm{d}t \gtrsim 10^{12}\ \mathrm{Vs^{-1}}$) [3.22, 3.23 and 3.50]. This was indeed observed in computer simulations [3.2] and also experimentally [3.3b, 3.50]. In the computer calculations it was found [3.11] that the number of the dipole domains which are formed simultaneously roughly equals the factor by which the $n_0 L$ product exceeds the 'critical value' (see Section 4.4.1.2, $[n_0 L]_{\mathrm{crit}} = 10^{11}$ to $10^{12}\ \mathrm{cm^{-2}}$). After the last dipole (of the multiple ones) has reached the anode in the first cycle, only *one* dipole is normally formed during the following cycles at the most pronounced nucleation centre since the electric field throughout the sample can now only rise at a slow rate governed by the discharge of the domains at the anode.

By having a high-Q resonant load circuit instead of a resistive or low-Q load, the rate of the bias change may at every cycle be sufficiently high to initiate multiple dipole domains. The important point is now that in multiple-domain operation a number of domains with their associated capacitances lie in series, which means that the overall capacitance of the RC circuit determining the speed of quenching is

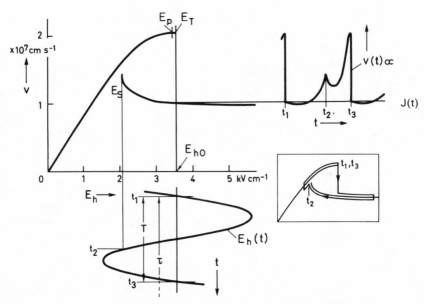

Fig. 3.25 Illustration of quenched dipole-domain operation ($n_0 L = 10^{12}$ cm^{-2}).

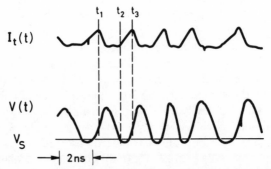

Fig. 3.26 Measured total-current waveform and associated terminal voltage for quenched dipole operation of n-type GaAs device ($n_0 L \approx 1\cdot2 \times 10^{13}$ cm^{-2}). After Ikoma *et al.* [3.49].

correspondingly reduced. A higher frequency limit than for single-domain quenching should, therefore, be expected [3.50] as discussed in more detail in Section 5.1.2.

The quenched multiple-domain mode is illustrated in Fig. 3.27. Calculated spatial field distributions at four consecutive time instants are shown for an oscillation frequency of about ten times the transit-time frequency [3.11]. The multiple dipoles, which begin to develop at $t = t_1$, have grown to their mature state at $t = t_2$. They have been quenched at $t = t_3$, and the remaining field variations are due to the assumed random doping fluctuations. Most of the local sample field is at above-threshold values for an appreciable fraction of the cycle. Consequently, the negative differential conductance of almost the whole sample is available during that time (this is already a feature of hybrid operation as discussed in Section 3.5.4 below).

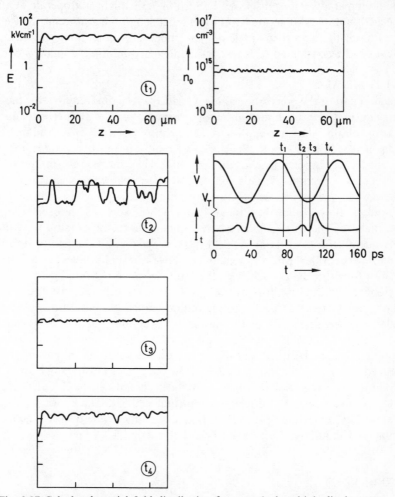

Fig. 3.27 Calculated spatial field distribution for quenched multiple-dipole operation at four consecutive time instants during oscillation cycle. The corresponding doping profile and waveforms of terminal voltage and total current are indicated. [$L = 70\ \mu$m, $f = 14\cdot3$ GHz, $(n_0/f) = 4 \times 10^4$ s cm^{-3}, $n_0 L = 4\cdot2 \times 10^{12}$ cm^{-2}, $(\delta n_0/n_0)_{r.m.s.} = 11\%$.] After Thim [3.11].

Quenched multiple-dipole operation certainly appears to be an interesting and effective way for achieving TE oscillator operation not limited to (or near) the transit-time frequency of the element, c.f. [3.59]. Although non-transit-limited operation has been achieved experimentally (for further modes free of the transit-time limitation see Sections 3.5.4 and 4.2), the particular *quenched multiple-dipole* mode has not so far been positively identified. This might be due to the fact that samples with a pronounced nucleation centre or with localized inhomogeneities (e.g. doping gradients) as discussed in Section 3.3.2 might not operate at all in the quenched multiple-dipole mode. On the other hand, when a number of elements operating in the quenched single-domain mode are series-connected, the overall system can be considered to work in the quenched-multiple domain mode [3.71].

3.5.4 Hybrid Mode
In the last part of the preceding section we saw that the time required for the domain quenching may play an important role in determining certain properties of quenched-mode operation. The same applies to the domain formation time in all the resonant modes if it is not negligible. Space-charge build-up and decay rates are, in fact, entirely decisive in the quenched accumulation-layer mode (LSA mode) which will be discussed in Section 4.2. This applies also to a possible intermediate-state mode, constituting a *hybrid* between the mature-dipole domain modes and the quenched accumulation-layer mode. In this hybrid mode the duration of the oscillation period, determined by the resonant circuit, is comparable to the dipole formation time. Consequently the drop in outside field E_1 ($\propto J$, see Fig. 3.18) is greatly delayed, and for most of the cycle the field E_1 remains at above-threshold values as, to some extent, it is also found in the quenched multiple-dipole mode (Fig. 3.27). In this case the developing field inhomogeneity distorts the static $v(E)$ characteristic somewhat but negative conductivity according to the falling branch of the $v(E)$ curve still remains available at the terminals of the device for sustaining oscillations. The (single) dipole, which is not fully formed, may in this mode either reach the anode and there be annihilated, or be quenched during its transit if the a.c. swing of the space-averaged field E_h is sufficient to extend below the domain sustaining field E_s.

Hybrid-mode operation was discovered during computer simulations [3.52] of TE devices and subsequently proved experimentally [3.53]. The hybrid mode has the advantage of permitting operation at frequencies above the transit-time frequency combined with relatively high efficiencies (see Section 5.2.4.1).

REFERENCES

3.1   Bonch-Bruevich, V. L. and Kogan, S. M., The formation of domains in semiconductors with negative differential resistance. *Fizika Tverdogo Tela* **7** (1965), p. 23. [In Russian, English transl.: *Sov. Phys. Solid-State* **7** (1965), p. 15.]

3.2   Kroemer, H., Nonlinear space-charge domain dynamics in a semiconductor with negative differential mobility. *IEEE Trans. Electron Dev.* **ED–13** (1966), p. 27.

3.3a Hasty, T. E., Stratton, R. and Jones, E. L., Effect of nonuniform conductivity on the behaviour of Gunn effect samples. *J. Appl. Phys.* **39** (1968), p. 4623.

3.3b Ohtomo, M., Nucleation of high-field domains in n-GaAs. *Jap. J. Appl. Phys.* **7** (1968), p. 1368.

3.4   McCumber, D. E. and Chynoweth, A. G., Theory of negative-conductance amplification and of Gunn instabilities in 'two-valley' semiconductors. *IEEE Trans. Electron Dev.* **ED–13** (1966), p. 4.

3.5a Kroemer, H., The Gunn effect under imperfect cathode boundary conditions. *IEEE Trans. Electron Dev.* **ED–15** (1968), p. 819.

3.5b Böer, K. W. and Döhler, G., Influence of boundary conditions on high-field domains in Gunn diodes. *Phys. Rev.* **186** (1969), p. 793.

3.5c Conwell, E. M., Boundary conditions and high-field domains in GaAs. *IEEE Trans. Electron Dev.* **ED–17** (1970), p. 262.

3.6a Shaw, M. P., Solomon, P. R. and Grubin, H. L., The influence of boundary conditions on current instabilities in GaAs. *IBM J. Res. Dev.* **13** (1969), p. 587.

3.6b Shaw, M. P., Solomon, P. R. and Grubin, H. L., A comparison of current instabilities in n-GaAs and n-CdS. *Solid-State Commun.* **7** (1969), p. 1619.

3.7 Gunn, J. B., Properties of a free, steadily travelling electrical domain in GaAs. *IBM J. Res. Dev.* **10** (1966), p. 300.

3.8 Yamashita, A. and Nii, R., Trapping of high field domain in n-type GaAs. *Jap. J. Appl. Phys.* **5** (1966), p. 263.

3.9 Denker, S. P., Rational design of Gunn- and L.S.A.-diode electrodes. *Electron. Lett.* **4** (1968), p. 294.

3.10 Copeland, J. A., Theoretical study of a Gunn diode in a resonant circuit. *IEEE Trans. Electron Dev.* **ED–14** (1967), p. 55.

3.11 Thim, H. W., Computer study of bulk GaAs devices with random one-dimensional doping fluctuations. *J. Appl. Phys.* **39** (1968), p. 3897.

3.12 Butcher, P. N., Theory of stable domain propagation in the Gunn effect. *Phys. Lett.* **19** (1965), p. 546.

3.13 Butcher, P. N., Fawcett, W. and Hilsum, C., A simple analysis of stable domain propagation in the Gunn effect. *Brit. J. Appl. Phys.* **17** (1966), p. 841.

3.14 Butcher, P. N. and Fawcett, W., Stable domain propagation in the Gunn effect. *Brit. J. Appl. Phys.* **17** (1966), p. 1425.

3.15 Butcher, P. N., Fawcett, W. and Ogg, N. R., Effect of field dependent diffusion on stable domain propagation in the Gunn effect. *Brit. J. Appl. Phys.* **18** (1967), p. 755.

3.16 Allen, J. W., Shockley, W. and Pearson, G. L., Gunn domain dynamics. *J. Appl. Phys.* **37** (1966), p. 3191.

3.17 Knight, B. W. and Peterson, G. A., Domain velocity, stability, and impedance in the Gunn effect. *Phys. Rev. Lett.* **17** (1966), p. 257.

3.18 Knight, B. W. and Peterson, G. A., Theory of the Gunn effect. *Phys. Rev.* **155** (1967), p. 393.

3.19 Heindle, W., Phenomenological theory of the Gunn effect (in German). *Verhandl. Deutsche Physikal. Gesellsch.* **2**/VI (1967), p. 52. (European Meeting 'Semiconductor Device Research', Bad Nauheim, April 19–21, 1967.)

3.20 Heinle, W., Basic equations of Gunn domain dynamics. *Phys. Lett.* **24A** (1967), p. 533.

3.21a Heinle, W., Principles of a phenomenological theory of Gunn-effect domain dynamics *Solid-State Electron.* **11** (1968), p. 583.

3.21b Heinle, W., Theory of the Gunn effect (in German). *Internat. Elektron. Rundschau* **22** (1968), p. 546.

3.22 Kurokawa, K., Transient behaviour of high-field domains in bulk semiconductors. *Proc. IEEE* **55** (1967), p. 1615.

3.23 Kurokawa, K., The dynamics of high-field propagating domains in bulk semiconductors. *Bell Syst. Techn. J.* **46** (1967), p. 2235.

3.24 Copeland, J. A., Electrostatic domains in two-valley semiconductors. *IEEE Trans. Electron Dev.* **ED–13** (1966), p. 189.

3.25 Copeland, J. A., Stable space-charge layers in two-valley semiconductors. *J. Appl. Phys.* **37** (1966), p. 3602.

3.26 Heinle, W., Gunn domain propagation in non-uniformily doped samples. *Brit. J. Appl. Phys.* **18** (1967), p. 1537.

3.27 Gunn, J. B., Electronic transport relevant to instabilities in GaAs. *Proc. Int. Conference Phys. Semiconductors Kyoto 1966*, in *J. Phys. Soc. Japan* **21** (1966), Suppl., p. 505.

3.28 Foyt, A. G. and McWhorter, A. L., The Gunn effect in polar semiconductors. *IEEE Trans. Electron Dev.* **ED–13** (1966), p. 79.

3.29 Kuru, I., Robson, P. N. and Kino, G. S., Some measurements of the steady-state and transient characteristics of high-field dipole domains in GaAs. *IEEE Trans. Electron Dev.* **ED–15** (1968), p. 21.

3.30 U.S. Patent No. 3, 365, 583 (inventor: J. B. Gunn, priority: June 12, 1964).

3.31 Heeks, J. S., Woode, A. D. and Sandbank, C. P., Coherent high-field oscillations in long samples of GaAs. *Proc. IEEE* **53** (1965), p. 554.

3.32 Heeks, J. S., Some properties of the moving high-field domain in Gunn effect devices. *IEEE Trans. Electron Dev.* **ED–13** (1966), p. 68.

3.33 Carroll, J. E., Nonuniform motion of high-field domains in the Gunn effect. *Electron. Lett.* **2** (1966), p. 194.

3.34 Heinle, W., AEG-Telefunken Research Institute, private communication (1969).

3.35 Heinle, W., On the equivalent circuit of a Gunn diode. *J. Electronics* **23** (1967), page 541.

3.36a Hobson, G. S., Small-signal admittance of a Gunn-effect device. *Electron. Lett.* **2** (1966), p. 207.

3.36b Hobson, G. S., The equivalent circuit of a Gunn-effect device. *Proc. 1966 MOGA Conference*, IEE Conf. Publ. No. 27, p. 314.

3.37 Matino, H. and Kuru, I., Reactance of GaAs bulk oscillator. *Proc. IEEE* **54** (1966), p. 291.

3.38a Carroll, J. E., Mechanisms in Gunn effect microwave oscillators. *Rad. Electron. Engng* **34** (1967), p. 17.

3.38b Carroll, J. E. and Giblin, R. A., Low frequency analog for a Gunn-effect oscillator. *IEEE Trans.* **ED–14** (1967), p. 640.

3.39a Robrock, II, R. B., A lumped model for characterizing single and multiple domain propagation in bulk GaAs. *IEEE Trans.* **ED–17** (1970), p. 93.

3.39b Robrock, II, R. B., Extension of the lumped bulk device model to incorporate the process of domain dissolution. *IEEE Trans.* **ED–17** (1970), p. 103.

3.39c Robrock, II, R. B., Effect of device parameters on domain dynamics in bulk GaAs. *Proc. IEEE* **58** (1970), p. 804.

3.40 Levinstein, M. E. and Shur, M. S., Behaviour of the high-field domains below the voltage of nucleation threshold. *Phys. Stat. Sol.* **28** (1968), p. 827.

3.41a Getsinger, W. J., The packaged and mounted diode as a microwave circuit. *IEEE Trans. Microwave Theory Tech.* **MTT–14** (1966), p. 58.

3.41b Getsinger, W. J., Mounted diode equivalent circuits. *IEEE Trans. Microwave Theory Tech.* **MTT–15** (1967), p. 650.

3.41c Spiwak, R. R., A low-inductance millimeter wave semiconductor package. *IEEE Trans. Microwave Theory Tech.* **MTT–19** (1971), p. 732.

3.42 Gunn, J. B., Effect of domain and circuit properties on oscillations in GaAs. *IBM J. Res. Dev.* **10** (1966), p. 310.

3.43 Carroll, J. E., Oscillations covering 4 Gc/s to 31 Gc/s from a single Gunn diode. *Electron. Lett.* **2** (1966), p. 141.

3.44 Robson, P. N. and Mahrous, S. M., Some aspects of Gunn effect oscillators. *Symp. Microwave Applications of Semiconductors, London*, June 30 to July 2, 1965, and *Rad. Electron. Engng* **30** (1965), p. 345.

3.45 U.K. Patent No. 1, 108, 372 (inventors: B. G. Bosch and H. Pollmann, priorities: June 9 to Dec. 4, 1965).

3.46   Shevchenko, T. G. and Tsvirko, Y. A., Gunn-effect cavity-controlled generator. *Electron. Lett.* **5** (1969), p. 107.

3.47   Bosch, B. G. and Pollmann, H., Frequency synchronisation of Gunn-effect oscillators. *IEEE Trans. Electron Dev.* **ED–13** (1966), p. 194.

3.48a   Heinle, W., Determination of current waveform and efficiency of Gunn diodes. *Electron. Lett.* **3** (1967), p. 52.

3.48b   Levinstein, M. E. and Shur, M. S., Calculation of the parameters of the microwave Gunn generator. *Electron. Lett.* **4** (1968), p. 233.

3.49a   Ikoma, T., Torizuka, H. and Yanai, H., Observations of current waveforms of the transferred-electron oscillators. *Proc. 1968 MOGA Conference, Nachrichtentechn. Fachberichte* **35** (1968), p. 388.

3.49b   Ikoma, T., Torizuka, H. and Yanai, H., Observations of voltage and current waveforms of the transferred electron oscillators. *Proc. IEEE* **57** (1969), p. 340. Erratum: *Proc. IEEE* **57** (1969), p. 1466.

3.50   Thim, H. W. and Barber, M. R., Observation of multiple high-field domains in n-GaAs. *Proc. IEEE* **56** (1968), p. 110.

3.51   Barber, M. R., High-power quenched Gunn oscillators. *Proc. IEEE* **56** (1968), p. 752.

3.52   Bott, I. B. and Fawcett, W., Theoretical study of the effect of temperature on X band Gunn oscillators. *Electron. Lett.* **4** (1968), p. 207.

3.53   Huang, H. C. and MacKenzie, L. A., A Gunn diode operated in the hybrid mode. *Proc. IEEE* **56** (1968), p. 1232.

3.54   Mantena, N. R. and Wright, M. L., Circuit model simulation of Gunn effect devices. *IEEE Trans. Microwave Theory Techn.* **MTT–17** (1969), p. 363.

3.55   Robinson, G. Y., White, R. M. and MacDonald, N. C., Probing of Gunn effect domains with a scanning electron microscope. *Appl. Phys. Lett.* **13** (1968), p. 407.

3.56   Copeland, J. A., Growth of two- and three-dimensional space charge from negative differential resistivity. *J. Appl. Phys.* **39** (1968), p. 5101.

3.57   Wessel-Berg, T., Conduction processes, normal modes, and drift instabilities in bulk semiconductors. Microwave Lab., Stanford University, Stanford, Calif., ML Report No 1315, April 1965.

3.58   Pötzl, H. W. and Richter, K., Microwave conductivity anisotropy of hot electrons in n-InSb at 77° K. *Proc. IEEE* **55** (1967), p. 1497.

3.59   Harrison, R. I., Denker, S. P. and Berger, H., A theoretical appraisal of the quenched multiple domain mode in GaAs microwave diodes. *Rad. Electron. Engng* **37** (1969), p. 11.

3.60   Lundström, I., Initial growth of domains in Gunn oscillators. *Electron. Lett.* **4** (1968), p. 120.

3.61   Harrison, R., Simple transient and nonlinear analysis of high-field domains in GaAs. *Brit. J. Appl. Phys.* **1** (1968), p. 973.

3.62   Pokorny, J., Initial time constants of the Gunn-domain space-charge growth. *Electron. Lett.* **5** (1969), p. 452.

3.63   Hobson, G. S., The external negative conductance of Gunn oscillators. *Solid-State Electron.* **12** (1969), p. 711.

3.64   Ohmi, T., Takeoka, Y. and Nishimaki, M., Observations of high-field domain widths in bulk GaAs oscillators. *Proc. IEEE* **56** (1968), p. 2188.

3.65   Gelmont, B. L. and Shur, M. S., Analytical theory of stable domains in high-doped Gunn diodes. *Electron. Lett.* **6** (1970), p. 385.

3.66   Engelmann, R. and Heinle, W., Pulse discrimination by Gunn-effect switching. *Solid-State Electron.* **14** (1971), p. 1.

3.67a   Yanai, H., Suzuki, N., Sugeta, T. and Tanimoto, M., Effect of electrode structure in dipole-domain formation. *Proc. Int. Symp. Gallium Arsenide and Related Compounds, Aachen, Oct. 1970*, p. 153 (Institute of Physics and Physical Society, London 1971).

3.67b Suzuki, N., Yanai, H. and Ikoma, T., Simple analysis and computer simulation on lateral spreading of space charge in bulk GaAs. *IEEE Trans. Electron Dev.* **ED–19** (1972), p. 364.

3.68   Shoji, M., Theory of transverse extension of Gunn domains. *J. Appl. Phys.* **41** (1970), p. 774.

3.69   Freeman, K. R., Sozou, C. and Hartnagel, H. L., Two-dimensional Gunn-domain growth in bulk GaAs. *Phys. Lett.* **34A** (1971), p. 95.

3.70   Kawashima, M. and Kataoka, S., Measurement of transverse spreading velocity of a high-field domain in a 3-terminal Gunn device. *Electron. Lett.* **6** (1970), p. 781.

3.71   Thim, H. W., Bell Telephone Labs., private communication, 1968.

3.72   Yang, P. K. and Sugano, T., Experimental observation of accumulation mode operation of bulk Ge oscillators. *IEEE Trans. Electron Dev.* **ED–18** (1971), p. 383.

3.73   Ladbrooke, P. H., Some possible consequences of a transverse doping gradient in bulk-effect oscillators. *J. Phys. D: Appl. Phys.* **3** (1970), p. 437.

3.74   Chang, K. K. N., Liu, S. G. and Prager, H. J., Infrared radiation from bulk GaAs. *Appl. Phys. Lett.* **8** (1966), p. 196.

3.75   Liu, S. G., Infrared and microwave radiations associated with a current-controlled instability in GaAs. *Appl. Phys. Lett.* **9** (1966), p. 79.

3.76a Southgate, P. D., Laser action in field-ionized bulk GaAs. *Appl. Phys. Lett.* **12** (1968), p. 61.

3.76b Southgate, P. D., Stimulated emission from bulk field-ionized GaAs. *IEEE J. Quantum Electron.* **QE–4** (1968), p. 179.

3.77   Southgate, P. D. and Mazzochi, R. T., Stimulated emission in field-ionized bulk InP. *Phys. Lett.* **28A** (1968), p. 216.

3.78   Fallmann, W. F., Hartnagel, H. L. and Mathur, P. C., Experimental Gunn-domain amplitudes vs. bias voltages of planar devices. *Phys. Stat. Sol.* **A 6** (1971), p. K77.

3.79   Kak, A. C., Gunshor, R. L. and Jethwa, C. P., Equivalent-circuit representation for stably propagating domains in bulk GaAs. *Electron. Lett.* **6** (1970), p. 711.

3.80   Tsukada, T. and Hamasaki, J., Small-signal admittance of Gunn diodes. *Electron. Commun. Japan* **52–B** (1969), p. 77.

4
Dipole-Domain Free Operation

4.1 INTRODUCTION

As stated in Chapter 1, the aim in the search for a semiconductor bulk-effect device capable of generating and amplifying power at microwave frequencies was to find a suitable effect producing negative differential conductivity (n.d.c.). A bulk effect—the Gunn effect—was indeed eventually discovered but the n.d.c. basically responsible for this effect was generally masked by the occurrence of severe field distortions (domains) and by associated transit-time relaxation oscillations as discussed in the preceding chapter.

The Gunn-effect transit-time oscillations have proved to be extremely useful by themselves and—as we have seen in the last two sections—the n.d.c. may be available directly, at least for a certain fraction of the oscillation cycle, even if dipole domains develop. Nevertheless, continuing attempts have been made to find methods of completely preventing the formation of space-charge dipole domains and, in this way, of making use of the n.d.c. as manifested by the static $v(E)$ characteristic. Such an operation seemed to promise freedom from the power and frequency limitations imposed by the transit-time effects. A number of useful domain-free operational modes have, indeed, been devised. These modes, however, do not necessarily exhibit a negative terminal conductance. For example, microwave (transmission) amplification has proved to be possible in domain-free devices by utilizing the growth of space-charge waves which results from the negative-mobility property of the carriers.

In particular, it has been established that the formation of space-charge dipoles can be prevented in a TE element if (1) the time interval during which the electric field stays in the negative-mobility region of the $v(E)$ characteristic is kept short enough, (2) the product of sample length and doping concentration is sufficiently small, or alternatively (3) the product of at least one of the sample dimension perpendicular to the carrier drift direction and, again, the doping concentration remains below a certain level. Furthermore it has been proposed that the amount of space-charge build-up can be restricted by (4) controlled carrier injection into the TE semiconductor, like injection across a p-n junction.

These methods of achieving dipole-domain-free operation, and the active modes based thereupon, will now be investigated in some detail.

4.2 QUENCHED ACCUMULATION LAYER (LSA MODE)

4.2.1 Basic Concept

In analogy to the mechanism of tunnel-diode relaxation oscillators, cf. [4.1], Atalla *et al.* proposed the operation of a GaAs element, connected to an external resonant circuit, in a 'homogeneous-field mode' [4.2]. This mode is characterized by a switching process between spatially homogeneous high-field ($E > E_p$) and low-field ($E < E_p$) states, with the switching time set approximately at the negative dielectric relaxation time for preventing excessive space-charge growth. On the other hand, when performing computer simulations on GaAs samples connected to a resonant circuit, Copeland discovered the existence of a mode with *l*imited *s*pace-charge *a*ccumulation ('LSA mode'), which also resulted in a substantially homogeneous electric field over the sample length [4.3]. Copeland's calculations constitute a refinement of the concept of the homogeneous-field mode; in particular, he was able to derive general conditions that must be fulfilled to prevent the formation of dipole domains. In the LSA mode the conversion of the d.c. power conveyed to the semiconductor into a.c. power is effected by the negative differential mobility of the individual electron rather than by the movement across the sample of space-charge bunches as in the case of the dipole-domain mode.

Consider an electron drifting, with a velocity v, in a homogeneous electric field

$$E(t) = E_0 + \hat{E} \sin (2\pi f t)$$

where $\hat{E}$ denotes the amplitude of an a.c. component. During one oscillation cycle of duration T the electron absorbs from the d.c. part of the field the mean power (d.c. Fourier coefficient)

$$P_{dc} = \frac{eE_0}{T} \int_0^T v(t) \, dt \qquad (4.1)$$

and from the a.c. part the power (fundamental Fourier coefficient)

$$P_{ac} = \frac{e\hat{E}}{T} \int_0^T v(t) \sin (2\pi f t). \qquad (4.2)$$

P_{ac} is negative, i.e. the carrier delivers power to the a.c. field, if the oscillatory part of $v(t)$ and the a.c. field are in antiphase. Such a condition can be achieved if the $v(E)$ characteristic has a range of $\mu' = dv/dE < 0$ and the quantities E_0, $\hat{E}$ are suitably chosen. Figure 4.1 shows the calculated d.c.-to-a.c. conversion efficiency $\eta = (-P_{ac})/P_{dc}$ and the a.c. resistance R_{ac} plotted against the a.c. amplitude $\hat{E}_h$ for a constant bias field of $E_{no} = 10 \text{ kV cm}^{-1}$ and using the $v(E)$ curve given in Fig. 4.2

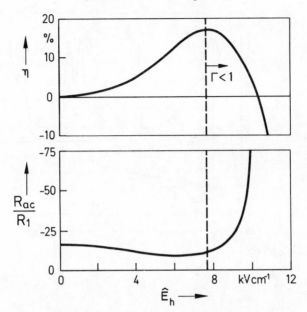

Fig. 4.1 Efficiency η and a.c. resistance R_{ac} (normalized to low-field resistance R_1) versus a.c. electric-field amplitude $\hat{E}_h$ for LSA operation of a GaAs sample biased at $E_{h0} = 10 \, \text{kV cm}^{-1}$. After Copeland [4.9].

[see eqn (3.4)]*. These efficiency values apply to an n-type GaAs sample without spatial doping fluctuations operated in the homogeneous-field mode, particularly the LSA mode (for further details see Section 5.2). The negative a.c. resistance R_{ac} is obtained from the square of the r.m.s. voltage $(\hat{E}_h L)^2/2$ divided by the total power $n_0 L A P_{ac}$ which the device under consideration delivers, where A denotes the cross-sectional area. In Fig. 4.1 the quantity R_{ac} is normalized to the positive low-field resistance $R_1 = L/(en_0\mu_1 A)$. The limitation of the space charge as necessary in this mode can, however, be achieved only over the range indicated in Fig. 4.1 by $\Gamma < 1$. This will be discussed in Section 4.2.3.

Soon after its existence was predicted, several workers claimed to have verified the LSA mode experimentally [4.3–4.6]. 'Anomalous' oscillatory behaviour of GaAs elements operated with a resonant load had been observed already prior to Copeland's discovery [4.7, 4.8] and might also be attributed to homogeneous-field operation. However, calculations carried out later on [3.11] showed that in a real sample with random spatial doping fluctuations it can be difficult to differentiate between the LSA mode, i.e. the quenched accumulation-layer mode, and the quenched multiple-dipole mode mentioned in Section 3.5.3.

The principle of preventing the formation of any substantial space-charge during

---

*The subscript h, denoting the space-average values, is used here with the electric field quantities because a minor field inhomogeneity exists in LSA-operated samples as we shall discuss below; generally one has $E \simeq E_h$.

79

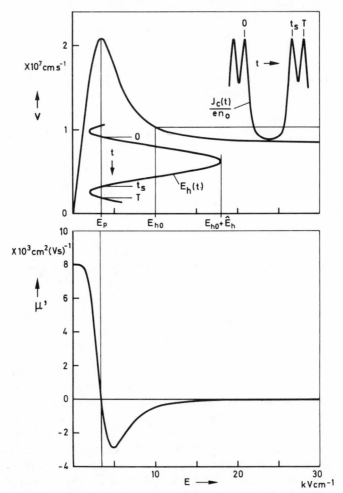

Fig. 4.2 Drift-velocity/field characteristic $v(E)$, differential mobility $\mu'(E)$, and super-imposed a.c. field $\Delta E_h(t)$, with resulting normalized drift-current density $J_c(t)/en_0$, to illustrate the principle of LSA operation. Adapted from Copeland [4.9].

LSA operation can be understood by referring to Fig. 4.2. The GaAs sample is assumed to be supplied with a sufficiently large d.c. bias field E_{h0} and connected to a resonant circuit. The quality factor of the circuit has a value such that an amplitude $\hat{E}_h$ of the a.c. field is obtained as shown, and this causes the field to swing back into the positive-mobility range ($E_h < E_p$) of the characteristic for a certain fraction of the oscillation cycle. The parameters E_{h0}, $\hat{E}_h$, $T = 1/f$, and the doping concentration n_0 now have to be adjusted appropriately. An electron accumulation detaches from the stationary accumulation layer, which according to Section 3.3 is always present at the cathode, and begins to move towards the anode. During the time interval

0 to t_s ($E_h > E_p$, $\mu' < 0$) this layer must not grow so much that it causes an appreciable distortion of the electric field. As we have seen in Section 3.2, growth of space charge is determined by the (negative) dielectric relaxation time $\tau_R = \varepsilon/en\mu'$, where $n \simeq n_0$ in LSA operation since here space-charge accumulation is supposedly small. Furthermore, the remaining fraction of the period from t_s to T ($E_h < E_p$, $\mu' > 0$) has to be long enough to ensure that the space-charge of the travelling layer is quenched during this interval to a level below that existing at the time $t = 0$. In this way, a periodic formation and quenching of a relatively weak accumulation layer is obtained merely in the vicinity of the cathode, while the major part of the sample remains free of space charge.

The calculated time and space dependence of the field distribution associated with a developing and, then, quenched accumulation layer near the cathode of an LSA-operated GaAs element can be seen from Fig. 4.3 [4.10]. Specifically, the figure shows

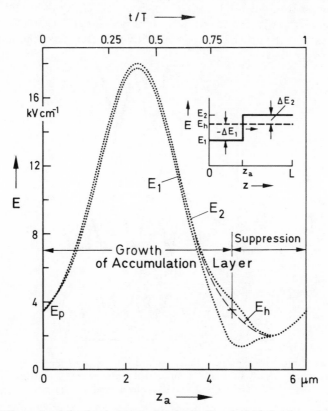

Fig. 4.3 Example of the development of cathode-side accumulation layer in LSA-operated GaAs sample. The field distributions E_1 and E_2 are plotted as a function of accumulation-layer position z_a and of time t ($E_{h0} = 10\ \mathrm{kV\ cm^{-1}}$, $\hat{E}_h = 8\ \mathrm{kV\ cm^{-1}}$, $n_0 = 10^{15}\ \mathrm{cm^{-3}}$, $L = 20\ \mu\mathrm{m}$, $f = T^{-1} = 20\ \mathrm{GHz}$, $E_2 = E_h = E_p$ and $E_2 - E_1 = 40\ \mathrm{Vcm^{-1}}$ at $t = 0$). After Heinle [3.34].

the constant electric fields in front of and behind the accumulation layer, E_2 and E_1 respectively (see inset), as a function of the location z_a of the drifting layer which, due to the neglect of diffusion, is of infinite charge density. The layer is quenched after about $0.9\ T$ when it has traversed a distance of approximately $0.55 v_p/f$, cf. [4.9]. In Fig. 4.3 an initial accumulation layer (at $z_a/L = 0$) has been assumed as would be associated with, for example, a doping inhomogeneity. In a real sample the accumulation layer at the cathode contact will, rather, be the result primarily of carrier injection during the time interval when $E_1 > E_p$. The influence of the relatively small field distortion due to the accumulation layer (of either origin) can, however, be neglected to a first degree when calculating the properties of an LSA oscillator [4.10].

Figure 4.4 shows the measured current waveforms of a GaAs element oscillating in the LSA mode [3.49]. To confirm that LSA operation had indeed been achieved, the waveform of the displacement current (middle trace) was measured separately and subtracted from the total current (upper trace). The drift current (lower trace) obtained in this way is seen to follow fairly closely the $v(E)$ characteristic (Fig. 4.2) hence indicating little field distortion which is a characteristic of this mode.

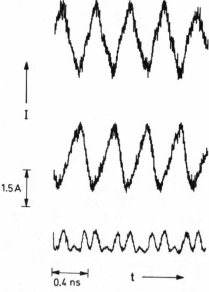

Fig. 4.4 Current waveforms of LSA-operated GaAs sample. From top to bottom: total current, negative of displacement current, drift current ($L = 100\mu$m, $n_0 \approx 2 \times 10^{14}$ cm^{-3}, $f = 2.4$ GHz). After Ikoma *et al.* [3.49].

Steady-state LSA operation constitutes a pure *large*-signal mode since E_{ho} and $\hat{E}_h$ must be of the same order of magnitude to fulfil the requirements for space-charge limitation. One possibility of initiating the LSA oscillation is based on a dipole-domain oscillation set up, at first, near the (lower) transit-time frequency [4.3, 4.6b].

A suitable higher-harmonic component of this domain oscillation, in connection with the negative domain conductance $-G_D$, may then excite an a.c. voltage across the LSA resonant circuit and thus establish a small-signal oscillation at the LSA frequency. A transition to the LSA oscillation will become possible and further domain formation will be suppressed, when the a.c. voltage has reached a value such that the total electric field in the sample drops below E_p for a sufficiently long fraction of the period T, in spite of the relatively large d.c. bias field E_{h0}. Figure 4.5 shows an oscillogram of the transient voltage waveform of such a mode transition. First a

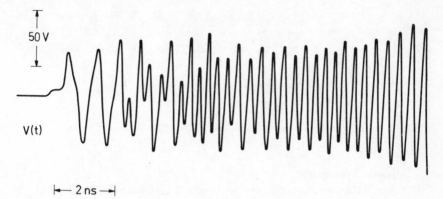

Fig. 4.5 Transient waveform of GaAs-sample voltage when starting the LSA oscillation. After Ikoma *et al.* [3.49].

domain oscillation of 0·8 GHz is dominant. After a few cycles it excites an LSA oscillation at 2·4 GHz and this grows to suppress the domain formation [3.49]. However, when using samples that are long compared to v_{h0}/f_{LSA} as desirable with LSA operation, it is advisable to avoid any domain formation because of the destructively high domain peak fields E_2 which may develop, and a different way of triggering the LSA oscillation must be employed (see Section 7.1.1).

The equivalent circuit of a TE sample operated in the LSA mode is shown in Fig. 4.6. It contains an a.c. resistance R_{ac} (see also Fig. 4.1) which is negative due to the negative value of the time-averaged mobility μ_{LSA} (defined in Section 4.2.3). In series with R_{ac} is a positive resistance R_I representing the inactive semiconducting region of length L_I at the cathode in which the electrons are accelerated. For an n-type GaAs element biased at $E_{h0} = 10$ kV cm^{-1}, this region is considered to be constant at approximately 0·4 μm [4.11], whence $R_I = 0.4 \times 10^{-4}$ cm R_1/L, i.e. it is negligible as compared to R_{ac} for most purposes. The series connection of R_I and R_{ac} is shunted by the geometrical capacitance C_h. If the sample is encapsulated, the usual parasitic elements must be added to this equivalent circuit, cf. [3.41].

Samples oscillating in the LSA mode can exhibit a negative differential conductance with respect to the mean device current at d.c. and sufficiently low frequencies [4.9, 4.12].

This negative conductance has an upper frequency limit $f_{2,\max} \approx f/Q$ analogous with that mentioned in Section 3.5.1 concerning the negative dipole-domain conductance. Here f denotes the LSA frequency and Q the quality factor of the resonant circuit (see also Section 5.1.4.1).

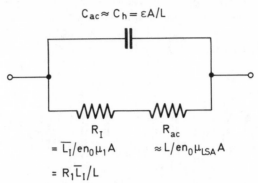

Fig. 4.6 Equivalent circuit of TE sample operated in LSA mode.

4.2.2 Analytic Approach

As with dipole-domain dynamics, the operation of TE elements in the LSA mode was treated on the one hand by comprehensive digital-computer calculations yielding exact numerical results [4.9, 4.13–4.15]*. On the other hand LSA operation was investigated in an analytic way which gives semi-quantitative, or merely qualitative, results because of the necessary simplifications, but which has the advantage of providing a better insight into the underlying physical mechanism.

A number of analytic investigations describing the dynamic behaviour of LSA devices have been carried out [4.10, 4.17, 4.18]. In the following we will sketch Heinle's approach for an n-type semiconductor starting from the basic integral eqn (3.8). Thus the same simplifications are made as those stated in Section 3.4.1. As in the treatment of dipole dynamics given in Section 3.4, a system of ordinary differential equations has to be solved. In this case, however, the equations are linear. An initial small electron accumulation layer is assumed at the cathode contact of the sample. The large-signal effect of carrier injection from the cathode contact during the maximum-field portion of each cycle is not taken into account.

Since it is assumed that there are no large deviations from the homogeneous field distribution, we need consider only first-order terms with regard to the field deviation ΔE as defined by

$$E(z, t) = E_h(t) + \Delta E(z, t) \qquad (4.3a)$$

---

*LSA operation of a GaAs element has also been simulated on an analogue computer [4.61]. The accuracy of the results obtained by an analogue computation will naturally depend on the degree of sophistication which can be retained in the modelling.

where

$$E_h = \frac{1}{L} \int_0^L E(z, t) \, dz$$

is the field in the homogeneous state. Then, the drift velocity becomes

$$v = v_h + \mu'_h \Delta E. \tag{4.3b}$$

Inserting eqns (4.3) into the basic eqn (3.8a) yields the partial differential equation

$$\frac{\partial \Delta E}{\partial t} + \frac{en_0}{\varepsilon} \mu'_h \Delta E + \frac{\partial \Delta E}{\partial z} v_h = \frac{v_h}{L} [E(L, t) - E(0, t)]. \tag{4.4}$$

This equation can be solved by separation of the variables. Here, we treat the special field distribution indicated by the inset in Fig. 4.3, i.e. $E_i(t) = E_h(t) + \Delta E_i(t)$; E_i is independent of z and $i = 1,2$. From eqn (4.4) it follows

$$\frac{d \Delta E_i}{dt} = - \frac{en_0}{\varepsilon} \mu'_h \Delta E_i - \frac{v_h}{L}(\Delta E_1 - \Delta E_2). \tag{4.5}$$

Introducing the substitutions

$$\Delta E_- = \Delta E_2 - \Delta E_1 \equiv E_2 - E_1 \quad \text{and} \quad \Delta E_+ = \Delta E_2 + \Delta E_1$$

one obtains for eqns (4.5) the solutions, cf. [4.9],

$$\Delta E_-(t) = \Delta E_-(t_0) \exp\left(- \frac{en_0}{\varepsilon} \int_{t_0}^t \mu'_h \, dt \right) \tag{4.6a}$$

$$\Delta E_+(t) = \Delta E_-(t) \left(\frac{\Delta E_+(t_0)}{\Delta E_-(t_0)} + \frac{2}{L} \int_{t_0}^t v_h \, dt \right) \tag{4.6b}$$

which describe the time dependence of the field deviations and, thus, also that of the charge per unit area $\varepsilon \Delta E_- \equiv \varepsilon(E_2 - E_1)$, during the interval t_0 to t. Together with the voltage balance condition

$$E_h L = E_1 z_a + E_2(L - z_a)$$

eqns (4.6) completely determine the dynamic behaviour of the accumulation layer. A graphical representation, for a particular set of operating parameters such that the space-charge control conditions (Section 4.2.3) are met, is given in Fig. 4.3 (with $t_0 = 0$). There, the time scale is converted to distance using the formula (3.26) for the velocity of the accumulation layer.

The total current density is (with $i = 1$ or 2)

$$J_t = \varepsilon \frac{dE_i}{dt} + en_0 v_i$$

from which follows by consideration of eqns (4.5) and (4.6):

$$J_t = \varepsilon \frac{dE_h}{dt} + en_0 v_h \left(1 + \frac{\varepsilon \Delta E_-(t)}{en_0 L}\right). \tag{4.7}$$

The second term in the bracket constitutes the charge in the accumulation layer relative to the fixed charge of all the donors in the sample and may be considered negligibly small as compared to unity in this LSA approach. Thus, the total current density is determined by the terminal voltage $V(t) = E_h(t)L$ not only in the zero-order approximation [4.9] but also in this first-order approximation. Using eqn (4.7) and carrying out a Fourier analysis, the efficiency and output power of an LSA device can be determined [4.10] if the space-charge control conditions, as described in the following section, are taken into account (see Section 5.2).

4.2.3 Space-charge Control

In this section we investigate the conditions that have to be fulfilled in order to avoid the formation of appreciable space charge. We follow Copeland's small-signal, kinematical treatment which he supplemented by a rigorous computer analysis [4.9].

Making use of eqn (4.6a) we define a growth factor Γ_N for the density of the drifting charge per unit area $\rho_A = \varepsilon \Delta E_-$. For an electric-field swing as given in Fig. 4.2, with a negative-mobility portion from $t_0 = 0$ to $t = t_s$, this growth factor is

$$\Gamma_N \equiv \frac{\rho_A(t_s)}{\rho_A(0)} = \exp\left(-\frac{e}{\varepsilon} \frac{n_0}{f} \mu_N\right) \tag{4.8a}$$

where the definition

$$\mu_N = \frac{1}{T} \int_0^{t_s} \mu'_h \, dt \tag{4.8b}$$

has been used. Since always $\mu_N < 0$, one has $\Gamma_N > 1$. The factor Γ_N must remain below a critical limit Γ_c in order to prevent small space-charge perturbations from growing too strongly, which would result in marked field distortions. Hence one has

$$\frac{n_0}{f} < h_N \ln \Gamma_c \tag{4.9a}$$

where

$$h_N = \frac{\varepsilon}{e(-\mu_N)} \tag{4.9b}$$

i.e. there is an *upper* limit for the ratio of doping concentration to operating frequency. The space-charge control parameter h_N, depending only on E_{h0}, $\hat{E}_h$ and μ', is shown in Fig. 4.7 as a function of the a.c. field amplitude $\hat{E}_h$. For calculating this curve the $v(E)$ characteristic of Fig. 4.2, as expressed by eqn (3.4), has been used.

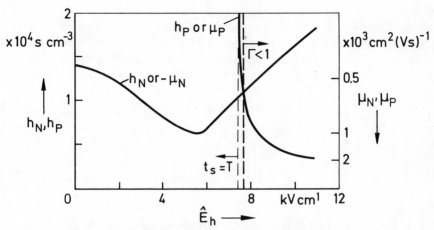

Fig. 4.7 Space-charge control factors h_N, h_P or time-averaged mobilities $- \mu_N$, μ_P, as a function of the a.c. field amplitude $\hat{E}_h$ for GaAs sample biased at $E_{h0} = 10 \, \text{kV cm}^{-1}$ on abscissa. Adapted from Copeland [4.9].

In addition it must be ensured that the space charge does not grow during a complete r.f. cycle to prevent progressive growth over many cycles:

$$\Gamma \equiv \frac{\rho_A(T)}{\rho_A(0)} = \exp\left(-\frac{e}{\varepsilon}\frac{n_0}{f}\mu_T\right) < 1. \tag{4.10}$$

This inequality requires a positive time average of the mobility:

$$\mu_T = -\frac{1}{T}\int_0^T \mu_h' \, dt > 0. \tag{4.11}$$

It ought not to be concluded from eqn (4.11) that the mean a.c. conductance at the sample terminals is positive. This quantity is rather obtained by a different time-averaging process [see eqn (4.2)] which yields a *negative* mean value

$$\mu_{LSA} = \frac{n_0 P_{ac}}{e n_0 \hat{E}_h^2/2} = \frac{2}{\hat{E}_h T}\int_0^T v \sin\left(2\pi f t\right) dt. \tag{4.12}$$

From the above, merely kinematical considerations one only obtains the two requirements given by inequalities (4.9), and (4.10) or (4.11). Copeland [4.9] additionally introduced a requirement for the *decay factor* $\Gamma_P < 1$, which characterizes space-charge quenching during the time interval $t_s \leq t \leq T$ of positive differential mobility. He postulated

$$\Gamma_P \equiv \frac{\rho_A(T)}{\rho_A(t_s)} = \exp\left(-\frac{e}{\varepsilon}\frac{n_0}{f}\mu_P\right) < \Gamma_d \tag{4.13a}$$

with

$$\mu_P = \frac{1}{T}\int_{t_s}^T \mu_h' \, dt. \tag{4.13b}$$

87

As a result of inequality (4.10) one has for the critical limit $\Gamma_d < 1/\Gamma_c$. This third condition is necessary most probably because of the existence of the initial accumulation layer due to carrier injection from the cathode [4.10], which is not taken into account in the above kinematical approach but was included in the exact computer calculations of [4.9]. Inequality (4.13a) defines a *lower* limit in n_0/f given by

$$\frac{n_0}{f} > h_P |\ln \Gamma_d| \qquad (4.14a)$$

where

$$h_P = \frac{\varepsilon}{e\mu_P}. \qquad (4.14b)$$

The dependence of the control factor h_P on $\hat{E}_h$ is also shown in Fig. 4.7. From inequality (4.10) it follows that $\Gamma = \Gamma_N \Gamma_P < 1$, hence $h_P < h_N$. A lower limit for n_0/f has been observed experimentally (see, e.g., [4.3, 4.4 and 4.6]).

The magnitude of the growth factor Γ_N that can be tolerated, depends on the size of the initial space-charge perturbations. For the critical limits Copeland [4.9] has chosen—somewhat arbitrarily as he remarks—the values $\Gamma_c = \exp(5)$ and $\Gamma_d = \exp(-6)$ to find a range of n_0/f values suitable for LSA operation. This leads to the relations $\Gamma < \exp(-1)$ and $6h_P < n_0/f < 5h_N$. Based on the results of the computer calculations (see Figs 4.7 and 5.13) and on experimental evidence, e.g. [4.3, 4.4, 4.6], it appears that for safe LSA operation of TE elements made of good-quality* n-type GaAs and operated at room temperature, the permissible range of the ratio of doping concentration to circuit resonant frequency is given by

$$2 \times 10^4 \text{ s cm}^{-3} < n_0/f < 2 \times 10^5 \text{ s cm}^{-3}. \qquad (4.15)$$

This range can, however, be fully utilized only if E_{h0} and $\hat{E}_h$ have their required values which change across the range. Figure 4.8 shows where, for a GaAs device, the LSA regime is situated in the $\hat{E}_h$-E_{h0} plane. Modifications to relation (4.15) are necessary if the transverse dimensions of the semiconductor sample become small because of r.f. field leakage into the external medium [4.64].

The n_0/f range depends strongly on the waveform of the a.c. electric field in the device, and relation (4.15) is valid for a sinusoidal waveform only. It has been shown [4.13] that by providing an a.c. square-wave field in the sample, with a suitable multi-resonant circuit, the upper n_0/f limit is *raised* to approximately 1×10^6 s cm^{-3}. This results from the fact that the field then stays only for a negligibly short time in the region of the $v(E)$ characteristic where the negative slope is at its steepest (see also Section 5.2.3.1).

In Section 4.4.3 a three-terminal device configuration is described in which the rate of carrier injection from the cathode contact can be controlled (reduced). This device is

---

*The influence of spatial doping fluctuations on LSA operation is investigated in Section 5.2.4.2.

thought to be capable of exhibiting a (small-signal) negative terminal conductance as a result of the domain-free operation which should be achievable [4.77]. In principle, devices with controlled carrier injection can also be operated in the (large-signal) LSA mode and, as discussed in Section 5.1.2.1, will show a *reduced* lower n_0/f limit, down to approximately 10^3 s cm^{-3}.

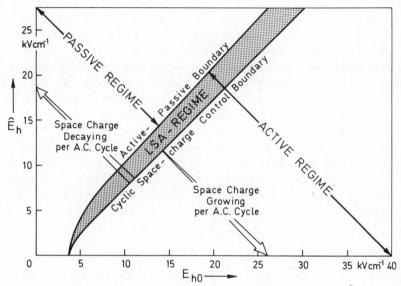

Fig. 4.8 Location of LSA regime in the plane of a.c. field amplitude ($\hat{E}$) versus bias field (E_{ho}) for GaAs sample. After Harrison *et al.* [4.19].

4.3 GROWING SPACE-CHARGE WAVES

4.3.1 Introductory Remarks

In Section 3.2 it was shown that a small imbalance in charge distribution in a semiconducting sample exhibits a time dependence proportional to $\exp(-t/\tau_R)$, with $\tau_R = \varepsilon/en\mu'$ denoting the dielectric relaxation time.

In this section we investigate the linear temporal and spatial response of a neutralized system* of drifted electrons in a TE semiconducting specimen to a small electromagnetic excitation which is periodic in time. As a result of the excitation considered, a charge imbalance in the form of space-charge bunches is created. These bunches propagate with approximately the electron drift velocity, thus constituting space-charge waves (carrier waves, electro-kinetic waves). The system is able to respond collectively only to excitations with wavelengths $\lambda > L_{\mathrm{Db}}$ (the

---

*A neutral ensemble of charge carriers in a semiconductor is often termed a 'solid-state plasma' because of the analogy to the electron-ion plasma of the gaseous state. In our case the plasma is composed of mobile negative charge carriers (electrons) and fixed positive charge carriers (ionized impurities).

Debye length L_{Db} was defined in Section 3.2). This natural wavelength limit follows from electron diffusion since an electron traverses a distance of the order of a Debye length L_{Db} merely because of its thermal velocity.

In the subsequent treatment a collective-mode description is adopted, i.e. the electrons are considered as a single ensemble with average properties as defined for instance by eqns (2.7) and (2.9). The general space-charge-wave analysis is based on the system of Maxwell's field equations (including Poisson's equation) and an additional particle equation (viz. the equation of motion), from which solutions in the form of propagating waves have to be obtained.

Since in the initial stage of dipole-domain formation the small-signal space-charge range must be passed through, an investigation of space-charge waves is a pre-requisite for stability considerations. Furthermore the treatment of domain-free travelling-wave amplifiers ([7.156]; Section 7.2.2) demands such an analysis. First we consider the situation for a homogeneous sample of infinite cross-section, i.e. the one-dimensional case. A two-dimensional approach then follows which yields approximate solutions and is necessary when investigating particular device con-figurations in which the influence of the conditions at the boundaries parallel to the electron drift direction plays a decisive role in obtaining dipole-domain free operation.

4.3.2 One-Dimensional Approach

Here, we shall be interested in how small time-periodic disturbances influence the steady-state electric-field distribution in a TE semiconductor element of infinite lateral extension. In the usual way, this is investigated using a linearized theory (small-signal theory), cf. [3.4, 3.21, 4.20, 4.21]. To be consistent with the preceding large-signal domain case, here we follow again Heinle [3.21] with his particular simplifying assumptions such as the neglect of diffusion (see Section 3.4.1). The influence of diffusion is briefly discussed at the end of this section.

After separating the variables of eqn (3.5) into d.c. terms, which are simplified by assuming a homogeneous d.c. field distribution within the sample, and into small deviations from it, i.e. $E(z,t) = E_{h0} + \Delta E(z,t)$, one obtains the approximate d.c. current density

$$J_0 = en_0 v_{h0}$$

and for the small-signal perturbation terms the linearized partial differential equation

$$\Delta J_t(t) = \varepsilon \frac{\partial \Delta E}{\partial t} + en_0 \, \mu'_{h0} \Delta E + \varepsilon v_{h0} \frac{\partial \Delta E}{\partial z} \qquad (4.16)$$

with the drift current density

$$\Delta J = en_0 \mu'_{h0} \Delta E + \varepsilon v_{h0} \frac{\partial \Delta E}{\partial z},$$

where $v_{h0} \equiv v(E_{h0})$ and $\mu'_{h0} \equiv \mu'(E_{h0})$. We are interested in time-periodic solutions of eqn (4.16), setting

$$\Delta E(z,t) = \tilde{E}(z) \exp(j\omega t), \quad \Delta J_t(t) = \hat{J}_t \exp(j\omega t).$$

Inserting these quantities into eqn (4.16) leads to an inhomogeneous differential equation

$$\hat{J}_t = j\omega\varepsilon\tilde{E} + \tilde{J} = j\omega\varepsilon\tilde{E} + en_0\mu'_{h0}\tilde{E} + \varepsilon v_{h0}\frac{d\tilde{E}}{dz} \tag{4.17}$$

which has the general solution

$$\tilde{E}(z) = \hat{E}_a + \tilde{E}_b(z) = \frac{\hat{J}_t}{(\omega_R + j\omega)\varepsilon} + \hat{E}_b \exp(-jkz) \tag{4.18}$$

with the dispersion relation

$$v_{h0}k = \omega - j\omega_R \tag{4.19a}$$

where

$$\omega_R = 1/\tau_R = en_0\mu'_{h0}/\varepsilon \tag{4.19b}$$

is the dielectric 'relaxation frequency'.

Equation (4.18) states that a periodic perturbation in current with frequency ω will produce two components in the electric-field oscillation. In the first the field oscillates uniformly over the entire sample length with an amplitude $\hat{J}_t/(jv_{h0}k\varepsilon)$. The LSA mode (Section 4.2) results as a large-signal case from this homogeneous field oscillation. On the spatially constant perturbation an electric wave is superimposed as the second component that varies as $\exp[j(\omega t - kz)]$. This travelling wave leads to domain formation when reaching the large-signal range (Section 3.4).

The complex quantity k is called the wave propagation number, $k = \beta + j\alpha$, its real part β being the phase constant $2\pi/\lambda$ and its imaginary part α the amplitude constant ($\alpha > 0$ indicates wave growth, $\alpha < 0$ decay). From eqn (4.19a) it follows that $\beta = \beta_e$, where

$$\beta_e = \omega/v_{h0} \tag{4.20a}$$

is the electronic phase constant, and $\alpha = \alpha_R$, where

$$\alpha_R = -\omega_R/v_{h0} \tag{4.20b}$$

is a growth coefficient related to dielectric relaxation. With definitions (4.20) the dispersion relation (4.19) simply reads

$$k = \beta_e + j\alpha_R. \tag{4.21}$$

Since $\beta = \beta_e = \omega/v_{h0}$, the phase velocity $v_{ph} = \omega/\beta$ and the group velocity $v_{gr} = d\omega/d\beta$ of the wave are equal, in magnitude and direction, to the d.c. carrier drift velocity v_{h0}. Thus, a space-charge wave, caused by the modulation of electron

91

density by the r.f. electric field, propagates synchronously with the d.c. drift current. From eqns (4.18–4.21) the amplitude of the electric wave is

$$|\tilde{E}_b(z)| = \text{const} \cdot \exp(\alpha_R z).$$

For a positive differential mobility μ'_{h0} one has $\alpha_R < 0$ and the space-charge wave is attenuated during its propagation in the sample. The amplitude of the wave will grow if the working point lies on the negatively-sloped part of the $v(E)$ characteristic where α_R is positive (convective amplification). In the latter case amplification is possible as long as the system remains stable, i.e. as long as it does not break into self-supporting oscillations. The stability behaviour will be investigated in Sections 4.4.1.2. and 4.4.2.3.

From vacuum-tube electronics it is well known that a time-periodic perturbation impressed on an *electron beam* sets up the Hahn-Ramo [4.22] fast and slow space-charge waves which carry, respectively, positive and negative energy with regard to the average beam energy and propagate with constant amplitude ($\alpha = 0$) as long as no interaction process is involved, cf. [4.23]. The dispersion diagram $k(\omega) \equiv \beta(\omega)$ for this passive electron-beam case is schematically shown in Fig. 4.9.

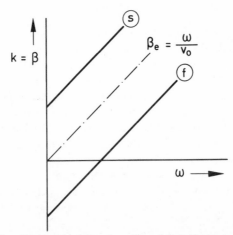

Fig. 4.9 Dispersion diagram $k(\omega)$ for slow (s) and fast (f) space-charge waves on electron beam in vacuum. v_0 denotes the d.c. beam velocity.

Since the carriers drifting in a *semiconductor* sample are subjected to scattering processes (collisions), the corresponding fast and slow wave modes are 'resistively coupled' [4.24]. Considering now a zero-diffusion semiconductor, cf. [4.24, 4.25], and first the case of $\mu'_{h0} > 0$, both propagating space-charge waves will be attenuated at the same rate if $\nu < 2\omega_{p1}$, where $\nu = \tau_p^{-1}$ is the collision frequency and $\omega_{p1} = (ne^2/\varepsilon m^\star)^{\frac{1}{2}}$ the plasma frequency of the electrons. For $\nu > 2\omega_{p1}$, i.e. for stronger collision damping, the attenuation of what originally was the fast wave increases still further while the attenuation of what used to be the slow wave decreases. At the same time the phase velocities of both waves become equal and identical to the average carrier-drift velocity. For most practical purposes, particularly in the usual collision-dominated case with $\nu \gg 2\omega_{\mathrm{pl}}$, we can then assume that a degenerate state exists with only one wave left which is the synchronous wave we have obtained in the above analysis. The dispersion diagram $k(\omega)$ for the collision-dominated zero-diffusion semiconductor is schematically shown in Fig. 4.10. In our case of the TE

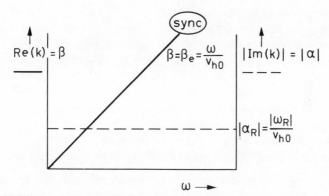

Fig. 4.10 Dispersion diagram $k(\omega)$ for synchronous space-charge wave in collision-dominated zero-diffusion semiconductor.

sample biased for $\mu'_{\mathrm{h0}} < 0$, the remaining synchronous space-charge wave will *grow*, as already discussed, because of the inherent negative-mobility semiconductor properties.

Taking diffusion into account by an appropriate modification of the starting equations (Section 3.2) leads to the dispersion relation

$$\left(v_{\mathrm{h0}} - \frac{en_0}{\varepsilon}\frac{\mathrm{d}D}{\mathrm{d}E}\bigg|_{E_{\mathrm{h0}}}\right) k = \omega - j(\omega_\mathrm{R} + k^2 D_{\mathrm{h0}}) \tag{4.22a}$$

instead of eqn (4.19a), cf. [4.20, 4.26, 4.27]. The symbol D denotes the field-dependent diffusion coefficient with $D_{\mathrm{h0}} \lesssim D(E_{\mathrm{h0}})$. Various curves of $D(E)$ for GaAs have been given in Fig. 2.10. Since eqn (4.22a) is quadratic in k an additional wave solution is now obtained. This constitutes a backward travelling (i.e. opposite to electron drift) space-charge wave which is heavily damped even for $\mu'_{\mathrm{h0}} < 0$. Also the phase velocity of the growing forward wave, already known from eqn (4.20), is no longer identical to the carrier drift velocity but deviates generally by a few percent.

   If the field dependence of the diffusion coefficient D is neglected [3.57, 4.21] the dispersion relation simplifies to

$$v_{\mathrm{h0}}k = \omega - j(\omega_\mathrm{R} + k^2 D_{\mathrm{h0}}). \tag{4.22b}$$

In the case of GaAs this approximation is possible for $n_0 \lesssim 10^{14}$ cm^{-3}. Equations (4.22) indicate that, owing to diffusion, wave growth is inhibited beyond an upper cut-off frequency

$f_{\text{s.c.w.}}$ which is defined by $\alpha = 0$ or $k = \beta$ yielding $\omega_R + \beta^2\,D_{h0} = 0$. From eqn (4.22b) it follows that $\beta = \beta_e$ and

$$f_{\text{s.c.w.}} = \frac{v_{h0}}{2\pi}\sqrt{\left|\frac{\omega_R}{D_{h0}}\right|} \quad \text{or} \quad \lambda_{\min} = 2\pi L_{Db} \qquad (4.23)$$

[see definitions (3.3b) and (4.20)]. A further discussion of the diffusion cut-off frequency will be found in Section 5.1.1.3.

4.3.3 Two-Dimensional Approach

If the lateral extension of the TE element is finite, the one-dimensional approach obviously fails and the influence of the surrounding material on space-charge propagation and growth has to be taken into account. The space charge in this case constitutes to some degree surface charge for which the dielectric relaxation time is no longer simply given by ε/σ', but is a function of the spatial distribution of the electric field as well as of the conductive, dielectric, and magnetic properties of the material adjacent to the semiconductor surface. In treating the laterally limited case we follow here the two-dimensional small-signal theory given by Kino and Robson [4.28] and its extension [4.29, 4.30] to include the effect of high-permeability material at the surface. In this analysis the dispersion relation $\vec{k}(\omega)$ for the drifting-carrier system is investigated. As in the one-dimensional case an amplifying wave occurs due to the spatial wave growth along the carrier drift direction (z direction) if a positive $\alpha = Im(k_z)$ exists. (Compare the 'convective instability', [4.31, 1.27].)

We consider a thin n-type semiconductor layer (permittivity ε_I) with thickness $2a$ and infinite extension in the x and z directions (Fig. 4.11). It is sandwiched between

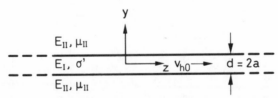

Fig. 4.11 Geometrical arrangement for two-dimensional analysis of space-charge waves.

an insulating material of permittivity ε_{II} and permeability* μ_{II}. The latter is a tensor if the insulating material is ferrimagnetic (ferrite) and subjected to a d.c. magnetic field. The net doping density n_0 in the semiconductor is assumed to be uniform while diffusion as well as trapping effects are neglected. As in Section 4.3.2 the variable quantities are separated into a homogeneous d.c. and a small-signal r.f. term, e.g. $E = E_{h0} + \Delta E$ with $\Delta E = \tilde{E}\,\exp(j\omega t)$. Then the first two (small-signal)

*To avoid confusion, note that the symbols μ, μ_I, μ_{II}, and μ_0 denote the permeability whereas the differential electron mobility in the semiconductor appears here only in the form of its spatial components μ_x', μ_y', and μ_z'.

Maxwell equations in the formalism of the Nabla operator $\vec{\nabla} = (\partial/\partial x,\ \partial/\partial y,\ \partial/\partial z)$ become

$$\vec{\nabla} \times \vec{E} = -j\omega\mu\overset{\leftrightarrow}{\vec{H}} \tag{4.24}$$

and

$$\vec{\nabla} \times \vec{H} = \vec{J}_t = j\varepsilon\omega\vec{E} + \vec{J}_c \tag{4.25a}$$

or

$$\vec{\nabla} \cdot \vec{J}_t = \vec{\nabla} \cdot (j\varepsilon\omega\vec{E} + \vec{J}_c) = 0 \tag{4.25b}$$

with $\vec{H}$ denoting the r.f. magnetic field vector and $\vec{J}_c$ the r.f. carrier current density, i.e.

$$\vec{J}_c = \vec{J} = \overset{\leftrightarrow}{\sigma}'\vec{E} + e\tilde{n}\vec{v}_{h0} \tag{4.26}$$

if diffusion is neglected. The differential conductivity $\overset{\leftrightarrow}{\sigma}'$ is now a tensor because of the anisotrophy resulting from the special direction $\vec{v}_{h0}$ of electron drift. The r.f. charge $e\tilde{n}$ is determined by Poisson's equation (third Maxwell equation)

$$\vec{\nabla} \cdot \varepsilon\vec{E} = e\tilde{n}. \tag{4.27}$$

The divergence of the second Maxwell equation (4.25a) leads to the current continuity equation (4.25b) which, with eqns (4.26) and (4.27), is a generalization of the one-dimensional small-signal eqn (4.18). In the two-dimensional geometry of Fig. 4.11 no current flows in the x direction, hence

$$\left.\begin{aligned} \vec{E} &= (0,\tilde{E}_y,\tilde{E}_z), \\ \vec{J} &= (0,\tilde{J}_y,\tilde{J}_z), \\ \vec{H} &= (\tilde{H}_x,0,0). \end{aligned}\right\} \tag{4.28}$$

The electric drift-field is applied in the z direction, i.e.

$$\left.\begin{aligned} \vec{E}_{h0} &= (0,0,E_{h0}), \\ \vec{v}_{h0} &= (0,0,v_{h0}). \end{aligned}\right\} \tag{4.29}$$

Thus the conductivity tensor can be written in the following diagonal form, cf. [3.57, 3.58]

$$\overset{\leftrightarrow}{\sigma}' = en_0\begin{bmatrix} \mu'_x & 0 & 0 \\ 0 & \mu'_y & 0 \\ 0 & 0 & \mu'_z \end{bmatrix} = \varepsilon\begin{bmatrix} \omega_{Rx} & 0 & 0 \\ 0 & \omega_{Ry} & 0 \\ 0 & 0 & \omega_{Rz} \end{bmatrix} = -\varepsilon v_{h0}\begin{bmatrix} \alpha_{Rx} & 0 & 0 \\ 0 & \alpha_{Ry} & 0 \\ 0 & 0 & \alpha_{Rz} \end{bmatrix} \tag{4.30}$$

where

$$\mu'_x = \mu'_y$$

$$\mu'_y = (\partial v_y/\partial E_y)_{E_{h0}} = v_{h0}/E_{h0} = \mu_{h0}$$

and

$$\mu'_z = (\partial v_z/\partial E_z)_{E_{h0}} = \mu'_{h0}.$$

The transverse mobility μ'_y is equal to the positive total mobility at the bias point as already stated in Section 3.3.2.

The fundamental eqns (4.24) to (4.27) have now to be specified for the semiconductor (region I) and the external medium (region II) using relations (4.28) and (4.29) and noting that $\overleftrightarrow{\sigma}'_{\mathrm{I}} = \overleftrightarrow{\sigma}'$ as given by eqn (4.30), $\overleftrightarrow{\sigma}'_{\mathrm{II}} = 0$, $\tilde{n}_{\mathrm{II}} = 0$ ($\vec{\mathscr{J}}_{c\mathrm{II}} = 0$). As a trivial solution one finds a homogeneous field oscillation $\vec{\tilde{E}}_{\mathrm{I}} = \vec{\tilde{E}}_{\mathrm{II}} = (0, 0, \hat{E}_a)$ in analogy to the first term of the one-dimensional solution (4.18). The wave solutions for each region can be written in the form $\vec{\tilde{E}}_i = \vec{\tilde{E}}_i \exp[-j(k_{yi}y + k_z z)]$ with a characteristic equation connecting k_{yi}, k_z and $\omega (i = \mathrm{I}, \mathrm{II})$. The boundary condition at the semiconductor surface ($y = \pm a$) establishes a relationship between $k_{y\mathrm{I}}$ and $k_{y\mathrm{II}}$ and hence, at least in principle, the dispersion relation can be evaluated in the usual form $k_{yi}(\omega)$ and $k_z(\omega)$.

The boundary conditions at the interface $y = \pm a$ follow directly from the fundamental eqn (4.24) to (4.27) if they are specified for the interface region. For this purpose Stoke's theorem* is applied to the first Maxwell equation (4.24) and Gauss's theorem** both to the current continuity eqn (4.25b) with $\vec{\mathscr{J}}_c$ given by eqn (4.26), and to Poisson's equation (4.27). In this procedure an appropriate infinitesimal geometry is used with an extension $\delta y = \delta y_{\mathrm{I}} + \delta y_{\mathrm{II}}$ perpendicular to the interface. The limit $\delta y \to 0$ yields a relationship between the electric field values at the interface $y = \pm_a$:

$$\tilde{E}_{z\mathrm{I}} - \tilde{E}_{z\mathrm{II}} = 0 \tag{4.31}$$

$$j\beta_e \varepsilon_{\mathrm{II}} \tilde{E}_{y\mathrm{II}} + (\alpha_{\mathrm{R}y} - j\beta_e)\varepsilon_{\mathrm{I}} \tilde{E}_{y\mathrm{I}} = jk_z \tilde{\rho}_s \tag{4.32}$$

$$\varepsilon_{\mathrm{II}} \tilde{E}_{y\mathrm{II}} - \varepsilon_{\mathrm{I}} \tilde{E}_{y\mathrm{I}} = \tilde{\rho}_s. \tag{4.33}$$

---

*Stoke's theorem reads

$$\int (\vec{\nabla} \times \vec{F}) \cdot d\vec{A} = \oint \vec{F} \cdot d\vec{s}$$

i.e. the area integral of the curl of the field vector $\vec{F}$ is equal to the integral of $\vec{F}$ along the circumference of that area.

**Gauss' theorem reads

$$\int \vec{\nabla} \cdot \vec{F} \, dV = \oint \vec{F} \cdot d\vec{A}$$

i.e. the volume integral of the divergence of the field vector $\vec{F}$ is equal to the integral of $\vec{F}$ across the total surface of that volume.

Here $e\tilde{n}_1 \delta y_1 \to \tilde{\rho}_s$ is the mobile r.f. surface charge density which accumulates from the semiconductor at the interface, and β_e is the electronic phase constant as defined in eqn (4.20a). The appearance of $\tilde{\rho}_s$ is a result of the neglect of diffusion.

The boundary conditions (4.31) to (4.33) are identical to the Hahn boundary conditions for the surface of an electron beam in vacuum [4.33], there $\tilde{\rho}_s$ being the surface charge distribution necessary to transform the actually scalloped surface of the modulated electron beam into a boundary surface independent of z, cf. [4.23, 4.32].

The derivation of the general dispersion relation is rather tedious [4.29. 4.30]. Disregarding the magnetic properties of the media involved by setting $\mu_I = \mu_{II} \approx 0$ greatly simplifies the analysis [4.28]. This simplification is justified in the case of a non-magnetic semiconductor like GaAs with purely dielectric material at its surface where $\mu_I = \mu_{II} = \mu_0$ (permeability of free space). We shall investigate this case in detail below. The resulting dispersion relation has a particular clear form in the limit of small semiconductor thickness 2a. For the case of 'magnetic surface loading' (μ_{II} large, but still $\mu_I \approx 0$) the relatively simple modifications of that specialized dispersion relation are then introduced at the end of this section.

Putting $\mu_I = \mu_{II} = 0$ the first Maxwell equation reads $\vec{\nabla} \times \vec{\tilde{E}} = 0$, i.e. $\vec{\tilde{E}}$ can be derived from a scalar potential Φ,

$$\vec{\tilde{E}} = \vec{\nabla}\Phi. \tag{4.34}$$

Equations (4.25) to (4.29) remain unchanged. Writing the wave solution in the two media, $i = $ I, II, as

$$\tilde{\Phi}_i = \hat{\Phi}_i \exp[-j(k_{yi}y + k_z z)] \tag{4.35}$$

leads to a system of algebraic equations for the wave amplitudes $\hat{E}_{yi}$ and $\hat{E}_{zi}$ in both media with the characteristic equation

$$k_{yI}^2 = k_y^2 \tag{4.36a}$$

where k_y is defined as

$$k_y = k_z \left[-\frac{\alpha_{Rz} + j(k_z - \beta_e)}{\alpha_{Ry} + j(k_z - \beta_e)} \right]^{\frac{1}{2}}, \tag{4.36b}$$

and

$$k_{yII}^2 = -k_z^2 \tag{4.37}$$

respectively. Thus two solutions are obtained in each medium, viz. $k_{yI} = \pm k_y$ and $k_{yII} = \pm jk_z$, yielding

$$\tilde{\Phi}_I = (\hat{\Phi}_s \cos k_y y + \hat{\Phi}_A \sin k_y y) \exp(jk_z z) \tag{4.38}$$

$$\tilde{\Phi}_{II} = [\hat{\Phi}_+ \exp(-k_z y) + \hat{\Phi}_- \exp(k_z y)] \exp(jk_z z). \tag{4.39}$$

As, in most cases of interest, a *small* amplitude constant $\alpha = \text{Im}(k_z)$ is considered yielding a phase constant $\beta = \text{Re}(k_z) \approx \beta_e$ we may discuss the solution, eqns (4.36) to (4.39), by approximating $k_z \approx \beta_e$. This leads to a real k_y, viz. in a rough estimate

$$k_y \approx \beta_e \sqrt{\frac{\alpha_{Rz}}{-\alpha_{Ry}}} = \beta_e \sqrt{\frac{-\mu'_z}{\mu'_y}}, \tag{4.40}$$

since in an appropriately biased TE semiconductor $\mu'_z < 0$ (growth) and $\mu'_y > 0$ (decay). As a consequence the solution (4.38) corresponds to a wave penetrating into the semiconductor volume. For *isotropic* mobility, $\mu'_z = \mu'_y$ or $\alpha_{Rz} = \alpha_{Ry}$, k_y would become imaginary and solution (4.38) be a surface wave [4.28]. $\hat{\Phi}_S$ is called the symmetric part and $\hat{\Phi}_A$ the antisymmetric part of the wave (with respect to the lateral or y direction). In the symmetric geometrical arrangement of Fig. 4.11 each wave part is able to fulfil the boundary conditions (4.31) to (4.33) separately and thus becomes an independent wave with individual properties. Each part, on the other hand, can be considered as a superposition of two elemental waves causing the propagation in the y direction to have standing-wave character [compare formulations (4.35) and (4.38)!] and thus leaving effective only the propagation in the z direction.

In the dielectric (medium II) the solution (4.39) constitutes a surface wave (k_{yII} imaginary) which, by its very nature, propagates only in the z direction if, as above, $k_z \approx \beta_e$ is regarded as a real quantity. Since y extends to infinity, $\hat{\Phi}_+$ has a physical meaning for $y > 0$ only, and $\hat{\Phi}_-$ for $y < 0$ only.

If we now specify the boundary conditions (4.31) to (4.33) for $y = a$ (thus $\hat{\Phi}_- = 0$) using solutions (4.38) and (4.39) and relation (4.34) we obtain for the symmetric wave ($\hat{\Phi}_A = 0$) an additional relation between k_y and k_z, viz.

$$\varepsilon_I k_y \tan k_y a = \varepsilon_{II} k_z \frac{j(k_z - \beta_e)}{\alpha_{Ry} + j(k_z - \beta_e)}. \tag{4.41}$$

Because of symmetry the boundary condition at $y = -a$ leads to the same expression. Combining eqns (4.41) and (4.36b) by eliminating α_{Ry} eventually leads to the general dispersion relation

$$k_z = \beta_e + j\gamma \tag{4.42a}$$

with

$$\gamma = \frac{\alpha_{Rz}}{1 - \dfrac{\varepsilon_{II}}{\varepsilon_I} \dfrac{k_y}{k_z} \cot k_y a} \tag{4.42b}$$

where k_y is given by eqn (4.36b). Similarly one finds for the antisymmetric wave ($\hat{\Phi}_S = 0$) the expression

$$\gamma = \frac{\alpha_{Rz}}{1 - \dfrac{\varepsilon_{II}}{\varepsilon_I} \dfrac{k_y}{k_z} \tan k_y a}. \tag{4.42c}$$

98

The permittivity ε_{II} in eqn (4.42b) has just to be replaced by $\varepsilon_{\text{II, eff}} = \varepsilon_{\text{II}} \coth k_z d_{\text{II}}$ if the dielectric II is of finite width d_{II} and backed by a metal sheet [4.28, 4.67].

Equation (4.42a) is written in the form of the dispersion relation of the one-dimensional case, eqn (4.21), with α_R replaced by γ. This form was chosen because γ becomes approximately real, i.e. $\alpha \simeq \gamma$, for small wave growth $\alpha \ll \beta_e$ and then can be considered as an effective dielectric growth coefficient. A general analysis for this case based on eqns (4.36b) and (4.42) with $\gamma \simeq \alpha$ was carried out in analytic form [4.34], and Fig. 4.12a shows the behaviour of wave growth as a function of

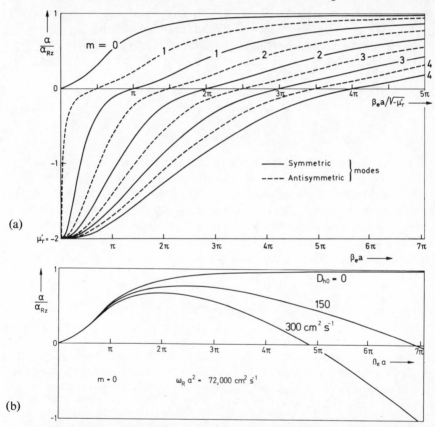

Fig. 4.12 Space-charge-wave mode chart: growth coefficient α versus frequency times semiconductor thickness ($\beta_e a$) for the condition $|\alpha| \ll \beta_e$; permittivity ratio $\varepsilon_{\text{II}}/\varepsilon_{\text{I}} = 10$; $\mu_r' = \mu_y'/\mu_z' = -2$. (a) Diffusionless case showing modes up to order $m = 4$. (b) Influence of diffusion on mode $m = 0$.

semiconductor thickness and/or frequency obtained in this way. The periodicity of the cotangent in eqn (4.42b) or the tangent in (4.42c) leads to an infinite number of lateral mode orders for both the symmetrical and the antisymmetrical wave. The mode order m indicates the number of nodes in the r.f. potential within the semiconductor slab from its centre $y = 0$ to its surface $y = \pm a$. All *higher* mode orders $m \geq 1$ show a *lower* cut-off frequency for growth: below it the damping influence of the positive transverse mobility becomes predominant. This lower cut-off frequency increases with the mode order m.

At relatively *small slab thickness* and/or frequency, i.e. if

$$\beta_e a \leq \frac{\pi}{2} \sqrt{\frac{-\mu_z'}{\mu_y'}}, \tag{4.43}$$

only the symmetric fundamental mode ($m = 0$) stays in a growth regime, even though with a strongly reduced growth coefficient. For GaAs one has $\mu_y'/\mu_z' \approx -2$, and for the mode $m = 0$ condition (4.43) falls more or less into the range $|k_y|a < 0.5$ [4.34, 4.35] which is the range of the 'quasi plane wave' [4.28]. Since $\cot k_y a$ can then be replaced by $1/(k_y a)$ eqn (4.42b) yields

$$\gamma \simeq \frac{\alpha_{Rz}}{1 + \dfrac{\varepsilon_{II}}{\varepsilon_I} \dfrac{1}{k_z a}}. \tag{4.44a}$$

For small wave growth $\alpha \ll \beta_e$ one may substitute $k_z \simeq \beta_e$ in eqn (4.44a) and γ becomes real, i.e. $\gamma \simeq \alpha$. In this case eqn (4.44a)* shows quantitatively the growth reduction caused by the external dielectric at small semiconductor thickness $2a$, as already inferred from Fig. 4.12, the physical reason being the leakage of r.f. electric field energy from the semiconductor into the adjoining dielectric medium. If additionally $\beta_e a \varepsilon_I/\varepsilon_{II} \ll 1$ (heavy 'dielectric surface loading' $\varepsilon_{II} \gg \varepsilon_I$, or very thin semiconductor layer $\beta_e a \ll 1$) the growth coefficient α becomes directly proportional to frequency:

$$\alpha \simeq \gamma \simeq \frac{\varepsilon_I}{\varepsilon_{II}} \beta_e a \alpha_{Rz}.$$

In the limit of *infinite slab thickness* $2a \to \infty$ the growth coefficient $\alpha = \gamma$ of all modes up to $m \to \infty$ approaches the one-dimensional value α_{Rz}, and their lateral change after eqn (4.38) is characterized by $k_y a \to (1 + 2m)\pi/2$ for the symmetric modes. A superposition of the symmetric modes then leads to the plane wave of the one-dimensional case [4.35].

If field independent *diffusion* is included into the two-dimensional theory [4.36] one finds as the most important result that in addition to the *lower* limiting frequency for growth (Fig. 4.12a), all modes now show an *upper* limiting frequency which decreases with the

---

*In the approximation of eqn (4.44a) the quantity γ is independent of the transverse mobility μ_y' since k_y has dropped out. As a consequence relation (4.44a) still holds if the r.f. electron flow is forced to be laminar in the z direction, e.g. by a suitable magnetic d.c. field. This case can formally be obtained by setting $\mu_y' = 0$ ($\alpha_{Ry} = 0$) yielding a zero surface charge density, i.e. $\rho_s = 0$, in the boundary conditions (4.32) and (4.33); cf. [4.62, 4.63].

mode order m. As a consequence, the modes beyond a certain order will show no growth at any frequency. For the lowest symmetric mode order $m = 0$ the situation is particularly simple. Here, in the limit of small growth with $|\gamma| \ll \beta_e$, the dispersion relation is modified approximately in the same way as in the one-dimensional case, i.e. in eqn (4.42a) γ is replaced by $\alpha = \gamma - \beta_e^2 D_{no}/v_{no}$ [compare with eqns (4.19a) and (4.22b)!]. The modified growth coefficient α is plotted in Fig. 4.12b for two different diffusion constants D_{no}.

If the *magnetic properties* of medium II are taken into account, a relatively simple result follows for the fundamental wave mode within the range of quasi-planarity at small semiconductor thickness, i.e. $|k|ya < 0.5$, as assumed for deriving eqn (4.44a). In the case of large isotropic permeability μ_{II} one obtains [4.29]

$$\gamma \simeq \frac{\alpha_{Rz}}{1 - \dfrac{\varepsilon_{II}}{\varepsilon_I[1 - (v_{no}/c_{II})^2]^{\frac{1}{2}}} \dfrac{1}{k_z a}} \tag{4.44b}$$

with $c_{II} = (\varepsilon_{II}\,\mu_{II})^{-\frac{1}{2}}$ being the phase velocity of free electromagnetic waves in medium II. In the case of magnetic loading with a ferrite subjected to a d.c. magnetic field $\vec{H}_0$ [4.30] the permeability becomes a tensor $\overset{\leftrightarrow}{\mu}_{II}$. For obtaining the strongest effect on the space-charge-wave behaviour $\vec{H}_0$ has to be applied along the z direction. Then

$$\overset{\leftrightarrow}{\mu}_{II} = \mu_0 \begin{bmatrix} m_1 & -jm_2 & 0 \\ jm_2 & m_1 & 0 \\ 0 & 0 & 1 \end{bmatrix}$$

where the relative permeability components m_1 and m_2 depend on H_0 and ω, with their maximum (both positive and negative) values reached near the ferrite resonance. Thus, contrary to the case of large *isotropic* μ_{II} above, now only the *transverse* directions exhibit a large permeability. Instead of eqn (4.44b) the following expression is derived

$$\gamma \simeq \frac{\alpha_{Rz}}{1 - \dfrac{\varepsilon_{II}\sqrt{m_1}}{\varepsilon_I} \dfrac{1}{k_z a}}. \tag{4.44c}$$

For small wave growth $\alpha \ll \beta_e$ one can replace again k_z by β_e in eqn (4.44c) as long as $c_{II} > v_{no}$ or $m_1 > 0$. Then γ is still real and hence $\gamma \simeq \beta$. Thus, both cases of magnetic surface loading lead to a further reduction of wave growth. This is expected since the r.f. electric field energy leaking from the semiconductor into medium II is now associated with additional energy of the magnetic field. In the case of $c_{II} < v_{no}$ or $m_1 < 0$ the situation is more complicated because γ remains complex. Small enough $|k_z|a$ even leads to a purely imaginary γ with $k_z \simeq \beta$ and thus growth is suppressed completely. Additionally, the phase velocity $v_{ph} = \omega/\beta$ of the space-charge wave deviates in this case from the drift velocity v_{no}. On the other hand the magnetic material II acts now as a slow wave circuit and coupling between an externally excited slow volume wave in medium II and the space-charge wave in the semiconductor may lead again to wave growth [4.30]. This type of growth is directly related to the wave growth in the electron-tube travelling-wave amplifier.

101

4.4 SMALL-SIGNAL NEGATIVE CONDUCTANCE IN STABLE DEVICES

The space-charge-wave analysis of Section 4.3 will now be applied to the question of stability against the formation of travelling space-charge dipole domains in a TE semiconductor specimen. For this purpose the small-signal admittance of the specimen of length L is formulated and then its poles are investigated in the complex frequency plane. Qualitatively, stability is expected if the gain of the space-charge waves along L stays below a critical value so as to prevent domains from developing. With a stable TE device amplification of r.f. signals, both of the reflection and transmission type, becomes possible.

4.4.1 Subcritical Doping-Length Product

4.4.1.1 *General Remarks*

If semiconductor samples with relatively large lateral extension 2a are considered, r.f. stray fields external to the semiconductor can be neglected and stability requires a sufficiently small product of net doping and sample length. A rough estimate of the critical product below which dipole domains cannot form is obtained in the following way [4.37]. If a mature domain is to develop its formation must be completed within the domain transit time $\tau \approx L/v_\mathrm{D}$. Since domain build-up is governed by the dielectric relaxation time $\tau_\mathrm{R} = \varepsilon/\sigma'$ one might expect stability for

$$\tau \lesssim |\tau_\mathrm{R}| \quad \text{or} \quad \frac{L}{v_\mathrm{D}} \lesssim \frac{\varepsilon}{|\sigma'|}. \tag{4.45a}$$

Assuming for simplicity that the domain travels with the average carrier drift velocity $v_\mathrm{D} \simeq v_\mathrm{h0}$ and using an average differential conductivity $\sigma' \simeq en\mu'_\mathrm{h0}$ relation (4.45a) becomes

$$n_0 L \lesssim (n_0 L)_\mathrm{crit} = \frac{\varepsilon v_\mathrm{h0}}{e|\mu'_\mathrm{h0}|} \tag{4.45b}$$

if one takes $n = n_0$. Inserting numerical values for n-type GaAs ($\varepsilon \approx 12\cdot6\,\varepsilon_0$, $v_\mathrm{h0} \approx 1\cdot5 \times 10^7$ cm s^{-1} and $\mu'_\mathrm{h0} = -2\,000$ cm^2 (Vs)$^{-1}$) yields a critical product of $(n_0 L)_\mathrm{crit} \approx 5 \times 10^{10}$ cm^{-2}. The existence of such a critical $n_0 L$ value was first pointed out by Kroemer [4.38]. It will be shown in Section 4.4.1.3 that samples with subcritical $n_0 L$ products and bias fields in the n.d.c. range exhibit a small-signal external negative sample conductance [4.38] at transit angles near

$$\theta = \omega\tau = \beta_\mathrm{e}L \approx 2i\pi, \quad \text{with } i = 1,2,\ldots. \tag{4.46}$$

Achievement of reflection-type microwave amplification with GaAs devices [4.39] relies on this fact. By providing a suitable load impedance such a device with subcritical $n_0 L$ can naturally be made to oscillate near frequencies given by eqn (4.46) cf. [4.40].

Equation (4.46) indicates that the small-signal negative conductance is limited to frequencies near the transit-time frequency and some higher harmonics. However, by series-connecting (stacking) several elements the effective length of the overall device can be increased [4.41]. This readily increases the *effective* transit time relative to the duration of the oscillation period and thereby overcomes the handicap of the transit-time limitation which will be discussed in detail in Section 5.2.5. Such a series-connection of small-n_0L-product elements is an attractive alternative to LSA operation of a large-n_0L-product device.

Starting from the general solution (4.19) for one-dimensional space-charge waves we now investigate the behaviour of a TE element, having ohmic contacts at the cathode and anode, with particular respect to the sub-critical doping length product.

4.4.1.2 *Stability Analysis*

A detailed investigation of the stability criteria of TE elements was first carried out by McCumber and Chynoweth [3.4] cf. [3.17, 3.18]. To determine whether the element is stable the usual method of examining the singularities of its (small-signal) impedance $Z(p)$ or admittance $Y(p) = 1/Z(p)$ for complex frequencies $p = j\omega$ is employed. The system is small-signal stable if any existing poles of $Z(p)$ and of $Y(p)$ are situated in the left half of the p plane. Poles on the right half lead to instabilities, specifically, those of $Z(p)$ in the voltage at constant current (infinite load resistance R_L) and those of $Y(p)$ in the current at constant voltage ($R_L = 0$).

The ohmic-contact boundary at the cathode $z = 0$ of the element requires the condition $\tilde{E}(0) = 0$ and thus, making use of eqn (4.20), the general space-charge wave solution (4.19 becomes)

$$\tilde{E}(z) = \frac{\hat{J}_t}{jkv_{no}\varepsilon} [1 - \exp(-jkz)]. \tag{4.47}$$

The boundary condition $\tilde{E}(0) = 0$ is a consequence of the high net-doping concentration at the contact. The same condition applies to the anode $z = L$ but is not used here since it would lead to the trivial solution $\hat{J}_t = 0$. This results from the neglect of diffusion. With diffusion taken into account the anode boundary condition determines the amplitude of the additional heavily damped backward wave (p. 93). The latter, however, does not contribute much to the r.f. terminal voltage of the element and thus neglect of the anode boundary condition is justified in calculating the sample admittance [4.20, 4.26].

Integrating eqn (4.47) over the entire sample length L yields the r.f. voltage $\hat{V}$, and, setting $jkL = \vartheta(j\omega)$, the sample impedance becomes [3.4, 3.21]

$$Z(j\omega) = \frac{V(j\omega)}{\hat{J}_t A} = R_\tau \frac{\vartheta - 1 + \exp(-\vartheta)}{\vartheta^2} \tag{4.48}$$

where

$$R_\tau = \frac{\tau}{C_h} = \frac{L^2}{\varepsilon v_{ho} A}. \tag{4.49}$$

103

According to dispersion relation (4.19) one has

$$\vartheta(j\omega) = (\omega_R + j\omega)\tau \tag{4.50a}$$

where $\tau = L/v_{h0}$. Using the definitions (4.20) for α_R and β_e, and the definition (4.46) for the transit angle θ the quantity ϑ can also be expressed as

$$\vartheta = (-\alpha_R + j\beta_e)L = -\alpha_R L + j\theta \tag{4.50b}$$

where $\alpha_R L$ determines the total wave gain along the sample length L. Generalizing now the complex frequencies $j\omega \to p$ the behaviour of the impedance function $Z(p)$ in the complex p plane is investigated. It is convenient to analyse first the singular points in the complex ϑ-plane and then find their corresponding positions in the p-plane by using eqn (4.50) in the form

$$\tau p = \vartheta + \alpha_R L. \tag{4.50c}$$

No *poles* of $Z(p)$ exist and, hence, the TE element is always voltage stable. The infinitely many *zeros* which $Z(p)$ possesses, on the other hand, may lead to current instability (cyclic dipole domain formation as discussed in Section 3.4). The two zeros p_1 lying farthest to the right in the complex p-plane are given [3.4] by $\vartheta_1 = \vartheta(p_1) = \xi_1 + j\eta_1 = -2\cdot09 \pm j7\cdot46$. Thus, current stability is obviously obtained if $\text{Re}(p_1) < 0$, i.e., using eqn (4.50c), if

$$\tau\text{Re}(p_1) = \alpha_R L + \xi_1 < 0 \quad \text{or} \quad \alpha_R L < 2\cdot09. \tag{4.51a}$$

Inequality (4.51a) is always satisfied for $\alpha_R < 0$ [$\mu'_{h0} > 0$, see definitions (4.19b) and (4.20b)!]. For a working point on the negatively-sloped part of the $v(E)$ characteristic one has $\alpha_R > 0$ ($\mu'_{h0} < 0$), and relation (4.51a) states that stability requires the space-charge-wave gain to stay below a critical limit. The criterion can be written in the form

$$n_0 L < (n_0 L)_{\text{crit}} = 2\cdot09 \frac{\varepsilon v_{h0}}{e|\mu'_{h0}|} \tag{4.51b}$$

which is very similar to inequality (4.45b) with its simple physical interpretation. Noting that $\mu'_{h0} \equiv \mu'(E_{h0})$, relation (4.51b) defines a boundary curve for stability as shown in Fig. 4.13 (solid curve with $r_L = 0$) where the critical $n_0 L$ product is calculated for n-type GaAs as a function of the d.c. bias field E_{h0} [3.21, 4.42]. This boundary determines the threshold field E_T above which domain oscillations set in. The lowest-order oscillation frequency according to the small-signal analysis is $\omega_1 = |\text{Im}(p_1)| = 7\cdot46/\tau$, i.e. somewhat larger than the transit-time frequency $\omega_\tau = 2\pi/\tau$. The latter is established only for points far enough away from the stability boundary (domain build-up time small compared to oscillation period!). For large $n_0 L$ products the threshold field E_T and the peak field E_p practically coincide. No dipole domains are possible at all below a certain minimum $n_0 L$ value, $(n_0 L)_{\text{min}} \approx 10^{11} \text{ cm}^{-2}$.

So far a zero load resistance has been considered in the discussion of current

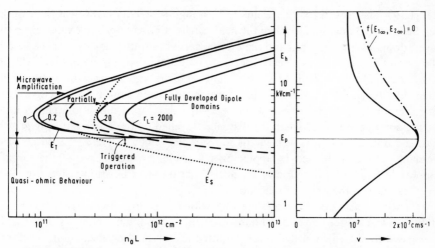

Fig. 4.13 Operational ranges of n-type GaAs samples calculated as functions of d.c. bias field E_{h0} and doping-length product n_0L. Normalized load resistance $r_L = R_L/R_\tau$ with R_τ defined by eqn (4.49). After Heinle [3.34]. Dashed curve: stability boundary measured on a GaAs device of $L = 340\ \mu$m and using cooling for n_0 variation, $r_L \approx 0$ [4.44a]. (N.B. On the ordinate it should read E_{h0} instead of E_h.)

stability. If a real load resistance $R_L > 0$ is connected in series to the sample impedance, the zeros of $[Z(p) + R_L]$ have to be investigated [4.42, 4.84, 4.85]. As it turns out, the boundary in Fig. 4.13 moves to the right with increasing load resistance R_L. In the figure $r_L = R_L/R_\tau$ is used as the parameter, with R_τ defined by eqn (4.49). The quantity R_τ is related to the low-field sample resistance R_1 by $R_\tau/R_1 = \tau/\tau_{R1}$ with $\tau_{R1} = \varepsilon/\sigma_1$ being the dielectric relaxation time at low or zero bias field. For GaAs with a low-field mobility of $\mu_1 \approx 8\,000$ cm^2 (Vs)$^{-1}$ one obtains $R_\tau/R_1 \approx 8 \times 10^{-11}\ n_0L/$cm^{-2}; hence the shift of the boundary curves is relatively small for practical load resistances. At any rate, an originally stable system can break into oscillations if the load resistance is reduced. Considering the more general case of a complex load impedance, it is found that for n_0L products only slightly above the critical value, domain instabilities can be inhibited by providing a sufficiently capacitive imaginary load [4.43, 4.85].

In Fig. 4.13 the range of possible small-signal microwave amplification and the domain operating ranges including triggered operation (Section 3.4) are indicated. The exact location of the stability boundary, particularly the minimum value of n_0L for domain formation, depends strongly on the shape of the $v(E)$ characteristic. For the curves of Fig. 4.13 the analytical relationship of eqn (3.4) was used [3.34]. A measured stability curve (dashed) for $v_L \approx 0$ is also shown which gives a somewhat larger $(n_0L)_{min} \approx 2 \times 10^{11}$ cm^{-2} but exhibits the same general shape as the calculated curves [4.44]. The measurement indicates lower threshold fields E_T than theory. Apparent threshold fields below the velocity peak value E_p and decreasing with increasing sample length L, are frequently observed in relatively long GaAs samples, c.f. [1.19]. The reason for this seems to be a high-resistivity layer near the cathode (cf. Sections 3.3 and 6.2.2). When the field reaches the velocity-peak value E_p in this layer, domains are triggered and may travel across the entire length L if the field in the rest of the sample is already beyond the domain sustaining value E_S [3.28]. The measured threshold field is consistent with this model since it is still higher than the 'threshold' E_S for triggered domain operation (lower dotted curve). The limiting curve between fully and only partially developed dipole domains (upper dotted curve) is obtained when the distance travelled by the domain during growth plus the domain width b just equals the sample length L. Here the duration of domain growth is defined by replacing the actual growth curve of the domain amplitude E_2-E_1 as a function of the accumulation layer position z_a (cf. Fig. 3.18) by its tangent in the point of inflection [3.21a].

When comparing experimental results with the simple small-signal theory for stability according to eqn (4.51) it has to be kept in mind that the theory postulates a *homogeneous* d.c. field distribution within the device. For ideal ohmic cathode contacts the d.c. field is expected to be very inhomogeneous (cf. Fig. 4.16); the associated space-charge accumulation makes the stability range at high bias field E_{h0} for devices with larger n_0L products questionable. However, the thin high-resistivity layer near the cathode frequently encountered in real devices (see Section 6.2.2) ensures a fairly homogeneous d.c. field distribution and a high-bias-field stability range is observed also in large-signal simulations [4.87].

Diffusion has a stabilizing effect on the TE element [3.4], since wave growth is reduced, i.e. $\alpha < \alpha_R$. However, for low frequencies and small net doping n_0 the influence on the critical n_0L product is only slight [4.26, 4.45]. Instead of relation (4.51a) the stability criterion $\mathrm{Re}(p_1) < 0$ becomes

$$\tau\mathrm{Re}(p_1) = \alpha_R L + \xi_1(1 + \alpha_R L\delta') + (\xi_1^2 - \eta_1^2)\delta < 0$$

or

$$\alpha_R L < \frac{2\cdot09 + 51\cdot3\delta}{1 - 2\cdot09\,\delta'}$$

$$(4.52)$$

where $\delta = D_{h0}/(v_{h0}L)$ and $\delta' = (\mathrm{d}D/\mathrm{d}v)_{E_{h0}}/L$ are the diffusion parameters. With $\delta = 0\cdot02$ and $\delta' = 0\cdot02$ one obtains $\alpha_R L < 3\cdot25$. In GaAs $[D_{h0} \approx 300\ \mathrm{cm}^2\ \mathrm{s}^{-1},\ (\mathrm{d}D/\mathrm{d}E)_{E_{h0}} \approx -0\cdot4\ \mathrm{cm}\ (\mathrm{Vs})^{-1},\ \mu'_{h0} = -2\,000\ \mathrm{cm}^2\ (\mathrm{Vs})^{-1}$, and $v_{h0} = 1\cdot5 \times 10^7\ \mathrm{cm}\ \mathrm{s}^{-1}$ at $E_{h0} = 6\ \mathrm{kV}\ \mathrm{cm}^{-1}]$ this corresponds to a device with about $L \approx 10\mu\mathrm{m}$ and $n_0L < 1\cdot7 \times 10^{11}\ \mathrm{cm}^{-2}$, thus $n_0 < 1\cdot7 \times 10^{14}\ \mathrm{cm}^{-3}$. The lowest-order oscillation frequency $\omega_1 = |\mathrm{Im}(p_1)|$ at the stability boundary is shifted somewhat away from its value without diffusion as a result of the changed phase velocity of the growing forward wave:

$$\omega_1\tau = |\eta_1|(1 + \alpha_R L\delta' + 2\xi_1\delta)$$

$$= 7\cdot46(1 + \alpha_R L\delta' - 4\cdot18\delta).$$

$$(4.53)$$

Using again $\delta = \delta' = 0\cdot02$ and $\alpha_R L = 3\cdot25$ one obtains $\omega_1 = 7\cdot40/\tau$.

Additionally, diffusion provides stability even at large ('supercritical') n_0L products

106

under certain conditions [4.46]. This stable state is characterized by a stationary high-field layer adjacent to the anode with the field in the rest of the sample remaining fairly homogeneous somewhat below E_p. According to a small-signal analysis [4.47a], based on the appearance of an absolute instability in the dispersion relation (4.22b), the condition for a stable high-field layer at the anode reads

$$\alpha_R > \frac{v_{h0}}{4D_{h0}} \quad \text{or} \quad n_0 > \frac{v_{h0}^2 \varepsilon}{4D_{h0}\, e |\mu_{h0}'|}. \tag{4.54}$$

Using the numerical values of GaAs stated above one finds $n_0 \gtrsim 6 \times 10^{14}$ cm^{-3}. Since the d.c. current flowing in a diffusion-stabilized TE device is smaller than the threshold current, it was predicted that such devices can be used as bistable subnanosecond switching elements [4.48, 4.86], see also Section 8.3.1.3.

4.4.1.3 *Negative Dynamic Conductance*

If a TE element is biased in the negatively-sloped part of the $v(E)$ characteristic, i.e. if $\mu_{h0}' < 0$, the space-charge wave grows, according to Section 4.3.2. This causes the small-signal admittance $Y(\theta) = 1/Z(\theta)$, eqn (4.48), of the stable sample with subcritical $n_0 L$ product to exhibit pronounced negative-conductance peaks at those transit angles that ensure inphase operation at the two sample contacts. At antiphase operation, on the other hand, the conductance becomes positive even though $\mu_{h0}' < 0$. This undulating behaviour is a result of the r.f. power contributed by the bunching process of the carriers as manifested by the last term of the current density eqn (4.17). Depending on the transit angle θ this power can be either negative or positive thus enhancing or overcompensating the negative power generated by the differential mobility [second term of eqn (4.17); cf. [4.82, 5.53]]. The transit-time dependence can easily be seen from eqns (4.48) and (4.50b) in the limiting case of zero differential mobility at $E_{h0} = E_p$, i.e. $\alpha_R = 0$, whence

$$\text{Re}[Z(\theta)] = R_{sr}(\theta) = \text{const} \cdot \frac{1 - \cos \theta}{\theta^2}. \tag{4.55}$$

The sample resistance thus falls to zero for $\theta = 2i\pi$, $i = 1, 2, \ldots$, i.e. at the transit-time frequency and its harmonics. For a particular sample biased in the n.d.c. region and exhibiting a gain parameter $\alpha_R L = 1 \cdot 6$, the conductance $G(\theta) = \text{Re } Y(\theta)$ and capacitance $C(\theta) = \tau \text{Im } Y(\theta)/\theta$, given by eqns (4.48) and (4.50b), are plotted in Fig. 4.14. The undulation caused by the transit-time effect of the space-charge waves is apparent. The negative-conductance peaks by far exceed the value $G_{h0} = en_0\mu_{h0}'L/A$ of a sample with no space-charge wave present. The 'resonant behaviour' near $\theta = 2i\pi$ as shown by the $Y(\theta)$ function can formally be attributed to an equivalent 'RCL' series circuit in parallel to the (negative) homogeneous sample conductance G_{h0} and the geometric sample capacitance C_h. In this series resonant circuit the frequency-independent elements are imaginary, however; the 'resistance' and 'inductance' are negative imaginary whereas the 'capacitance' is positive imaginary.

107

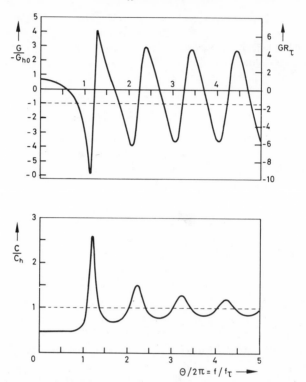

Fig. 4.14 Theoretical small-signal admittance $1/Z = Y = G + j\omega C$ as a function of transit angle θ; obtained from eqn (4.48) for $\alpha_R L = 1\cdot 6$ (diffusion neglected).

The effects of *diffusion* on the small-signal admittance have been investigated by a number of authors [3.4, 4.20, 4.21, 4.26 and 4.45]. Instead of eqn (4.48) one obtains in the case of $\alpha_R \ll v_{ho}/D_{ho}$ (the backward wave is neglected, [4.20]).

$$Z(\theta) = r \frac{\vartheta - 1 + \exp(-\vartheta)}{\vartheta(j\theta - \alpha_R L)} \tag{4.56}$$

where $\vartheta(\theta) = jkL$ corresponds to the growing forward-wave solution of eqn (4.22a). For the special diffusion parameters $\delta' = 0\cdot 02$ and $\delta = 0\cdot 02$ and for $\alpha_R L = 2$ the small signal admittance $Y(\theta) = 1/Z(\theta)$ as evaluated from eqn (4.56) is shown in Fig. 4.15. A similar undulation is obtained as in Fig. 4.14 where diffusion was neglected. This undulation dies out, however, near the diffusion cut-off frequency for space-charge wave growth as given by eqn (4.53). Above it the transit-time effect of the space-charge wave is eliminated and a wide-band negative conductance appears [4.45b]. However, there, the negative conductance stays below the homogeneous-field value G_{ho}, since a damped space-charge wave is still excited at the cathode. This eventually leads to an upper frequency limit as discussed in Section 5.1.1.

A *parasitic series resistance* to the active element, even if small compared to the absolute value of the intrinsic negative resistance $R_{h0} = 1/G_{h0}$, might drastically lower or even completely wipe out the terminal negative conductance of the device [4.49]. Thus, for realizing the high-frequency wide-band negative conductance special care has to be taken when forming the ohmic contacts (cf. Section 6.2.2).

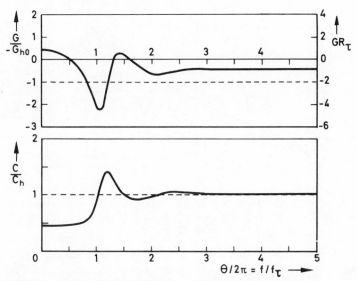

Fig. 4.15 Theoretical small-signal admittance $1/Z = Y = G + j\omega C$ as a function of transit angle θ; obtained from eqn (4.56) for $\alpha_R L = 2$ and with diffusion parameters $\delta = D_{h0}/(v_{h0}L) = 0.02$ and $\delta' = (dD/dv)_{E_{h0}}/L = 0.02$ (corresponding to $L \approx 10\,\mu$m in GaAs).

In the *first order* (small-signal) approximation no negative terminal conductance is obtained at zero and low frequencies since electron injection from the cathode produces an internal space charge which grows with increasing small-signal voltage $\hat{V}$. According to a theorem of Shockley ([1.9]; cf. [4.38, 4.50]) the amount of this space-charge build-up is so large that the current grows in spite of the drop of electron drift velocity in the n.d.c. bias region. This effect ought to be included for calculating the *zero-order* d.c. solution of the field assuming the appropriate boundary conditions $E_0(0) = E_0(L) = 0$ with diffusion taken into account, cf. [3.4]. Contrary to the assumption of homogeneous d.c. field in Section 4.3.2 it leads to an *inhomogeneous* distribution $E_0(z)$ that becomes more pronounced with decreasing net-doping density n_0. Figure 4.16 shows a series of $E_0(z)$ distributions in n-type GaAs calculated for different bias values $E_{h0} = V_0/L$ [4.51]. Referring to such curves a simple explanation of Shockley's theorem is possible [4.50]. The field increases from zero at the cathode to high enough values in the semiconductor bulk as are required by the applied bias voltage $E_{h0}L$. Large enough bias causes the field to cross the velocity-peak value E_p somewhere in the bulk. Since at this point $dE_0/dz > 0$, Poisson's equation postulates $n - n_0 > 0$, i.e. an increased carrier density and hence a current

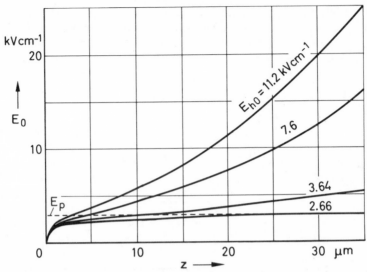

Fig. 4.16 Calculated d.c. electric-field distribution $E_0(z)$ of n-type GaAs sample for doping $n_0 = 4 \times 10^{13}$ cm^{-3} and various values of bias field E_{h0}. Cathode at $z = 0$. After Thim [4.51].

density $J_0 > J_p = e n_0 v_p$. This current density has to be constant within the entire bulk where, however, $v_0 < v_p$ except at the cross-over, $E_0 = E_p$, itself. Consequently, $n - n_0$ or dE_0/dz is smallest at the cross-over which thus has to be an inflection point. With increasing bias E_{h0} this point moves closer to the cathode and its slope becomes steeper so that a further increase in current density J_0 is required. A static negative conductance is predicted, however, if a special field dependence of the diffusion constant $D(E)$ is assumed [4.88].

For an accurate calculation of the small-signal admittance the non-uniform d.c. field distribution has to be taken into account. Closed-form analytical expressions were obtained with various degrees of sophistication: neglecting diffusion for the r.f. perturbations [4.52, 4.53], in particular in the special case of zero ion density when all electrons are injected from the cathode [4.27]; or taking diffusion partially into account and including trapping of the excess charge carriers [4.54]. Hot electron relaxation effects complicate the mathematics even more and only numerical computer simulations are possible for a rigorous calculation [3.4, 4.27]; analytical results have been obtained, however, for a homogeneous zero-order d.c. field distribution [3.4, 4.55]. The more general problem of a non-uniform doping profile has also been studied, but analytical results are obtainable only if diffusion and electron relaxation are neglected [4.56].

In a rather different approach [4.25] the TE element can be treated as a two-stream amplifier system, cf. [4.31], in which the two streams are made up of the fast central-valley electrons and of the slow satellite-valley electrons. Such a system is able to produce convective amplification under suitable conditions. If sufficient positive feedback is provided, leading to unity round-trip gain, the amplifier will be converted into an oscillator which, then, may be represented as a one-port device with a negative real part of its impedance. The feedback is positive for values of θ as given by eqn (4.46).

Figure 4.17 shows the *measured* frequency characteristic of the small-signal admittance $Y = G + j\omega C$ of a stable GaAs TE device with subcritical $n_0 L$ product

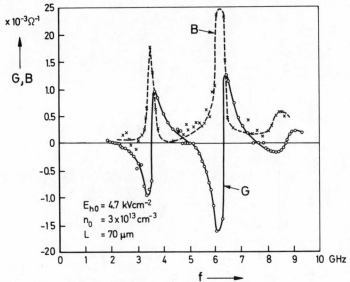

Fig. 4.17 Measured small-signal admittance $Y = G + jB$ for a GaAs TE device with $n_0 = 3 \times 10^{13}$ cm^{-3} and $L = 70\,\mu$m at a bias field $E_{h0} = 4\cdot7$ kV cm^{-1}. After Hakki [4.26].

[4.26]. The principal behaviour is identical to the theoretical curves of Figs 4.14 and 4.15. Negative conductance peaks up to the third transit-time harmonic were detected. The capacitance has sharp maxima at frequencies just above those corresponding to negative conductance peaks as expected from theory. The element provided reflection-type amplification when connected to the end of a dispersion-free transmission line (see Section 7.2.1). Results of another measurement near the first transit-time harmonic are plotted in Fig. 4.18 and compared to theoretical curves which the same authors obtained by a rather rigorous calculation taking the non-uniform d.c. electric-field distribution into account [4.53]. The good fit between theory and experiment gives confidence in the validity of the chosen (or adjusted) theoretical parameters, particularly of the $v(E)$ characteristic (cf. Fig. 2.8).

In Fig. 4.16 we have seen that near the cathode there is a range with $E_0 < E_p$ even for $E_{h0} > E_p$. For small $n_0 L$ values this static field distortion becomes rather pronounced. Then a negative conductance is no longer obtained since the region with $E_0 < E_p$ predominates, with $n_0 L = 10^{10}$ cm^{-2} appearing to be an absolute lower limit for the active properties of GaAs [4.26]. On the other hand, it was shown

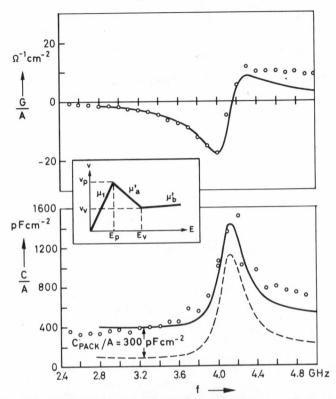

Fig. 4.18 Comparison of measured small-signal admittance $Y = G + j\omega C$ for a GaAs TE device (open circles) with calculations (solid lines) based on the $v(E)$ curve of the inset (dashed line: calculated capacitance of unpackaged chip; C_{PACK} = package capacitance). Sample data at the operating temperature of $-40°$C: $n_0 = 1\cdot6 \times 10^{13}$cm^{-3}, $\mu_1 = 6\,000$ cm^2 (Vs)$^{-1}$, $L = 50\,\mu$m, $A = 1\cdot75 \times 10^{-3}$cm^2, $E_{\text{h}0} = 5\cdot8$ kV cm^{-1}. The parameters $v_{\text{p}} = 2 \times 10^7$cm s^{-1} and $E_{\text{p}} = 4$ kV cm^{-1} were taken from threshold measurements on TE oscillators at $-40°$C, while v_{v} and the mobilities μ_{a}' and μ_{b}' were considered as adjustable parameters in the calculation; for the curves shown: $\mu_{\text{a}}' = -2\,500$ cm^2 (Vs)$^{-1}$, $\mu_{\text{b}}' = 0$, and $v_{\text{v}} = 0\cdot4v_{\text{p}}$. After McWhorter and Foyt [4.53].

theoretically [4.27] that a negative conductance should appear even for $n_0 \to 0$ ($jn_0L \to 0$), i.e. with samples in which all electrons are due to injection from the cathode. Thus, the experimentally observed lower n_0L limit is probably a consequence of trapping effects not considered in the theory.

4.4.2 Subcritical Doping-Thickness Product

4.4.2.1 *General Consideration*
As shown by the analysis of Section 4.3.3 for a TE semiconductor sample with at least one of the two transverse dimensions comparable to the space-charge wave-

length ($\lambda \approx 2\pi v_{ho}/\omega$, where ω is the radian frequency of the excitation), the r.f. electric field lines that extend into the surrounding material appreciably influence space-charge propagation and growth. In a simple physical interpretation, the 'quasi-stationary' field induced in the surrounding medium II leads there to power dissipation if $\sigma_{II} > 0$ and to reactive power being stored if $\sigma_{II} = 0$ (loss-less dielectric or magnetic medium) and hence causes the growth rate of the space-charge waves to be reduced. If this reduction is large enough to prevent the gain from exceeding a critical value, we expect that, similar to the one-dimensional case, dipole-domain formation is inhibited.

Experimentally TE elements have first been stabilized against space-charge domain formation in this way by covering the surface of GaAs samples with a high-permittivity material ('dielectric surface loading', [4.57, 4.58]). On the other hand, stabilization could be achieved by sufficiently reducing the transverse dimensions of the sample with air as surrounding medium [4.59].

So far *dielectric*, and not magnetic or resistive surface loading has mainly been used with, for example, polycrystalline barium titanate ($BaTiO_3$) as the surrounding material. In spite of the high permittivity of $BaTiO_3$, the phase velocity of a free electromagnetic wave in the dielectric $c_{II} = (\varepsilon_{II}\mu_{II})^{\frac{1}{2}}$, is still too high, compared with the electron drift velocity, to cause any appreciable influence on the growth rate. Rather it is the effect of the increase in capacitive reactance which has to be considered. To provide a rough idea of the physical process involved, this case is investigated here by using a highly simplified large-signal model [4.60, 4.61] which yields an approximate doping-times-thickness criterion for dipole-domain formation. A somewhat more rigorous derivation of such a criterion, based on the two-dimensional analysis of space-charge wave propagation presented in Section 4.3.3, will be given in Section 4.4.2.2, and Section 4.4.2.3 is devoted to experimental results obtained with dielectric-loaded GaAs elements.

We shall now consider the dielectric loaded TE element with a drifting dipole domain, as shown schematically in Fig. 4.19 where the relevant symbols are given. The influence of the stray electric fields in the dielectric surface layer may simply be

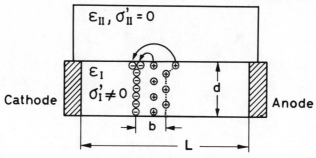

Fig. 4.19 Schematic representation of n-type dielectric-loaded TE semiconductor element with space-charge dipole domain. After Engelmann [4.60]. (Read σ_{II} instead of σ'_{II}.)

113

described by an increase in the overall dipole-domain capacitance C_D so that $C_D = C_{DI} + C_{DII}$. It is further assumed here that the domain capacitance C_{DI} within the semiconductor is unaffected by the dielectric surface loading. This is certainly justified for relatively light loading ($\varepsilon_{II} \approx \varepsilon_I$ and/or $d \gg b$). For heavy loading ($\varepsilon_{II} \gg \varepsilon_I$) or thin semiconductor layers ($d \lesssim b$), the stray fields tend to increase the domain width b due to the limited positive space charge available in the semiconductor (full depletion), as indicated in Fig. 4.19. However, C_{DI} might then be entirely negligible as compared to the stray-field capacitance C_{DII}, meaning that the model is still applicable.

For C_{DI} one has

$$C_{DI} = \varepsilon_I dw/b$$

where w is the second dimension of the semiconductor sample transverse to the electric bias field. The magnitude of C_{DII} is estimated from the penetration of the stray field into the dielectric material. For a dipole domain of width b the penetration is roughly equal to b if b is smaller than the thickness of the dielectric layer and the external permittivity ε_{II} is not excessively high. Therefore

$$C_{DII} = \varepsilon_{II} bw/b = \varepsilon_{II} w$$

being independent of b. If we assume a simplified equivalent circuit of the sample as given in Fig. 4.20 (neglecting in Fig. 3.21b the geometric capacitance C_h, the drift

Fig. 4.20 Large-signal equivalent circuit of TE semiconductor element with drifting dipole domain.

current I_D of the travelling dipole domain and the change of the domain capacitance C_D during growth; see [3.39, 8.14]), the domain charge growth is determined by the large-signal time constant

$$\tau_D \approx R_1 C_D = \frac{L}{en_0\mu_1}\left(\frac{\varepsilon_I}{b} + \frac{\varepsilon_{II}}{d}\right) \tag{4.57}$$

The ratio of the time constants with and without dielectric loading is

$$\frac{\tau_D}{\tau_{DI}} = 1 + \frac{\varepsilon_{II}b}{\varepsilon_I d} \tag{4.58}$$

where

$$\tau_{DI} \approx R_1 C_{DI} = \frac{L}{b}\tau_{RI} = \frac{L}{b}\frac{\varepsilon_I}{en_0\mu_1} \tag{4.59}$$

with τ_{RI} and μ_1 denoting low-field values [see eqn (3.18b) with $\mu = \mu_1' \approx \mu_1$].

Since the drift current I_D of the domain in parallel with C_D has been neglected, eqn (4.59) somewhat underestimates the growth time because the displacement current charging C_{DI} is overestimated. If we now assume heavy dielectric loading ($\varepsilon_{II} \gg \varepsilon_I$) or thin semiconductor layers ($d \ll b$), eqn (4.58) reduces to

$$\frac{\tau_D}{\tau_{DI}} = \frac{\varepsilon_{II}}{\varepsilon_I} \frac{b}{d}. \tag{4.60}$$

The formation of dipole domains is assumed to be suppressed if $\tau_D \gtrsim \tau = L/v_{h0}$, similarly to relation (4.45a). This yields

$$n_0 L \lesssim n_0 v_{h0} \tau_{DI} \frac{\tau_D}{\tau_{DI}}. \tag{4.61}$$

Inserting eqns (4.59) and (4.60) into eqn (4.61) eventually leads to a doping-times-thickness criterion for stability against dipole-domain formation, viz.

$$n_0 d \lesssim (n_0 d)_{\text{crit}} = \frac{\varepsilon_{II} v_{h0}}{e \mu_1} = \frac{\varepsilon_{II}}{\varepsilon_I} \frac{|\mu'_{h0}|}{\mu_1} (n_0 L)_{\text{crit}} \tag{4.62}$$

where $(n_0 L)_{\text{crit}}$ is taken from relation (4.45b) with $\varepsilon = \varepsilon_I$. If the dielectric material is also applied to the opposite surface of the sample (rectangular cross-section assumed), the r.h.s. of inequality (4.62) must be multiplied by two. It should be noted that the sample length L drops out of inequality (4.61) only if the dielectric loading is heavy or the semiconductor layer is thin.

As a *numerical example*, for an n-type GaAs sample of $n_0 = 10^{14}$ cm^{-3}, $v_{h0} = 10^7$ cm s^{-1}, $\mu_1 = 7\,000$ cm^2 (Vs)$^{-1}$ and subjected to one-sided surface loading with a medium having an effective relative permittivity of, say, $\varepsilon_{II}/\varepsilon_0 = 1\,000$, one obtains from inequality (4.62) for ensuring stability a permissible transverse dimension $d \lesssim 80$ μm ($n_0 d \lesssim 8 \times 10^{11}$ cm^{-2}). According to expression (4.62), it is possible to achieve stability, even without having to apply a special high-permittivity surface material, by a sufficient reduction of the dimension d, in agreement with experiments [4.59]. For example, a GaAs element with the active layer exhibiting the negative-mobility properties epitaxially grown on a semi-insulating GaAs substrate (see Section 6.1.2), i.e. with a 'surface' material of $\varepsilon_{II}/\varepsilon_0 \approx 12\cdot5$, would require $d \lesssim 1$ μm using the GaAs parameters above ($n_0 d \lesssim 10^{10}$ cm^{-2}). Employing thin-layered epitaxially-prepared TE elements for domain-stable operation is certainly most attractive.

4.4.2.2. *Stability Analysis*
In Section 4.3.3 the dispersion relation of the fundamental space-charge wave (mode order $m = 0$) in a semiconductor layer of thickness $d = 2a$ was investigated for the case of quasi wave-planarity, i.e. $|k_y|a < 0\cdot5$, see eqns (4.42a) and (4.44). This case generally corresponds to relatively small layer thickness d [4.35]. For deriving the small-signal impedance of such a device the wave can be approximated as being perfectly plane. This leads to an impedance expression of the same character as eqn (4.48) for the one-dimensional case, viz. [4.64]

$$Z(j\omega) = r\frac{\gamma}{\alpha_{Rz}}\frac{\vartheta - 1 + \exp(-\vartheta)}{\vartheta^2}. \tag{4.63}$$

Here, we have again set $jk_zL = \vartheta(j\omega)$ where k_z is the growing-wave solution of the combined eqns (4.42a) and (4.44). For small wave growth $\alpha \ll \beta_e$ this solution follows simply from setting $k_z \approx \beta_e$ in eqns (4.44), yielding $\gamma \approx \alpha$ and $k_z \approx \beta_e + j\alpha$.

The zeros of Z in the complex ϑ plane ($j\omega \to p$) remain at the positions $\vartheta_n = \xi_n + j\eta_n$ of the one-dimensional case and, thus, the criterion for current stability can be derived similarly by using the dispersion relation in the form $p = p(\vartheta)$. From eqns (4.42a) and (4.44a) one obtains for instance

$$\tau p = \vartheta\left(1 + \frac{\alpha_{Rz}}{\vartheta + j\dfrac{L}{a}\dfrac{\varepsilon_{II}}{\varepsilon_{I}}}\right) \tag{4.64}$$

As it turns out the most stringent stability criterion $\mathrm{Re}(p_n) < 0$ is now given by the lowest-order zero p_1 only if [4.80]

$$\frac{a}{L}\frac{\varepsilon_I}{\varepsilon_{II}} > 0{\cdot}144 \tag{4.65}$$

i.e. at relatively light dielectric loading and/or thick semiconductor layers. Then, for the zero with $\eta_1 > 0$ the condition $\mathrm{Re}(p) < 0$ yields

$$\xi_1 + \alpha_{Rz}a\frac{\eta_1}{\dfrac{\varepsilon_{II}}{\varepsilon_I} + \dfrac{a}{L}\eta_1} < 0 \quad \text{or} \quad \alpha_{Rz}a < 0{\cdot}28\frac{\varepsilon_{II}}{\varepsilon_I} + 2{\cdot}09\frac{a}{L} \tag{4.66a}$$

from eqn (4.65) by neglecting $\xi_1^2 \ll \eta_1^2$. The zero with $\eta_1 < 0$ requires a different dispersion relation and leads to the same condition. Thus an n_0d criterion can be formulated, viz.

$$n_0d < (n_0d)_{\mathrm{crit}} = \left(0{\cdot}27\frac{\varepsilon_{II}}{\varepsilon_I} + \frac{d}{L}\right)(n_0L)_{\mathrm{crit}} \tag{4.66b}$$

where $(n_0L)_{\mathrm{crit}}$ is given by relation (4.51b) with $\varepsilon = \varepsilon_I$. This criterion, however, still depends on the sample length L.

The lowest-order oscillation frequency near the stability boundary follows from $\omega_1 = \mathrm{Im}(p_1)$ with $\eta_1 = 7{\cdot}46$ and eqn (4.64). One finds (neglecting again $\xi_1^2 \ll \eta_1^2$)

$$\omega_1\tau = 7{\cdot}46 + \frac{2{\cdot}09\alpha_{Rz}a(\varepsilon_I/\varepsilon_{II})}{[1 + 7{\cdot}46(a/L)(\varepsilon_I/\varepsilon_{II})]^2}. \tag{4.67}$$

Thus the two-dimensional effect increases this frequency beyond the transit-time value $\omega_\tau\tau = 2\pi$ somewhat further than is the case in the one-dimensional approach.

The modifications necessary in the n_0d criterion for a loading material II with strong *magnetic* properties are obvious when comparing the appropriate eqns (4.44) for γ. Because the magnetic properties further decrease the growth coefficient in case of $\gamma \approx \alpha$, they cause the critical $n_0 d$ product to increase [4.29, 4.30].

If condition (4.65) is violated, i.e. for sufficiently thin semiconductor layers or heavy dielectric loading, *higher*-order zeros of the impedance function $Z(p)$ have to be selected for determining stability [4.80, 4.94]. This is due to the increase of the wave growth coefficient α with frequency (cf. Fig. 4.12a, mode number $m = 0$). Accordingly the frequency of the instability increases. However, the plane-wave approximation, i.e. $|k_y|a < 0.5$, is no longer applicable at the heavier dielectric loading, particularly for higher frequencies [4.35]. The limiting case of large $\varepsilon_{II}/\varepsilon_I$ values is characterized by a strong lateral change of the wave amplitude determined by $k_y a \approx \pi/2$ [4.67]. In this approximation one finds for $L > d$ and $n_0 d$ criterion that is independent of the sample length L and the permittivity of the outer medium II, viz.

$$(n_0 d)_{\mathrm{crit}} = \frac{K_1}{2.09}\,(n_0 L)_{\mathrm{crit}} \tag{4.68}$$

where $(n_0 L)_{\mathrm{crit}}$ is again given by relation (4.51b) with $\varepsilon = \varepsilon_I$. The factor K_1 depends only on the ratio of the differential mobilities μ'_y/μ'_z. For GaAs biased in the n.d.c. region with $\mu'_y/\mu'_z = -2$ one has $K_1 = 5.6$, i.e. the critical $n_0 d$ product is about 2.7 times the critical $n_0 L$ product. The frequency of the instability in this case is given by

$$\omega d/v_{h0} = K_2 \tag{4.69}$$

where K_2, like K_1, depends only on μ'_y/μ'_z ($K_2 = 8.5$ for $\mu'_y/\mu'_z = -2$). Since $d < L$ this frequency can be considerably higher than the transit-time frequency as already expected qualitatively from the discussion of the quasi plane-wave case above.

The criteria (4.66b) and (4.68) must have a smooth transition in the range of moderate dielectric loading, cf. [4.80, 4.89]. They are to be compared with criterion (4.62) that was derived from simple physical arguments. In general, there exists at least an order-of-magnitude agreement if $\varepsilon_{II}/\varepsilon_I$ is not excessively high.

In discussing the theoretical results, one has to keep in mind the approximations made. For instance, the boundary conditions will spatially disturb the homogeneous d.c. field E_0 assumed in the small signal theory of Section 4.3.3. Similarly as in the case of diffusion stability in supercritical $n_0 L$-product samples, a stationary high-field anode layer is theoretically predicted for thin-film TE devices [4.65, 4.66]. Such inhomogeneous d.c. fields are actually observed in experiments (see Section 4.4.2.3 below, particularly Fig. 4.22). If this disturbance is strong the stability condition might change drastically. On the other hand, a strong field disturbance can be suppressed by employing a suitable cathode boundary condition with limited carrier injection [4.92, 4.93].

As we have seen (Fig. 4.12b) *diffusion* reduces space-charge wave growth towards the higher-order oscillation frequencies, thus decreasing the numerical value in condition (4.65). By choosing $\delta \equiv D_{h0}/(v_{h0}L) \gtrsim 0.025$ (with $\delta' = 0$) this value approaches zero, i.e. only the fundamental frequency needs to be considered as in the one-dimensional case. The stability relation (4.66a) then becomes [4.96]

$$\alpha_{Rz} a < (-\xi_1 + \eta_1^2 \delta) \left(\frac{\varepsilon_{II}}{\varepsilon_1 \eta_1} + \frac{a}{L} \right). \tag{4.70}$$

Results of a rigorous computer analysis of two-dimensional domain dynamics with diffusion included have been reported [4.68].

A different stabilization scheme for TE devices was proposed [4.69] involving the two-stream interaction of an active TE semiconductor element and a passively biased semiconductor. The advantages of such a scheme are a nearly homogeneous d.c. field distribution and the possibility of electronic control of the growth rate by changing the drift-velocity difference in the two semiconductors. The additional d.c. power generated in the passive semiconductor is a disadvantage, however.

4.4.2.3 *Experimental Verification*

It has experimentally been proved that in GaAs samples with supercritical $n_0 L$ product capacitive surface loading increases the dipole-domain formation time [4.70] or has the desired effect of even completely inhibiting the formation of moving domains [4.57, 4.58, 4.71, 4.72 and 4.90]. Backing the dielectric material by a metal film generally enhances the effect as should be expected. Here, however, complications may arise with regard to the d.c. field and carrier distribution because of the field effect occurring between the metal-backed dielectric and the semiconductor. Domain inhibition could also be achieved purely by a reduction of the transverse sample dimensions without additional surface loading [4.59]. GaAs elements in which moving dipole domains have been completely suppressed are able to sustain oscillations at frequencies around the transit-time frequency $f_\tau \approx v_{h0}/L$ within a continuous range of one or two octaves [4.58a, 4.71, 4.73].

In the one-dimensional investigation of Section 4.4.1.3 we have seen that TE elements, stabilized by providing a subcritical $n_0 L$ product, show a negative terminal conductance near the transit-time frequency and possibly some harmonics of it where the positive feedback has the proper phase. In the two- (or three-) dimensional configuration of interest here, the resonant properties introduced by the finite *transverse* dimensions must also be taken into account. As seen in the preceding section these alter the optimum feedback condition from that valid in the one-dimensional case. Furthermore, existence of the additional feedback path via the outside dielectric material modifies the condition. These circumstances might also explain the observed *wide*-band behaviour of such devices, cf. [4.94].

There is some experimental indication [4.57, 4.74, 4.75] that in GaAs samples with subcritical $n_0 d$ product the static negative differential conductance due to the $v(E)$ characteristic, which actually is the property originally sought, is in fact available at the terminals for fields up to a few times the threshold field E_T, cf. [8.55]. Tapering of the sample, with the narrow end being at the cathode side, seems to favour this effect [4.57]. Shockley [1.9] has postulated that space-charge accumulation as a result of carrier injection from the cathode suppresses any static negative resistance (cf. Section 4.4.1.3). This conclusion is considered to hold also for an arbitrary three-dimensional geometry as long as the cathode contact establishes a zero electric field [4.76]. Then, imperfect cathode boundary conditions must be involved in producing a static negative conductance. In any case, the preliminary results obtained so far will have to be supplemented to allow a definite conclusion to be drawn on the negative-conductance properties of samples with subcritical n_0d product.

In samples where the dielectric surface loading does not extend over the cathode and anode contacts but covers only part (cf. [4.91a, b]) of the semiconductor as schematically indicated by Fig. 4.21(a), a strongly distorted d.c. field results and the stability which is obtained against moving dipole domains may not necessarily be due to satisfying the n_0d criterion [4.65, 4.66]. As Fig. 4.21(a) shows, the d.c. electric

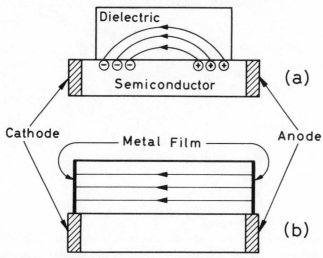

Fig. 4.21 Schematic representations of d.c. field distributions in dielectric surface material when dielectric (a) only partially covers the semiconductor element, (b) extends right across the contacts and is metallized on its sides. Adapted from Hofmann [4.65].

field lines within the dielectric have appreciable transverse components near the semiconductor surface. Thus a static accumulation layer is formed at the surface near the cathode, together with a depletion layer near the anode. This depletion reduces the effective width of the semiconductor in the anode region and, particularly if its transverse extension becomes comparable to the semiconductor thickness, a marked potential drop occurs at the anode (Fig. 4.22). Since the potential drop in the middle part of the semiconductor must decrease correspondingly, the electric field there may fall to a subcritical value $E_0 < E_p$. If for the remaining part of the sample, where $E_0 > E_p$, the product of doping times effective length is subcritical, no moving dipole domains can form. This distortion of the d.c. electric field could probably be avoided in a sample configuration as shown in Fig. 4.21(b). Relatively smooth potential profiles were obtained in thin-layer structures with suitable boundary conditions established at the cathode and anode contacts [4.93, 4.94, 4.95, 5.78, 7.162].

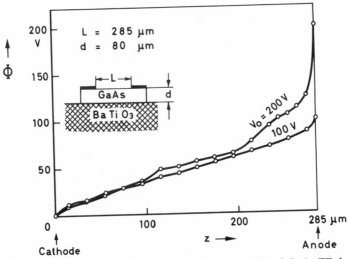

Fig. 4.22 Potential probe measurements on a thin, $n_0 d$-stabilized GaAs TE device: $d = 3·5$ μm, $L = 300$ μm (width $w = 250$ μm), $n_0 d \approx 5 \times 10^{11}$ cm^{-2} and $n_0 L = 5 \times 10^{13}$cm^{-2}. After Frey [4.79].

4.4.3 Controlled Carrier Injection

In an interesting proposal for a three-terminal structure, to be built from a TE semiconducting material like GaAs, the negative differential conductivity of the static $v(E)$ characteristic is exploited directly and—an important feature—can be controlled via the third electrode [4.77]. This device is essentially an npνn-transistor structure, i.e. a transistor which possesses a lightly-doped (ν) collector region, as illustrated in Fig. 4.23.

There are two important regions in the device, namely the essentially electron-free ν-region of the reserved-biased pνn junction, and the region providing means for controlled electron injection, cf. [3.5a] into the ν-region. The bias field in the pνn junction has to be such that the emitted electrons attain negative differential mobility, i.e. $E_0 > E_p$. Then the collector drift region is a TE-active zone which can be represented by a capacitor in parallel with a negative conductance as shown in the right-hand part of the small-signal equivalent circuit in Fig. 4.24. The equivalent circuit is completed by a parasitic series resistance R_s which essentially represents the emitter resistance. Due to the fairly high level of carrier injection, the emitter resistance is relatively low, and the influence of the emitter capacitance can thus be neglected. Also the base resistance is assumed to be negligible.

Swept-out n.d.c. Region

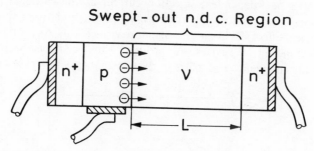

Fig. 4.23 Schematic representation of TE negative-resistance device with emitter-controlled carrier injection.

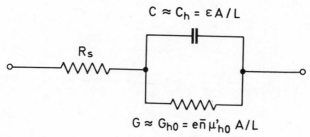

Fig. 4.24 Equivalent circuit of TE device with emitter-controlled carrier injection.

The mean electron density $\bar{n}$ and hence the magnitude of the small-signal negative conductance $-G$ can be controlled by the emitter-base voltage. The calculation of G is performed by a procedure similar to that of the ohmic-contact case (Section 4.4.1.3). Instead of the cathode boundary condition $\tilde{E}(0) = 0$ for an ohmic contact, the r.f. drift current density [cf. eqn (4.17)]

$$\hat{J}(z) = en\mu'_{ho}\tilde{E}(z) + \varepsilon v_{ho}\frac{\mathrm{d}\tilde{E}}{\mathrm{d}z}$$

must be set at zero at the cathode $z = 0$ (constant injection) using eqns (4.18) and (4.19). Calculations of G and of the dynamic capacitance C for such a GaAs device [4.77] assuming neutrality in the v region ($\bar{n} \approx n_0$) and taking into account the effects of carrier diffusion, have yielded the constant values $G \approx G_{ho} = e\bar{n}\mu'_{ho}A/L$ and $C \approx C_h = \varepsilon A/L$ for transit angles $\theta \gtrsim 3\pi$. As demonstrated in Fig. 4.25 $-G$ starts from zero at $\theta = 0$, shows a short undulation similar to that in Fig. 4.15 and then approaches the above value. Thus, as a result of the constant-injection boundary condition, the wide-band high-frequency conductance approaches the homogeneous-field value. In other words, constant carrier injection ensures minimum

121

excitation of the space-charge-wave part $\tilde{E}_b(z)$ in the general solution (4.19), making the homogeneous part $\hat{E}_a$ dominant. This conclusion was also obtained for the more general case of a field-dependent injection current at the cathode, including a depleted (reverse-biased) Schottky-barrier contact [4.78a] and for an ohmic cathode contact with a field-effect shield on a co-planar device [4.78b]. As for the wide-band negative conductance in TE devices with ordinary ohmic contacts (Section 4.4.1.3), an upper limiting frequency also exists in the constant-injection case [4.81].

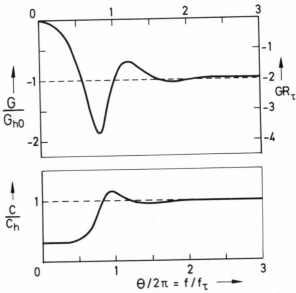

Fig. 4.25 Theoretical small-signal admittance $Y = G + j\omega C$ of an injection-controlled GaAs TE device; adapted from a calculation by Atalla and Moll [4.77]. Parameters are $\alpha_R L = 2$, $\delta = 0.02$ and $\delta' = 0.02$ (cf. caption of Fig. 4.15).

For achieving optimum performance, the injection level must be adequate to cause a reasonably homogeneous electric field in the collector drift region. This is obtained if the electron density approximately equals the fixed-impurity density $\bar{n} \approx n_0$, although it was shown that higher carrier injection, i.e. $\bar{n} > n_0$, improves the negative conductance properties [4.83]. In any case, it must be ensured that $\bar{n}L \lesssim 10^{11}$ cm^{-2} to avoid space-charge dipole formation (Section 4.4.1.2). Hence a fairly low doping (ν) is required for the collector drift region.

A TE device with emitter-controlled carrier injection—or alternatively with injection by photo generation of electron-hole pairs [4.77]—has still to await practical realization.

REFERENCES

4.1   Schuller, M. and Gärtner, W. W., Large-signal circuit theory for negative-resistance diodes, in particular tunnel diodes. *Proc. IRE* (1961), p. 1268.

4.2a   U.S. Patent No. 3,422,289 (inventors: M. M. Atalla, R. J. Archer, R. D. Hall and R. W. H. Engelmann, priority: Dec. 15, 1965).

4.2b   Engelmann, R. W. H., Circuit performance of field-controlled negative-conductivity devices. Informal Conf. on Active Microwave Effects in Bulk Semiconductors, New York City, Feb. 3–4, 1966. See also Techn. Report ECOM—01758—1 and —2, Gunn effect devices, Contract DA 28—043 AMC—01758 (E), US Army Electron. Command, March and June 1966.

4.3   Copeland, J. A., A new mode of operation for bulk negative resistance oscillators. *Proc. IEEE* **54** (1966), p. 1479.

4.4   Copeland, J. A., CW operation of LSA oscillator diodes—44 to 88 GHz. *Bell Syst. Techn. J.* **46** (1967), p. 284.

4.5   Copeland, J. A. and Spiwak, R. R., LSA operation of bulk n-GaAs diodes. *Int. Solid-State Circuits Conference Philadelphia Pa.*, Feb. 1967, Digest of Technical Papers, p. 26.

4.6a   Kennedy, W. K. and Eastman, L. F., High power pulsed microwave generation in gallium arsenide. *Proc. IEEE* **55** (1967), p. 434.

4.6b   Higgins, J. A., Grande, V. J. and Pearson, G. L., Signature of the LSA mode. *IEEE Trans. Electron Dev.* **ED–14** (1967), p. 719.

4.7   Shuskus, A. J. and Shaw, M. P., Current instabilities in gallium arsenide. *Proc. IEEE* **53** (1965), p. 1804.

4.8   Kennedy, W. K., Power generation in GaAs at frequencies far in excess of the intrinsic Gunn frequency. *Proc. IEEE* **54** (1966), p. 710.

4.9   Copeland, J. A., LSA oscillator diode theory. *J. Appl. Phys.* **38** (1967), p. 3096.

4.10   Heinle, W., Simple theory for LSA operation of Gunn-effect semiconductors. *Electron. Lett.* **3** (1967), p. 429.

4.11   Riley, T. J., Design calculations for cw millimetre wave L.S.A. oscillators. *Proc. 1968 Int. GaAs Symp.*, Inst. of Physics and Phys. Soc. Conf. Series No. 7, p. 173.

4.12   Spiwak, R. R., Frequency conversion and amplification with an LSA diode oscillator. *IEEE Trans. Electron Dev.* **ED–15** (1968), p.614.

4.13   Jeppesen, P. and Jeppsson, B., Computer simulation of LSA oscillators with high doping to frequency ratios. *Proc. IEEE* **57** (1969), p. 795.

4.14   Taylor, B. C. and Fawcett, W., Detailed computer analysis of LSA operation in CW transferred electron devices. *IEEE Trans. Electron Dev.* **ED–17** (1970), p. 907.

4.15   Curtice, W. R. and Purcell, J. J., Analysis of the LSA mode including effects of space charge and intervalley transfer time. *IEEE Trans. Electron Dev.* **ED–17** (1970), p. 1048.

4.16   Müller, R. R. and Nitz, B., Analogue-computer model for a gallium arsenide L.S.A. diode. *Electron. Lett.* **5** (1969), p. 498.

4.17   Bott, I. B. and Hilsum, C., An analytic approach to the LSA mode. *IEEE Trans. Electron Dev.* **ED–14** (1967), p. 492.

4.18   Olsson, K. O. I., LSA diode theory for long samples. *IEEE Trans. Electron Dev.* **ED–16** (1969), p. 202.

4.19   Harrison, R. L., Denker, S. P. and Hadley, M. L., Characteristic ranges for LSA oscillation. *IEEE Trans. Electron Dev.* **ED–15** (1968), p. 792.

4.20   Engelmann, R. W. H. and Quate, C. F., Linear, or 'small-signal', theory for the Gunn effect. *IEEE Trans. Electron Dev.* **ED–13** (1966), p. 44.

4.21   Suematsu, Y. and Nishimura, Y., Wave theory of the negative resistance element due to Gunn effect. *Proc. IEEE* **54** (1966), p. 322.

4.22   Ramo, S., Space charge and field waves in an electron beam. *Phys. Rev.* **56** (1939), p. 276.

4.23   Chodorow, M. and Susskind, C., *Fundamentals of Microwave Electronics* (New York, McGraw-Hill, 1964).

4.24   Vural, B. and Bloom, S., Streaming instabilities in solids and the role of collisions. *IEEE Trans. Electron Dev.* **ED–13** (1966), p. 57.

4.25 Steele, M. C. and Vural, B., *Wave Interactions in Solid State Plasmas* (New York, McGraw-Hill, 1969).

4.26 Hakki, B. W., Amplification in two-valley semiconductors. *J. Appl. Phys.* **38** (1967), p. 808.

4.27 Kroemer, H., Detailed theory of the negative conductance of bulk negative mobility amplifiers, in the limit of zero ion density. *IEEE Trans. Electron Dev.* **ED–14** (1967), p. 476.

4.28 Kino, G. S. and Robson, P. N., The effect of small transverse dimensions on the operation of Gunn devices. *Proc. IEEE* **56** (1968), p. 2056.

4.29 Hartnagel, H. L., Gunn instabilities with surface loading. *Electron. Lett.* **5** (1969), p. 303.

4.30 Masuda, M., Chang, N. S. and Matsuo, Y., Suppression of Gunn-effect domain formation by ferrimagnetic materials. *Electron. Lett.* **6** (1970), p. 605.

4.31a Nergaard, L. S., Analysis of a simple model of a two-beam growing wave tube. *RCA Rev.* **9** (1948), p. 585.

4.31b Tosima, S. and Horita, R., New type of two-stream plasma instability. *J. Appl. Phys.* **34** (1963), p. 2993.

4.32 Hahn, W. C., Small-signal theory of velocity-modulated electron beams. *Gen. Electric Rev.* **42** (1939), p. 258.

4.33a Kino, G. S., Carrier waves in semiconductors. Microwave Lab., Stanford University, Stanford, Calif., ML Rept. 1353, August 1965.

4.33b Kino, G. S., Growing surface waves in a semiconductor in the presence of a transverse magnetic field. *Appl. Phys. Lett.* **12** (1968), p. 312.

4.33c Kino, G. S., Carrier waves in semiconductors—I: Zero temperature theory. *IEEE Trans. Electron Dev.* **ED–17** (1970), p. 178.

4.34 Engelmann, R. W. H., Space-charge waves in semiconductor plates (in German). *Arch. Elektron. Übertr.* **25** (1971), p. 357.

4.35 Engelmann, R. W. H., Plane-wave approximation of carrier waves in semiconductor plates with nonisotropic mobility. *IEEE Trans. Electron Dev.* **ED–18** (1971), p. 587.

4.36a Dean, R. H., Optimum design of thin-layer GaAs amplifiers. *Proc. IEEE* **57** (1969), p. 1327.

4.36b Heinle, W., Inclusion of diffusion in the space-charge wave theory of Kino and Robson. *Electron Lett.* **7** (1971), p. 245.

4.37 Hilsum, C., A. simple analysis of transferred-electron oscillators. *Brit. J. Appl. Phys.* **16** (1965), p. 1401.

4.38 Kroemer, H., External negative conductance of a semiconductor with negative differential mobility. *Proc. IEEE* **53** (1965), p. 1246.

4.39 Thim, H. W., Barber, M. R., Hakki, B. W. and Uenohara, M., Microwave amplification in a d.c.-biased bulk semiconductor. *Appl. Phys. Lett.* **7** (1965), p. 167.

4.40 Engelmann, R. W. H. and Mathers, G. W., Oscillations in bulk GaAs due to an equivalent negative r.f. conductance. *Proc. IEEE* **54** (1966), p. 786.

4.41 Thim, H. W., Series-connected bulk GaAs amplifiers and oscillators. *Proc. IEEE* **56** (1966), p. 1245.

4.42 Heinle, W., Diagram of Gunn-effect phenomena. *Electron. Lett.* **2** (1966), p. 417.

4.43 Mahrous, S. and Hartnagel, H. L., Gunn-effect domain formation controlled by a complex load. *Brit. J. Appl. Phys.* **2** (1969), Ser. 2, p. 1.

4.44a Pollmann, H. and Engelmann, R. W. H., On supercritical reflection-type amplification and the stability criterion in bulk GaAs devices. *Proc. 8th Int. Conference MOGA Amsterdam, 1970,* p. 16/24 (Deventer, Kluwer, 1970).

4.44b Engelmann, R. W. H., On 'supercritical' transferred-electron amplifiers. *Arch. Elektron. Übertragungstechn.* **26** (1972), p. 357.

4.45a Suematsu, Y. and Nishimura, Y., Proposal on analysis of Gunn effect oscillator of negative resistance type. *J. Inst. Elec. Commun. Engrs Japan* **50** (1967), p. 94.

4.45b Suematsu, Y. and Nishimura, Y., Small-signal admittance of bulk semiconductor devices at higher frequencies. *Proc. IEEE* **56** (1968), p. 242.

4.46a Magarshack, J. and Mircea, A., Stabilization and wideband amplification using over-critically doped transferred-electron diodes. *Proc. 8th Int. Conference MOGA, Amsterdam,* 1970, p. 16/19 (Deventer, Kluwer, 1970).

4.46b Jeppesen, P. and Jeppsson, B., The influence of diffusion on the stability of the super-critical transferred electron amplifier. *Proc. IEEE* **60** (1972), p. 452.

4.47a Guéret, P., Convective and absolute instabilities in semiconductors exhibiting negative differential mobility. *Phys. Rev. Lett.* **27** (1971), p. 256.

4.47b Heinle, W., Instabilities in semiconductors with negative differential mobility. *J. Phys. D: Appl. Phys.* **5** (1972), p. 1344.

4.48 Thim, H., Stability and switching in overcritically doped Gunn diodes. *Proc. IEEE* **59** (1971), p. 1285.

4.49 Kroemer, H., Effect of a parasitic series resistance on the performance of bulk negative conductivity amplifiers. *Proc. IEEE* **54** (1966), p. 1980.

4.50 Kroemer, H., Negative conductivity in semiconductors. *Festkörperprobleme* **7** (1967), p. 264.

4.51 Thim, H. W., Temperature effects in bulk GaAs amplifiers. *IEEE Trans. Electron Dev.* **ED–14** (1967), p. 59.

4.52 Mahrous, S. and Robson, P. N., Small-signal impedance of stable transferred-electron diodes. *Electron. Lett.* **2** (1966), p. 107.

4.53 McWhorter, A. L. and Foyt, A. G., Bulk GaAs negative conductance amplifiers. *Appl. Phys. Lett.* **9** (1966), p. 300.

4.54 Mahrous, S., Robson, P. N. and Hartnagel, H. L., The stability and reflection gain of subcritically doped Gunn diodes. *Solid-State Electron.* **11** (1968), p. 965.

4.55 Hanson, D. C. and Rowe, J. E., Non-isothermal effects in bulk GaAs transit-time-mode oscillators. *Int. J. Electron.* **24** (1968), p. 415.

4.56 Shoji, M., Small-signal impedance of bulk semiconductor amplifier having a nonuniform doping profile. *IEEE Trans. Electron Dev.* **ED–14** (1967), p. 323.

4.57 *Electronics* **40** (1967), Dec. 11, p. 255.

4.58a Kataoka, S., Tateno, H., Kawashima, M. and Komamiya, Y., Microwave oscillation and amplification in a long bulk GaAs with $BaTiO_3$ sheets on the surface. *Proc. 1968 MOGA Conference, Nachrichtentechn. Fachber.* **35** (1968), p. 454.

4.58b Kataoka, S., Tateno, H. and Kawashima, M., Suppression of travelling high-field-domain mode oscillations in GaAs by dielectric surface loading. *Electron. Lett.* **5** (1969), page 48.

4.59 Kumabe, K., Suppression of Gunn oscillations by a two-dimensional effect. *Proc. IEEE* **56** (1968), p. 2172.

4.60 Engelmann, R. W. H., Simplified model for the domain dynamics in Gunn-effect semi-conductors covered with dielectric sheets. *Electron. Lett.* **4** (1968), p. 546.

4.61 Engelmann, R. W. H., On the transverse surface boundary effect in Gunn devices. *Proc. IEEE* **57** (1969), p. 818.

4.62 Giannini, F., Ottavi, C. M. and Salsano, A., Laminar electron flow in thin GaAs slabs. *Proc. IEEE* **58** (1970), p. 259.

4.63 Engelmann, R. W. H., Comment on laminar electron flow in thin GaAs slabs; F. Giannini, C. Ottavi and A. Salsano, Author's reply. *Proc. IEEE* **58** (1970), p. 1869. [Note printer's error: the last paragraph of "Author's Reply" is actually a "Further Comment" by R. W. H. Engelmann, see *Proc. IEEE* **59** (1971), p. 1136].

4.64 Guéret, P., Small-signal two-terminal impedance of a thin Gunn diode. *Electron. Lett.* **6** (1970), p. 213.

4.65 Hoffmann, K. R., Some aspects of Gunn oscillations in thin dielectric loaded samples. *Electron. Lett.* **5** (1969), p. 227.

4.66 Suga, M., Field distribution in a Gunn diode with a distributed capacitance electrode. *Proc. IEEE* **57** (1969), p. 253.

4.67   Hofmann, K. R., Stability criterion for Gunn oscillators with heavy surface loading. *Electron. Lett.* **5** (1969), p. 469.

4.68   Kataoka, S., Tateno, H. and Kawashima, M., Two-dimensional computer analysis of dielectric-surface-loaded GaAs bulk element. *Electron. Lett.* **6** (1970), p. 169.

4.69   Guéret, P., Stabilization of Gunn oscillations in layered semiconductor structures. *Electron. Lett.* **6** (1970), p. 637.

4.70   Vlaardingerbroek, M. T., Acket, G. A., Hofmann, K. and Boers, P. M., Reduced build-up of domains in sheet-type gallium arsenide Gunn oscillators. *Phys. Lett.* **28A.** (1968), p. 97.

4.71   Becker, R. and Bosch, B. G., An investigation of GaAs Gunn elements covered with $BaTiO_3$ (in German). *Verhandl. Deutsche Physikal. Gesellsch.* **4/VI** (1969), p. 209. (European Meeting "Semiconductor Devices Research", Munich, March 24–27, 1969).

4.72   Kuru, I. and Tajima, Y., Domain suppression in Gunn diodes. *Proc. IEEE* **57** (1969), p. 1215.

4.73   Kataoka, S., Tateno, H. and Kawashima, M., Improvements in efficiency and tunability of Gunn oscillators by dielectric surface loading. *Electron. Lett.* **5** (1969), p. 491.

4.74   Becker, R., AEG-Telefunken Research Institute, unpublished investigations (1969).

4.75   Chawla, B. R., Bartelink, D. J. and Coleman, D. J., Transverse electromagnetic wave amplification in n-GaAs. Symp. on Instabilities in Semiconductors, Yorktown Heights, N.Y., March 20–21, 1969 [*Bull. Amer. Phys. Soc.* **II/14** (1969), p. 747].

4.76a Kroemer, H., Generalized proof of Shockley's positive conductance theorem. *Proc. IEEE* **58** (1970), p. 1844.

4.76b Tateno, H., Kataoka, S. and Kroemer, H., Comments on "Generalized proof of Shockley's positive conductance theorem". *Proc. IEEE* **59** (1971), p. 1282.

4.77   Atalla, M. M. and Moll, J. L., Emitter-controlled negative resistance in GaAs. *Solid-State Electron.* **12** (1969), p. 619.

4.78a Hariu, T., Ono, S. and Shibata, Y., Wideband performance of the injection limited Gunn diode. *Electron. Lett.* **6** (1970), p. 666.

4.78b Holmstrom, R. and Mittleman, S. D., The shielded-cathode mode Gunn device: a proposed new mode of Gunn device operation. *Solid-State Electron.* **13** (1970), p. 513.

4.79   Frey, W., AEG-Telefunken Research Institute, unpublished investigations (1970).

4.80   Heinle, W. and Engelmann, R. W. H., Stability criterion for semiconductor plates with negative AC mobility. *Proc. IEEE* **60** (1972), p. 914.

4.81   Copeland, J. A., Bell Telephone Labs., private communication (1971).

4.82   Dascalu, D., Transit time effects in bulk negative-mobility amplifier. *Electron. Lett.* **4** (1968), p. 581.

4.83   Dascalu, D., Theory of space-charge effects in negative-mobility amplifiers operating in the emitter-current-limited injection mode. *Digest EUROCON 71, Lausanne*, Oct. 1971 (IEEE, Region 8, Lausanne, Switzerland, 1971).

4.84a Narayan, S. Y. and Sterzer, F., Stabilization of transferred-electron amplifiers with large n_0L products. *Electron. Lett.* **5** (1969), p. 30.

4.84b Sterzer, F., Stabilization of supercritical transferred-electron amplifiers. *Proc. IEEE* **57** (1969), p. 1781.

4.85   Maloberti, F. and Svelto, V., On the stability of transferred-electron amplifiers. *Alta Frequenza* **39** (1970), p. 1010.

4.86   Guéret, P. and Reiser, M., Switching behaviour of over-critically doped Gunn diodes. *Appl. Phys. Lett.* **20** (1972), p. 60.

4.87   Charlton, R., Freeman, K. R. and Hobson, G. S., Stabilization mechanism for "supercritical" transferred electron amplifiers. *Electron. Lett.* **7** (1971), p. 575.

4.88a Hauge, P. S., Static negative resistance in Gunn effect materials with field-dependent carrier diffusion. *IEEE Trans. Electron Dev.* **ED–18** (1971), p. 390.

4.88b Döhler, G., Shockley's positive conductance theorem for Gunn materials with field dependent diffusion. *IEEE Trans. Electron Dev.* **ED–18** (1971), p. 1190.

4.89 Hofmann, K. R., Stability theory for thin Gunn diodes with dielectric surface loading. *Electron. Lett.* **8** (1972), p. 124.

4.90 Hofmann, K. R. and 't Lam, H., Suppression of Gunn-domain oscillations in thin GaAs diodes with dielectric surface loading. *Electron. Lett.* **8** (1972), p. 122.

4.91a Teszner, J. L., Two-dimensional investigation of instabilities in a GaAs layer in contact with an outer semiconductor of positive resistance (in French). *Solid-State Electron.* **13** (1970), p. 1471.

4.91b Teszner, J. L., Tunable Gunn oscillator by semiconductor surface loading. *Electron. Lett.* **7** (1971), p. 146.

4.92 Dean, R. H. and Schwarz, P. M., Field profile in n-GaAs layer biased above transferred-electron threshold. *Solid-State Electron.* **15** (1972), p. 417.

4.93 Dean, R. H., A practical technique for controlling field profile in thin layers of n-GaAs. *IEEE Trans. Electron Dev.* **ED–19** (1972), p. 1144.

4.94 Dean, R. H., Reflection amplification in thin layers of n-GaAs. *IEEE Trans. Electron Dev.* **ED–19** (1972), p. 1148.

4.95 Frey, W. and Engelmann, R. W. H., On the potential profile of overcritically biased thin epitaxial GaAs layers. *Arch. Elektronik Übertragungstechn.* **27** (1973), p. 284.

4.96 Engelmann, R. W. H. and Heinle, W., Effect of diffusion on the small-signed stability of semi-conductor plates with negative a.c. mobility. *Arch. Elektronik Übertragungstechn.* **28** (1974), p. 66.

5
Comparative Description of Mode Properties

5.1 MODE CLASSIFICATION AND FREQUENCY LIMITATIONS

5.1.1 Introduction

As we have seen in the two previous chapters, numerous operational modes can be observed in a semiconductor device that exhibits the transferred-electron effect. Even though the basic electrical characteristic, viz. the drift-velocity/electric-field relationship, is of the general shape of a tunnel-diode characteristic, a much more complex behaviour is obtained because of an important difference: in a tunnel diode the negative resistance is associated with a *local* pn junction within the semiconductor, whereas in a TE device it is distributed over the whole semiconductor *volume*.

Which mode is excited depends largely on the doping concentration n_0 of the semiconductor (determining the dielectric relaxation time $\tau_R \approx \varepsilon/en_0\mu$ as introduced in Section 3.2), the amount of doping fluctuations and inhomogeneities, the contact conditions, the length L of the active region, the frequency f and the quality factor Q of the overall resonant circuit, and on the bias field. It should be stressed that the term 'mode' is used here in a rather loose sense, merely standing for 'class of behaviour' [5.1]. In the following a classification of the operational modes is attempted [5.2], as indicated in Table 5.1, by dividing into modes with mature space-charge dipoles in the semiconductor, modes with only partly-formed dipoles and modes with no dipoles occurring. In addition, it is necessary to differentiate between primary and secondary modes. The primary modes result from the TE effect as such, while the secondary modes occur only simultaneously with a primary one. Eventually, one has to distinguish between oscillatory and amplifying modes. The mode is termed oscillatory if it is basically unstable, and is called amplifying if stabilization is possible. The additional content of Table 5.1 is explained below.

5.1.2 Primary Modes

5.1.2.1 *Oscillatory Modes*

The modes which exhibit mature travelling dipole domains in the semiconductor are all oscillatory in character. The simplest case is pure *transit-time operation* in a

128

Table 5.1—Modes of TE Operation

Primary modes				Secondary modes
Mature dipole domains	Transit-time limited	(A) Pure transit-time mode (resistive or complex load) (B) Delayed dipole mode	} complex load: αL > 2·09, no stability possible: *oscillators*	(a) Negative conductance of dipole in transit or of d.c. current: modulation, amplification
		(C) Quenched single-dipole mode (D) Quenched multiple-dipole mode		(b) Non-linear capacitance and resistance of dipole domain: parametric effects
Partly-formed dipole domains	Not transit-time limited	(E) Hybrid mode		(c) Negative conductance at $f < f_{oscill}/Q$
No dipole domains, but some space-charge formation		(F) LSA mode (G) Controlled injection		(d) Free-electron parametric effects at $f \to f_{TE}$
		(H) Stable negative conductance (I) Growing space-charge waves	αL < 2·09 stability possible: *amplifiers* or *oscillators*	

resistive or resonant circuit in which the a.c. voltage swing does not interfere with the domain nucleation and extinction processes (total voltage always above threshold). As we have seen in Section 3.4.4 the oscillation frequency is determined by the drift length L and average drift velocity $\bar{v}_a \approx v_D$:

$$f_\tau = \frac{1}{\tau} \approx \frac{v_D}{L} \tag{5.1}$$

which constitutes the basic relationship observed by Gunn [1.18, 1.19]. Frequency tuning is possible only over a small range by changing v_D either with the d.c. bias field E_{h0} or, even more slightly, with the circuit by affecting the amplitude $\hat{E}_h$ of the space-averaged a.c. field. Because of the jitter in domain nucleation, drift, and extinction, this mode is the most noisy one of all oscillation modes as already indicated in Section 3.5.1 and, hence, is not suitable for microwave oscillator application. However, it can be applied advantageously for a whole series of functional and digital operations (Chapter 8).

Better microwave oscillation performance is obtained by modes in which the domain nucleation and/or extinction processes are controlled by the a.c. voltage swing $E_h L$ across a suitable resonant or non-resistive circuit. In this way delayed, and quenched, dipole-domain modes are obtained as discussed in Sections 3.5.2

129

and 3.5.3, respectively. Fairly large circuit tuning becomes possible, and because of the synchronizing effect of the circuit, domain-jitter noise is reduced. Performance is improved further by employing oscillatory modes in which the dipole domains are quenched before they have grown to their mature steady-state shape (hybrid operation; Section 3.5.4) or in which dipoles are prevented from growing at all (LSA operation; Section 4.2). For any of these circuit-controlled oscillation modes there exists no basic lower tuning limit if the circuit is able to support suitable *non-sinusoidal* voltage waveforms (relaxation oscillations; see e.g. [5.19] for delayed-domain mode and [4.2] and [4.13] for LSA mode). A lower limit results only for pure *sinusoidal* oscillations (single-frequency high-Q circuit). A basic upper frequency limit, however, is imposed in any case by the finite quenching times of the dipole domains or charge accumulations occurring in all of the modes. These frequency limits will now be estimated for each particular mode using simple physical arguments.

In the *delayed dipole-domain mode* (d.d.) only the formation process and not the quenching process is controlled by the circuit. Hence, the domain vanishes after its transit time, and the upper frequency limit is given by the transit-time frequency stated in eqn (5.1). The upper frequency limit of the *quenched single-dipole-domain mode* (q.s.d.) may be estimated as

$$f_{\text{q.s.d.}} \approx \frac{1}{2\pi\tau_D} \tag{5.2}$$

where τ_D is an RC time constant,

$$\tau_D = R_1\bar{C}_D = \tau_{R1}\frac{L}{\bar{b}} \tag{5.3}$$

which was introduced in Section 4.4.2.1 for domain growth [eqn (4.59)]. The bar indicates an average value during the quenching process. The domain width b is related to the domain excess potential Φ_D. By using eqns (3.19) and (3.24) one obtains

$$b = \left(\frac{2\varepsilon}{\varepsilon n_0}\Phi_D\right)^{1/2} = (2\tau_{R1}\mu_1\Phi_D)^{1/2}. \tag{5.4a}$$

Since the last phase of the quenching process, when b has already become quite small, takes up most of the time τ_D we may define $\bar{b}$ of eqn (5.3) by a value within this last phase. Quenching is completed when Φ_D reaches the order of the potential variations caused by doping fluctuations, say $\Phi_D \lesssim 0.1\,\text{V}$, $V = E_h L$ being the total voltage drop across the device. For simplification, we may assume $0 < E_h \approx E_p/2 < E_s$ during the quenching process (Fig. 3.25) and thus obtain from eqn (5.4a).

$$\bar{b} \approx (0.1\tau_{R1}\mu_1 E_p L)^{1/2}. \tag{5.4b}$$

By setting $\mu_1 E_p \approx 2v_D$ (cf. Fig. 3.15a) and by using eqn (5.1), substitution of expression (5.4b) into (5.3) yields

$$\tau_D \approx (5\tau_{R1}\tau)^{1/2}. \tag{5.5}$$

Thus, we obtain with eqns (5.1) and (5.5) for the upper limiting frequency normalized to the transit-time frequency

$$\frac{f_{\text{q.s.d.}}}{f_\tau} = \frac{1}{2\pi}\left(\frac{\tau}{5\tau_{\text{R1}}}\right)^{1/2} = \frac{1}{2\pi}\left(\frac{e\mu_1}{5\varepsilon v_{\text{D}}}n_0L\right)^{1/2} \tag{5.6}$$

an expression that is proportional to the square root of the n_0L product.

The upper frequency limit is extended if more than one dipole domain is formed in the sample (*quenched multiple-dipole-domain mode* [q.m.d.]; Section 3.5.3) [3.11, 3.50, 3.51]. Having a number of m identical dipoles in series reduces the dipole capacitance of eqn (5.3) to an effective value $\bar{C}_{\text{D, eff}} = \bar{C}_{\text{D}}/m$. Computer simulations [3.11] reveal that

$$m \approx n_0L/(n_0L)_{\text{crit}}$$

with $(n_0L)_{\text{crit}} \approx 0 \cdot 7 \times 10^{12}$ cm^{-2}. Thus, the upper limiting frequency becomes:

$$\frac{f_{\text{q.m.d.}}}{f_\tau} \approx \frac{f_{\text{q.s.d.}}}{f_\tau}m \approx \frac{1}{2\pi(n_0L)_{\text{crit}}}\left(\frac{e\mu_1}{5\varepsilon v_{\text{D}}}\right)^{1/2}(n_0L)^{3/2}. \tag{5.7}$$

As compared to eqn (5.6) a stronger dependence on the n_0L product results.

Next we consider the *hybrid mode* (hyb.) which also extends the frequency range of operation as outlined in Section 3.5.4. Since positive space charge (carrier depletion) still grows in this mode to form an immature dipole with the primary cathode accumulation layer, we may estimate the upper frequency limit from the average negative dielectric relaxation time τ_{RN}, viz.

$$f_{\text{hyb.}} \approx \frac{1}{2\pi|\tau_{\text{RN}}|}. \tag{5.8a}$$

With eqn (3.3a) ($\mu = \mu_{\text{N}}$ and $n = n_0$) one obtains

$$\frac{n_0}{f_{\text{hyb.}}} \approx \frac{2\pi\varepsilon}{e|\mu_{\text{N}}|} \tag{5.8b}$$

which is the n_0/f parameter [4.9] found to be of extreme importance in LSA operation (Section 4.2.3). With eqn (5.1) we thus have

$$\frac{f_{\text{hyb.}}}{f_\tau} \approx \frac{e|\mu_{\text{N}}|}{2\pi\varepsilon v_{\text{D}}}n_0L. \tag{5.9}$$

Here, μ_{N} is an average value of the negative differential mobility during space-charge growth, as defined by eqn (4.8b). The degree of dependence of $f_{\text{hyb.}}/f$ on the n_0L product falls just between those of eqns (5.6) and (5.7).

In the *LSA mode*, eventually, no dipole domains can form and the cathode only injects (negative) space charge into the device. This injected accumulation layer has to be quenched during each cycle. Thus we may estimate the upper frequency limit from the average positive dielectric relaxation time τ_{RP}, yielding

$$f_{LSA} \approx \frac{1}{2\pi\tau_{RP}}. \tag{5.10a}$$

With eqn (3.3a) ($\mu = \mu_P$ and $n = n_0$) it follows that

$$\frac{n_0}{f_{LSA}} \approx \frac{2\pi\varepsilon}{e\mu_P}. \tag{5.10b}$$

This leads to

$$\frac{f_{LSA}}{f_\tau} \approx \frac{\mu_P}{\mu_N} \frac{f_{hyb.}}{f_\tau} \approx \frac{e\mu_P}{2\pi\varepsilon v_D} n_0 L \tag{5.11}$$

where μ_P is an average value of the positive differential mobility during the time interval of quenching [eqn (4.13b)]. Equations (5.10) are almost identical to the frequency limit observed in computer simulations [4.9], see Section 4.2.3.

Since $\mu_P > |\mu_N|$ for GaAs and other n.d.c. materials, the LSA mode possesses the highest frequency limit in devices with ordinary ohmic contacts. A further extension is only possible if the cathode accumulation layer is avoided. This can be achieved if the ohmic contact is replaced by a contact with limited carrier injection leading to the *controlled-injection mode of operation* (c.i.) as discussed in Section 4.4.3 [4.77]. In this case no space-charge quenching is necessary as long as the injection establishes an almost uniform field, i.e. $\bar{n} \approx n_0$. Then, the upper frequency limit is solely given by the parasitic circuit losses (emitter resistance R_{em} + contact resistance R_c = series resistance R_s). The cut-off frequency can be calculated from the equivalent circuit of Fig. 4.24 by formulating the range of oscillation with

$$\text{Re}(Z_d) \leq 0 \tag{5.12}$$

where Z_d is the overall device impedance. The situation here is almost identical to the case of a tunnel diode. For

$$|G|R_s \ll 1 \tag{5.13}$$

one obtains from (5.12) a simple relation between the negative quality factor of the *parallel* circuit consisting of C and G,

$$Q_P = \frac{\omega C}{G} \tag{5.14}$$

and the positive quality factor of the *series* circuit of C and R_s,

$$Q_s = \frac{1}{\omega C R_s} \tag{5.15}$$

viz.

$$Q_P + Q_s \leq 0. \tag{5.16}$$

For $f/f_\tau \gtrsim 1$, i.e. transit angles $\theta \gtrsim 2\pi$, one has $G \approx G_{h0} = e\bar{n}\mu_{h0}A/L$ and $C \approx C_h = \varepsilon A/L$ (Fig. 4.25). Thus with $\omega_R = 1/\tau_R$ and eqn (3.3a) ($\mu = \mu'_{h0}$) eqn (5.14) becomes

$$Q_p \approx \omega/\omega_R = 2\pi f \tau_R. \tag{5.17}$$

Furthermore, introducing in eqn (5.15) the time constant

$$\tau_s = R_s C \approx R_s C_h \tag{5.18}$$

the upper limiting frequency is obtained from relation (5.16) as

$$f_{c.i.} \approx \frac{1}{2\pi(|\tau_R|\tau_s)^{1/2}} \tag{5.19a}$$

which can be written in the form

$$\frac{\bar{n}}{f_{c.i.}} \approx 2\pi \left(\frac{\varepsilon \bar{n} \tau_s}{e|\mu'_{h0}|} \right)^{1/2}. \tag{5.19b}$$

The requirement presumed in relation (5.13) follows as

$$\frac{\tau_s}{|\tau_R|} \ll 1. \tag{5.20}$$

Since the specific contact resistance R_c can be made as small as $10^{-4}\Omega\text{cm}^2$ [6.94] (see Section 6.2.2), R_s is dominated by the differential emitter resistance R_{em} in most cases of practical interest [4.77], i.e.

$$R_s A \approx R_{em} A \approx \frac{kT_L/e}{e\bar{n}v_{h0}}. \tag{5.21}$$

The differential emitter resistance defined by $R_{em} = \mathrm{d}V_{em}/\mathrm{d}J_{em}A$, follows from the emitter diode equation at large forward bias

$$J_{em} \approx \text{const.} \exp(eV_{em}/kT_L)$$

assuming that the injected bias current is equal to the emitter current, i.e. $J_0 = e\bar{n}v_{h0} \approx J_{em}$ (cf. textbooks on transistor operation). Inserting expressions (5.18) and (5.21) into eqn (5.19b) yields with $f_\tau \approx v_{h0}/L$

$$\frac{f_{c.i.}}{f_\tau} \approx \frac{e}{2\pi\varepsilon} \left(\frac{e}{kT_L} \frac{L|\mu'_{h0}|}{v_{h0}} \right)^{1/2} \bar{n}L. \tag{5.22}$$

At the upper frequency limit the negative Q of eqn (5.17) becomes

$$Q_p \approx - \left(\frac{|\tau_R|}{\tau_s} \right)^{1/2} = - \frac{e}{kT_L} \frac{Lv_{h0}}{|\mu'_{h0}|} \tag{5.23}$$

after employing relations (5.19) and (5.21).

In Fig. 5.1 *upper limiting frequencies* of the different large-signal oscillatory modes

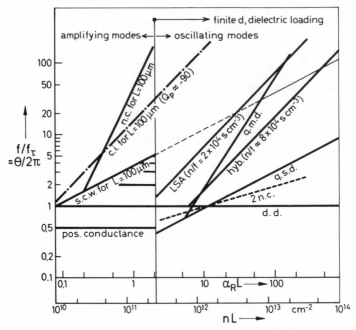

Fig. 5.1 Upper limiting frequencies of the different modes in units of the transit-time frequency as a function of the nL product (modified and supplemented mode chart after Copeland [5.1]), as calculated for GaAs.

are plotted in the plane of frequency (in units of the transit-time frequency f_τ) against the nL product, where $n = n_0$ for ohmic-contact modes and $n = \bar{n}$ for the controlled-injection modes, cf. [4.9]. The following numerical values have been chosen to prepare Fig. 5.1:

$$\left.\begin{array}{l} \varepsilon \approx 10^{-12}\text{F cm}^{-1} \\ \mu_1 = 8\,000 \text{ cm}^2\,(\text{Vs})^{-1} \\ \mu_P = 2\,000 \text{ cm}^2\,(\text{Vs})^{-1} \\ \mu_N \approx \mu'_{\text{ho}} = -500 \text{ cm}^2\,(\text{Vs})^{-1} \\ \bar{v}_a \approx v_D \approx v_{\text{ho}} = 10^7 \text{ cm s}^{-1}. \end{array}\right\} \quad \text{cf. Fig. 4.7} \left.\vphantom{\begin{array}{l} \\ \\ \\ \\ \\ \end{array}}\right\} \quad (5.24)$$

These values describe roughly the situation in GaAs at a bias field of about $E_{\text{ho}} = 10 \text{ kV cm}^{-1}$. In the case of the controlled-injection device we additionally chose $L = 100 \ \mu$m because the frequency ratio $f_{\text{c.i.}}/f_\tau$ depends not only on the nL product but also explicitly on the device length L [relation (5.22)].

Figure 5.1 demonstrates the superior high-frequency capabilities of the LSA, hybrid, and quenched-dipole-domain, modes for devices with large n_0L product. It should be noted, however, that n_0L is proportional to the d.c. power density $p_0 \approx e n_0 v_D A E_{\text{ho}} L$ which is limited because of heat removal problems (Section 6.3). In the domain modes large nL values additionally cause high domain peak fields

(Fig. 3.9) which must be avoided since they may lead to dielectric breakdown (cf. Figs 5.26 and 6.25). The upper frequency limits of all possible oscillatory modes in an ohmic-contact device tend to converge around the transit-time frequency for nL products approaching the critical value that separates the oscillatory and the amplifying modes (Section 4.4.1). There, a clear mode distinction is no longer possible.

5.1.2.2 *Amplifying Modes*

In the one-dimensional approach of small-signal space-charge-wave dynamics the nL product can be expressed in terms of the gain parameter $\alpha_R L$ which determines the space-charge growth along the total drift length L in a uniform field E_{ho} (see Section 4.3.2):

$$\alpha_R L = -\frac{\omega_R}{v_{ho}}L = -\frac{e\mu'_{ho}}{\varepsilon v_{ho}}nL. \tag{5.25}$$

For $\alpha_R L \leq 2\cdot09$ the gain becomes small enough for stable operation (Section 4.4.1.2). With the numerical values quoted in (5.24) this yields $nL \leq 2\cdot6 \times 10^{11}\,\mathrm{cm}^{-2}$ for suppression of moving domains. This range is marked by 'amplifying modes' in Fig. 5.1. Amplification of microwave power in stable devices is based on the small-signal negative conductance or on the excitation of travelling space-charge waves which grow in amplitude.

The upper limiting frequency of *space-charge-wave* (s.c.w.) growth is determined by carrier diffusion. From the small-signal dispersion relation one obtains (Section 4.3.2)

$$f_{\mathrm{s.c.w.}} = \frac{v_{ho}}{2\pi}\left(\frac{|\omega_R|}{D_{ho}}\right)^{1/2} = \frac{v_{ho}}{2\pi}\left(\frac{e|\mu'_{ho}|}{\varepsilon D_{ho}}n\right)^{1/2} \tag{5.26}$$

as already stated by eqn (4.23). The frequency $f_{\mathrm{s.c.w.}}$ constitutes also the ultimate limit for any of the oscillatory domain modes. Below the limit given by eqn (5.26) space-charge transit-time effects are important in determining the small-signal conductance of ohmic-contact devices, leading to negative values only near f_τ and multiples thereof. Above this limit the conductance becomes nearly frequency-independent and is negative for large enough nL values (Section 4.4.1.3). The upper limiting frequency for this *negative conductance* (n.c.) is [5.3]

$$f_{\mathrm{n.c.}} \approx \frac{(\omega_R L)^2}{4\pi D_{ho}} = \frac{(e\mu'_{ho})^2}{4\pi\varepsilon D_{ho}}(nL)^2. \tag{5.27}$$

Choosing again $L = 100\ \mu\mathrm{m}$, expressions (5.26) and (5.27) are plotted in Fig. 5.1 for the numerical values listed in (5.24) and a diffusion constant $D_{ho} = 200\ \mathrm{cm}^2\,\mathrm{s}^{-1}$ ($E_{ho} \approx 10\,\mathrm{kV\,cm}^{-1}$, [2.36b], see Fig. 2.10). Below $f_{\mathrm{s.c.w.}}$ we have a range of undulating negative conductance and of growing space-charge waves. Between $f_{\mathrm{s.c.w.}}$ and $f_{\mathrm{n.c.}}$ there exists a range of wide-band negative conductance. Below about $0\cdot5 f_\tau$ the device conductance becomes positive (space-charge-limited current regime cf. Sec. 4.4.1.3), and space-charge-wave amplification is impractical since the total device contains less than one half of a space-charge wavelength.

For the negative conductance of a stable device with *controlled carrier injection* (c.i.) the same considerations hold as outlined for above oscillatory modes.

The stable range can be extended to larger nL products if the lateral device dimension d (perpendicular to the electron drift) is reduced so as to cause two-dimensional effects to become important (Section 4.4.2). Owing to the lateral r.f. field changes then established, space-charge growth is reduced to $\alpha < \alpha_R$ and, hence, the condition $\alpha L \leq 2.09$ leads to a larger critical nL product. (Section 4.4.2.2.)

5.1.2.3 *Absolute Frequency Limits of the Modes*

Absolute values of the upper limiting frequencies for the non-transit-time limited modes are plotted in Figs. 5.2 and 5.3 against nL product and carrier density n, respectively. The numerical values of the parameters involved are the same as in Fig. 5.1 [eqns (5.24)]. In both figures the intrinsic frequency limit f_{TE} of the TE effect, as caused by the scattering processes between the central and satellite valleys of the conduction band, is indicated at 135 GHz (see Section 5.1.3 below).

Figure 5.2 shows the upper limiting frequency $f_{n.c.}$ for the small-signal negative conductance according to eqn (5.27), and, for two values of the drift length L, the

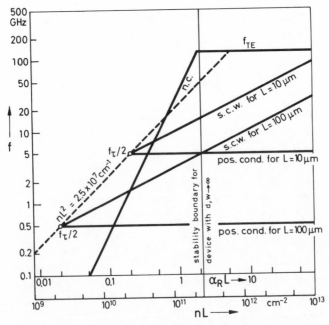

Fig. 5.2 Absolute upper frequency limits of the amplifying modes as a function of the nL product, calculated for GaAs.

corresponding value $f_{\text{s.c.w.}}$ for growing space-charge waves, according to eqn (5.26). The lower-frequency end points of the $f_{\text{s.c.w.}}$ lines were set at half the transit-time frequency for the reasons stated in Section 5.1.2.2 above. These points lie on the dashed line of constant nL^2 product:

$$nL^2 = \frac{\pi^2 \varepsilon D_{h0}}{e|\mu'_{h0}|} \tag{5.28}$$

or, with the numerical values given in eqn (5.24) for GaAs, $nL^2 = 2 \cdot 5 \times 10^7$ cm^{-1}. Equation (5.28) follows from eqn (5.26) and $f_{\text{s.c.w.}} = f_\tau/2 \approx v_{h0}/2L$ (see also [4.26, 5.4]). In ohmic-contact devices with an nL^2 product *below* this critical value no transit-time undulation occurs in the frequency dependence of the small-signal conductance which always remains positive if, additionally, the nL product is below about 8×10^{10} cm^{-2} (see Fig. 5.2). Thus with $nL^2 < 2 \cdot 5 \times 10^7$ cm^{-1} reflection-type amplification (Section 7.2.1) is limited to device lengths of $L \lesssim 3 \cdot 1$ μm and to frequencies of $f \gtrsim 16$ GHz. Space-charge-wave (travelling-wave) amplification (Section 7.2.2) would be possible only for frequencies smaller than $f_\tau/2$, i.e. in this case it is impractical as pointed out already.

In Fig. 5.3 the upper frequency limit of growing space-charge waves, which is also the ultimate limit for the dipole-domain modes, is compared to that of the LSA mode. The diagram indicates that at the lower carrier concentrations n, space-charge waves

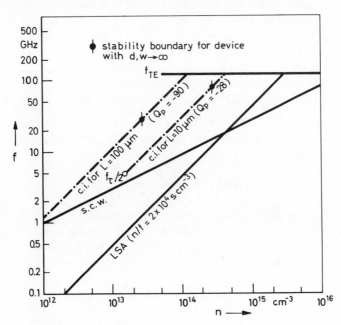

Fig. 5.3 Absolute upper frequency limits of space-charge wave growth and of LSA oscillation as a function of carrier density n, as calculated for GaAs.

have a higher cut-off frequency than the LSA mode. However, the use of controlled carrier injection raises the LSA cut-off frequency above the space-charge-wave value also in this range, as shown for the two drift lengths L of 10 and 100 μm.

It should be pointed out here that the frequency limitation is, of course, not the only aspect when comparing different modes. Other important factors are obtainable microwave power and d.c./a.c. power conversion efficiency which will be discussed in Section 5.2.

5.1.3 Intrinsic Cut-off Frequency of TE Effect

In estimating the upper frequency limits of the different modes we have assumed so far that the static drift-velocity/electric-field characteristic remains valid at the higher microwave frequencies. However, since the n.d.c. as manifested by the falling part of the characteristic results from electron scattering processes of finite duration, there must exist a cut-off frequency of this n.d.c. If the oscillation period reaches the order of the scattering times, the n.d.c. is expected to vanish. This behaviour can be taken into account by introducing an effective relaxation time [5.5], τ_{eff}, which describes, in a lumped way, the transfer of the electrons between the light and heavy-mass valleys of the conduction band. In a *small-signal* model two asymptotic cases can be distinguished.

1. For small frequencies ω, i.e. $\omega\tau_{\text{eff}} \ll 1$ the microwave mobility of the electrons is given by the slope of the static velocity/field characteristic for a.c. electric fields along the d.c. drift direction, viz. $\mu' = dv/dE$; it is given by the (always positive) total mobility $\mu = v/E$ in a direction perpendicular to the d.c. field (cf. Sections 3.3.2 and 4.3.3).
2. For frequencies beyond $\omega\tau_{\text{eff}} = 1$ experimental [5.6] and theoretical [5.7] investigations have revealed that the anisotropy of the conductivity tends to vanish; thus also the longitudinal mobility approaches roughly $\mu = v/E$.

In a descriptive way, then, one may express the frequency dependence of the (complex) microwave mobility of the electrons along the d.c. drift-field direction by the simple formula*

$$\mu'_m \approx \mu + \frac{\mu' - \mu}{1 + j\omega\tau_{\text{eff}}}. \tag{5.29}$$

Equation (5.29) is represented by an equivalent circuit (Fig. 5.4a) consisting of a conductance $G = en\mu'A/L$ and a capacitive shunt element with damping [5.8]. The time constant of this shunt element is $R_pC_p = \tau_{\text{eff}}$. The negative-mobility cut-off frequency follows from the condition $\text{Re}(\mu'_m) \le 0$ as

$$f_{\text{TE}} \approx \frac{1}{2\pi\tau_{\text{eff}}}\left(\frac{-\mu'}{\mu}\right)^{1/2}. \tag{5.30}$$

---

*The subscript m is used to distinguish the *microwave* differential mobility from the static differential mobility μ' treated previously.

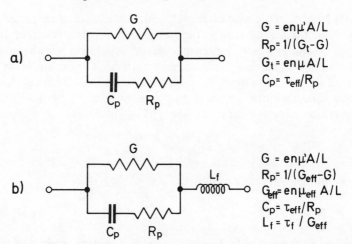

Fig. 5.4 Equivalent circuit of device admittance resulting from the microwave behaviour of the electron mobility μ'_m: (a) after simplified formula (5.29); (b) approximation for rigorous formula (5.32). After Zimmerl [5.8].

At $E_{h0} = 10\,\text{kV cm}^{-1}$, for example, one has for GaAs a mobility ratio $\mu/\mu' = -2\cdot2$ and, assuming $\tau_{\text{eff}} \approx 0\cdot8$ ps (cf. Fig. 5.5c below), one obtains $f_{\text{TE}} \approx 135$ GHz, the value used in Figs 5.2 and 5.3. From the field dependence of μ'/μ follows immediately that f_{TE} approaches 0 at either end of the negative-mobility range of the $v(E)$ characteristic. However, the largest cut-off frequency is not obtained at the maximum of the negative mobility because of a strong dependence of τ_{eff} on electron temperature and thus on the d.c. bias field E_{h0} [5.9, 5.10].

A more rigorous calculation of the microwave electron mobility involves the determination of the electron distribution function F from the *time-dependent* Boltzmann equation which reads symbolically for the vth valley

$$\frac{\partial F_v}{\partial t} = \left(\frac{\partial F_v}{\partial t}\right)_{\vec{E}} + \sum_i \left(\frac{\partial F_v}{\partial t}\right)_i. \qquad (5.31)$$

$$\underbrace{\qquad}_{\text{field term}} \quad \underbrace{\qquad}_{\text{collision term}}$$

The symbols were defined in Section 2.2 where the static case $\partial F_v/\partial t = 0$ had been discussed [eqn (2.3)]. Following the static treatment of Butcher and Fawcett [2.12, 2.26], several authors have solved eqn (5.31) by approximating each distribution function F_v by a displaced Maxwellian [5.8, 5.11–5.13]. Then, the time-dependent physical parameters of these specialized distribution functions (viz. electron number, momentum, and temperature or energy) simply follow from the balance equation for particle number, momentum and energy which are obtained, as the first three 'moments', from the Boltzmann equation (5.31). In this procedure the collision term of eqn (5.31) leads to six relaxation times*, which depend on electron temperature and hence field, for the physical parameters, viz. for carrier concentration (τ_n), momentum (τ_p) and energy (τ_e) in each valley [5.11]. The microwave mobility is calculated for small-signal deviations from the static equilibrium and is expressed in terms of these six relaxation times and their field derivatives [5.8, 5.12]. A considerably simpler formulation is obtained by formally introducing five** (generally complex) small-signal relaxation frequencies [5.13]. This leads to the expression

$$\mu'_m = \sum_{i=1}^{5} \frac{\mu^{(i)}}{1 + j\omega/\omega^{(i)}} \tag{5.32a}$$

where the mobilities $\mu^{(i)}$ may be complex, and where

$$\sum_{i=1}^{5} \mu^{(i)} = \mu' \tag{5.32b}$$

as follows directly from $\omega = 0$. The relaxation of a particular physical parameter (e.g. particle number, Δn_1, in central valley) is described by *all* five small-signal relaxation frequencies $\omega^{(i)}$, for example

$$\Delta n_1 = \sum_{i=1}^{5} a^{(i)} \exp(-\omega^{(i)}t) \tag{5.33}$$

and thus, basically, a single $\omega^{(i)}$ cannot be associated with a certain relaxation process as in the case of the physical relaxation times mentioned above. Both the $\mu^{(i)}$ and the $\omega^{(i)}$ are d.c.-field dependent. For GaAs the $\mu^{(i)}$ are plotted in Fig. 5.5a as a function of bias field E_{h0} and the $\omega^{(i)}$ are represented in the form $\tau^{(i)} = 1/\text{Re}\,(\omega^{(i)})$ and $1/f^{(i)} = -2\pi/\text{Im}\,(\omega^{(i)})$ in Fig. 5.5c. In calculating the $\mu^{(i)}$ and $\omega^{(i)}$ the same set of material parameters has been assumed as in the time-independent case which yielded the $v(E)$ characteristic of Fig. 2.7 (Section 2.3). As it turns out, all the $\omega^{(i)}$ have a *positive* real part as is necessary for a stable equilibrium state. Only two $\omega^{(i)}$ become complex-conjugate over certain bias regions; the others are always real, particularly the relaxation frequency leading to the largest relaxation time $\tau^{(1)} = 1/\omega^{(1)}$. The time $\tau^{(1)}$ most stringently determines the duration of the relaxation process of the electron population in the central and satellite valleys, $-\Delta n_1 = \Delta n_2$ [5.13].

If only the largest time constant is retained in eqn (5.32) and all others are set at

---

*It should be pointed out that these relaxation times are defined differently than the scattering relaxation times, τ_{sc}, introduced in Section 2.2 (Fig. 2.6). The τ_{sc} are functions of the *actual* carrier energy, the relaxation times introduced here depend on the *average* carrier energy (i.e. their temperature) and follow directly from the appropriately-averaged loss rates as they are, for example, plotted in Fig. 2.4. For the momentum relaxation time one obtains simply $\tau_p = \langle \tau_{sc} \rangle$.

**The number five follows from the five independent parameters of the two distribution functions for the central and satellite valleys: carrier concentration in central valley, carrier momentum and energy both in central and in satellite valleys.

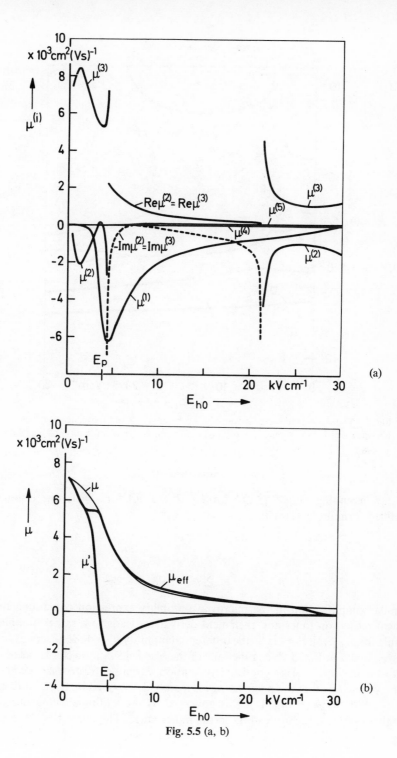

Fig. 5.5 (a, b)

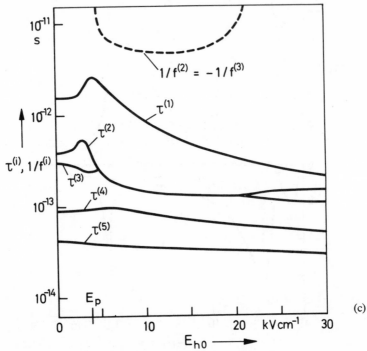

Fig. 5.5 Theoretical characteristic mobility constants $\mu^{(i)}$ and frequencies $\omega^{(i)}$ of the microwave electron mobility μ'_m of GaAs after eqn (5.32) as a function of bias field E_{h0}: (a) $\mu^{(i)}$; (b) $\mu' = \sum\limits_{i=1}^{5} \mu^{(i)}$, $\mu_{\mathrm{eff}} = \sum\limits_{i=2}^{5} \mu^{(i)}$; (c) $\tau^{(i)} = 1/\mathrm{Re}(\omega^{(i)})$, $1/f^{(i)} = -2\pi/\mathrm{Im}(\omega^{(i)})$. After Heinle [5.13].

zero, i.e. assuming $(\omega/|\omega^{(i)}|) \ll 1$ for $i \geq 2$ but not for $i = 1$, one obtains the simplified formula (5.29) with

$$\tau^{(1)} = \tau_{\mathrm{eff}}$$

and

$$\mu' - \mu^{(1)} = \sum_{i=1}^{5} \mu^{(i)} = \mu_{\mathrm{eff}} \approx \mu.$$

A similar simplification in the microwave-mobility expression containing the six physical relaxation times and their field derivatives leads to a rather complicated formula for τ_{eff} [5.8]. For GaAs the differential mobility μ' and the effective mobility μ_{eff} are plotted in Fig. 5.5b as functions of the bias field E_{h0}. As can be seen, in the range of n.d.c., $\mu_{\mathrm{eff}} \approx \mu$ is a good approximation. From a comparison with Fig. 5.5a one may conclude that the positive μ_{eff} is mainly determined by the quantities $\mu^{(2)}$ and $\mu^{(3)}$ whereas $\mu' = \mu_{\mathrm{eff}} + \mu^{(1)}$ is governed by the additional influence of the strongly negative $\mu^{(1)}$ thus leading to the n.d.c. range. The quantities $\mu^{(4)}$ and $\mu^{(5)}$ are almost negligible throughout the bias range shown.

For a d.c. bias field of $E_{h0} = 10.5$ kV cm^{-1} the course of the exact eqn (5.32) is plotted in the complex mobility plane of Fig. 5.6 and compared to the simplified eqn (5.29). It can be seen that in the range $\mathrm{Re}\mu'_{h0} \leq 0$ the two formulas yield roughly the same results. Thus the simple formula (5.29) suffices for approximately calculating the intrinsic cut-off frequency. At very high frequencies, however, the microwave mobility eventually tends to zero instead of approaching μ or μ_{eff}. In the equivalent circuit of Fig. 5.4a this can qualitatively be accounted for by adding a series inductance with an appropriate time constant $\tau_f = L_f G_{\mathrm{eff}}$ leading to the circuit of Fig. 5.4b, cf. [5.12].

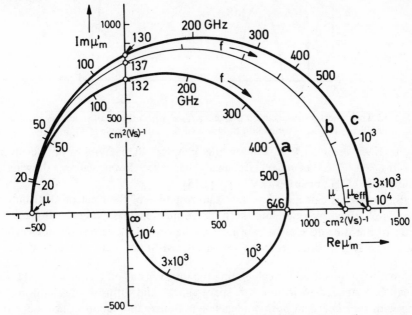

Fig. 5.6 Theoretical microwave electron mobility μ'_m of GaAs in the complex plane for a bias field of $E_{h0} = 10.5$ kV cm^{-1}: (a) rigorous formula (5.32); (b) simplified formula (5.29); (c) simplified formula (5.29) with μ replaced by $\mu_{\mathrm{eff}} = \sum_{i=2}^{5} \mu^{(1)}$. Adapted from Grasl and Zimmerl [5.12a]. (N.B. Left-hand notation should read μ' instead of μ.)

The intrinsic cut-off frequencies f_{TE} of GaAs and InP as obtained for $\mathrm{Re}\mu'_{h0} = 0$ from the exact formula (5.32) with the non-parabolicity of central conduction-band valley taken into account are plotted in Fig. 5.7 as a function of bias field [5.13, 5.14]. The maximum values are 200 GHz near 22 kV cm^{-1} for GaAs and 38 GHz near 12 kV cm^{-1} for InP. Assuming a parabolic band structure in the central valley leads to similar results with a reduced negative mobility range and somewhat smaller maximum values for f_{TE} [5.15]. Thus InP exhibits a much lower cut-off frequency (cf., however, more recent calculations for InP: Glover, G. H., *J. Appl. Phys.* **44**

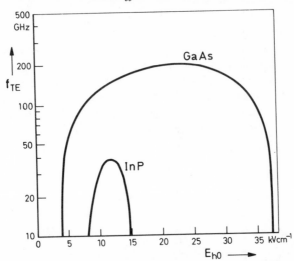

Fig. 5.7 Theoretical intrinsic cut-off frequency f_{TE} at which the small-signal negative
mobility vanishes, as a function of bias field E_{h0}. After Heinle [5.13, 5.14].

(1973), p. 1295; see also p. 152 ff. in this chapter) than GaAs. For the mixed
crystal $In_{0.2}Ga_{0.8}Sb$ the cut-off frequency has even lower values, showing a
maximum of 10 GHz near 0.97 kV cm^{-1} [5.15].

A rather rigorous calculation of the time response of the electron distribution
function is obtained by extending the iterative method of solving the Boltzmann
equation to the time-dependent case [5.16]. The results, however, do not deviate
strongly from those obtained with the displaced-Maxwellian calculation [5.17,
5.18].

The small-signal method for calculating the intrinsic cut-off frequency of the TE
effect can be extended to include r.f. space-charge and diffusion mechanisms by
starting with the *time-* and *position*-dependent Boltzmann equation necessary for
the description of space-charge *waves* [5.18]. The results for the upper frequency
limit justify using eqns (5.26) and (5.30) as asymptotic formulas for small and large
n_0L product, respectively. A somewhat smaller τ_{eff} is derived for eqn (5.30), particularly at the lower bias fields E_{h0}, indicating a larger cut-off frequency f_{TE} for space-
charge waves than for homogeneous-field oscillations. The values of diffusion constant D_{h0} from eqn (5.26) agree roughly with measured values of Fig. 2.10[2.36b].

The *large-signal* evaluation of the cut-off frequency as applicable to the LSA and
domain modes involves major modifications. Instead of the n.d.c. behaviour the
frequency dependence of the efficiency is calculated with the r.f. field amplitude as an
additional parameter (see Section 5.2.5). The resulting cut-off frequency is of the
same order of magnitude as estimated from the small-signal theory [5.12]. A substantially lower cut-off frequency results, however, for very short TE samples since
the finite heating time constant of the electrons causes an inactive region of about
$1\mu m$ length adjacent to the cathode contact [5.104].

144

Experimental verification of the intrinsic frequency limit of the TE effect is still scarce. Efficient LSA oscillations in GaAs samples have been obtained up to 88 GHz [4.4] with signals detectable at frequencies as high as 160 GHz [5.93]. The highest operating frequency reported so far for InP devices is 40 GHz [5.94].

5.1.4 Secondary Modes
The secondary modes associated with the TE effect are based on the external resistive or reactive properties of an oscillating TE semiconductor device.

5.1.4.1 *External Negative Conductance of Oscillator*
Most important is an external negative conductance (denoted subsequently by the suffix 2 n.c.) which can be used either for additional oscillation (self-modulation, [5.20 to 5.22a]) but also for reflection-type amplification at frequencies different from the primary oscillation frequency [4.12, 5.22, 5.23]. If the secondary voltage changes are slow the primary oscillation mode is able to respond to any such change keeping a steady-state condition. Thus the external negative conductance of the oscillator is simply given by the decrease of average current with increasing d.c. bias voltage [average terminal $I(V)$ characteristic]. However, for very fast secondary voltage changes that occur in a time smaller than $Q_L/(2\pi f_1)$, where Q_L is the loaded quality factor of the resonator oscillating at the primary frequency f_1, the primary oscillation remains more or less fixed in its initial state. The external conductance in this case may deviate strongly from the value derived from the average terminal $I(V)$ characteristic and may even become positive. Considering secondary voltages of nearly sinusoidal shape the transition between the two cases occurs at a secondary frequency of $f_2 \approx f_1/Q_L$.

The first case is always realized in the pure transit-time mode of the dipole-domain oscillator with zero primary-voltage amplitude, i.e. $Q_L \approx 0$. If the $n_0 L$ product of the device is not too small the current spikes of the oscillation (see Fig. 3.20) are narrow enough to be negligible in calculating the average current. The average terminal current/voltage characteristic is then approximately given by the dynamic characteristic of a dipole domain in transit (Fig. 3.15; parts above threshold). The frequency behaviour of the external negative conductance follows directly from the equivalent circuit of a device containing a travelling dipole domain as shown in Fig. 3.21. An upper frequency limit $f_{2n.c.}$ results because of the finite charging time of the domain capacitance C_D. In the small-signal case it is calculated by applying the condition

$$\mathrm{Re}\left(\frac{Y_r Y_D}{Y_r + Y_D}\right) = 0 \tag{5.34}$$

to the small-signal admittance given by relation (3.40). Neglecting the geometric rest capacitance C_r of the part outside of the dipole domain, i.e. postulating $|G_r| \gg \omega_2 C_r$ or

$$f_2 \ll \frac{1}{2\pi\tau_{R1}} \tag{5.35}$$

145

one obtains

$$f_{2\text{n.c.}} = \frac{1}{2\pi[\tau_{\text{R1}}|\tau_{\text{RD}}|(L/b_\infty - 1)]^{1/2}} \qquad (5.36)$$

where τ_{R1} and τ_{RD} are the dielectric relaxation times [eqn (3.3a)] of the mobilities $\mu'_{10} \approx \mu_1$ and μ'_{D}, respectively. With expressions (3.19) for b_∞ and (3.37) for μ'_{D}, and additionally using $f_\tau = \bar{v}_a/L$, where $\bar{v}_a \approx v_{\text{D}} = v_{1\infty}$, it follows

$$\frac{f_{2\text{n.c.}}}{f_\tau} = \frac{1}{2\pi} \left(\frac{1 - v_{2\infty}/v_{1\infty}}{1 - b_\infty/L} \cdot \frac{e\mu'_{1\infty}}{\varepsilon v_{1\infty}} n_0 L \right)^{1/2}. \qquad (5.37)$$

Equation (5.37) bears a certain similarity to the expression (5.6) for the cut-off frequency of the quenched-single-domain mode. The first square-root factor, however, depends also slightly on the $n_0 L$ product; for GaAs at a bias field of $E_{\text{h0}} = 4 \text{ kV cm}^{-1}$ it is, for example, equal to about $0.5\,(10^{12}\text{ cm}^{-2}/n_0 L)^{0.22}$ in the range of $3 \times 10^{11}\text{ cm}^{-2} < n_0 L < 3 \times 10^{13}\text{ cm}^{-2}$ [derived from the $v(E)$ characteristic of Fig. 3.9, cf. also Figs 3.10 to 3.12 and 3.15a]. With the numerical values $\mu'_{1\infty} = 8\,000\text{ cm}^2\,(\text{Vs})^{-1}$, $v_{1\infty} = 10^7\text{ cm s}^{-1}$ and $\varepsilon = 10^{-12}\text{ F cm}^{-1}$ one then obtains

$$\frac{f_{2\text{n.c.}}}{f_\tau} = 0.9 \left(\frac{n_0 L}{10^{12}\text{ cm}^{-2}} \right)^{0.28}$$

which is shown in Fig. 5.1 as the lower dashed line. In this case condition (5.35) is always satisfied.

The velocity difference $v_{1\infty} - v_{2\infty}$ increase with decreasing bias field E_{h0} after Figs 3.9 and 3.10, giving rise to an increasing cut-off frequency according to eqn (5.37). The maximum value is obtained for bias voltages below threshold [3.35]. This region, however, can be reached only if the primary r.f. voltage has a minimum amplitude, sweeping the device below threshold. This can be achieved by using either an external r.f. pump voltage (synchronous domain triggering) or a primary circuit with relatively high Q_{L} [5.24].

A detailed analysis of the external negative conductance of a dipole-domain oscillator with high-Q circuits [3.63] leads to the following results. If $f_2 \ll f_1/Q_{\text{L}}$ the average terminal $I(V)$ characteristic is practically independent of the frequency or mode of primary operation and only depends on the cavity load resistance R_{L}: the highest external negative conductance is obtained with the lightest loading (largest R_{L}) and occurs close to threshold as shown in Fig. 5.8. If $f_2 > f_1/Q_{\text{L}}$ the external conductance is not related any more to the average $I(V)$ characteristic. Negative values are predicted only for heavier loading in the delayed-dipole-domain mode close to the transit-time frequency. In both secondary frequency ranges the finite primary r.f. voltage swing across the device may enhance the negative conductance as compared to the constant-voltage case. The experiments are semi-quantitatively in agreement with the theoretical results [5.22a].

Similar considerations for the LSA mode yield negative external conductance only

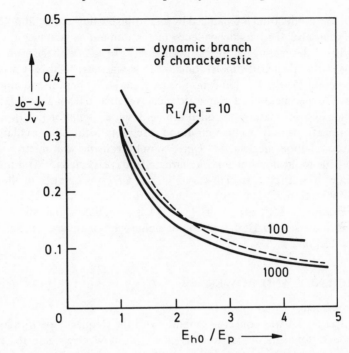

Fig. 5.8 Calculated average current-density/electric-field characteristic of resonator-controlled dipole-domain GaAs oscillator for different load resistances. Dashed line corresponds to dynamic branch of Fig. 3.15a ($E_h > E_p$). After Hobson [3.63].

if $f_2 < f_1/Q_L$ [4.9, 5.22b]. For $f_2 > f_1/Q_L$ the requirement for preventing build-up of space charge (Section 4.2.3) leads to a positive external conductance. A negative external conductance may occur again only if $f_2 \gg f_1$, since then the LSA oscillator is of a distributed nature for a secondary signal of frequency f_2 [5.22c].

5.1.4.2 Parametric Effects

The periodic dipole-domain nucleation and extinction in a pure transit-time TE oscillator, or the r.f. voltage swing of a resonator-controlled TE oscillator, causes a modulation of the instantaneous device capacitance and/or conductance because of the inherent non-linear properties involved. This modulation can lead to parametric effects (amplification or oscillation at a secondary frequency f_2) with the primary oscillation at f_1 acting as a pump signal ('self pumping'). The mechanism requires an idler circuit at the difference frequency $f_i = f_1 - f_2$. As usual in parametric systems, a degenerate case is obtained for $f_i = f_2 = f_1/2$.

Most obvious is a cyclic variation of the domain capacitance in the dipole-domain modes. The capacitance varies between a relatively high nucleation value (approaching infinity if no domain is present) and a finite steady-state value [5.25], cf. also

Fig. 3.21a. The modulation is enhanced by any voltage swing at the 'pump' frequency f_1 [5.26] because of the bias dependence of the domain capacitance $C_D = \varepsilon A/b_\infty$ (cf. Fig. 3.12). A degenerate parametric oscillator ($f_2 = f_1/2$) with the primary oscillation being in the quenched dipole-domain mode was described by Carroll [5.27]. Both degenerate [5.26c, 5.28] and non-degenerate [5.29] parametric amplification was observed experimentally. Parametric effects result also from a modulation of the conductance of a resonator-controlled dipole-domain oscillator [5.30].

Parametric effects may occur even in the case of an ideal LSA oscillator (space-charge accumulation neglected). Here, purely *resistive* parametric frequency conversion due to the non-linear $v(E)$ characteristic is expected [5.31]. Furthermore, if the 'pump' frequency f_1 is high enough, the *reactive* properties of the electrons discussed in Section 5.2.2 add to the parametric conversion [5.32]. Subharmonic signal generation in LSA oscillators [5.103] has been associated with this type of parametric conversion [5.31, 5.32]. An unambiguous experimental proof, however, is here still lacking.

5.2 EFFICIENCY AND POWER

5.2.1 Introduction

Whereas Section 5.1 was concerned mainly with the frequency performance of the different operational modes of the TE effect, we now investigate the d.c.-to-r.f. power conversion efficiency and the absolute power handling capability, first disregarding self-heating effects. Theoretical values of efficiency quoted in the literature vary considerably because of different definitions and different assumptions for the r.f. wave shapes and/or the $v(E)$ characteristic (cf. e.g., [4.17]). For this reason it is appropriate to give a more general analysis of efficiency values that can be expected from n.d.c. devices, before presenting a detailed description. We confine the discussion to the efficiency of oscillators since amplifier efficiencies in the negative-conductance mode are likely to approach similar values. An entirely different saturation behaviour is expected only for the unilateral travelling-wave amplifier described in Section 7.2.2 (see also concluding remarks in Section 5.2.4.1).

5.2.2 General Efficiency Estimates

The total d.c./r.f. conversion efficiency of an oscillator is defined as the ratio of the total (negative) r.f. power output P_t to the (positive) d.c. power input P_0,

$$\eta_t = \frac{-P_t}{P_0} = \frac{P_0 - P_{dis}}{P_0} \qquad (5.38)$$

where P_{dis} is the (positive) dissipated power which is converted to heat. In practice one is generally interested in a single frequency, mostly the fundamental r.f. component, and hence considers only the fundamental r.f. power P_1. Thus the practical efficiency becomes

$$\eta = \frac{-P_1}{P_0} = \frac{P_0 - P_{\text{dis}} + \sum\limits_{n=2}^{\infty} P_n}{P_0} \tag{5.39}$$

where the components P_n are the r.f. output powers at all higher harmonics. For a self-sustained oscillator the termination has to be passive at all frequencies, hence $P_n \leq 0$. To obtain high efficiency values the dissipated power should be made as small as possible. This means that either the current $I(t)$ or the voltage $V(t)$ should become as low as possible during part of one r.f. cycle. The optimum current and voltage waveforms are sketched in Fig. 5.9a showing the required phase relationship. In this ideal square-wave case the voltage drops to zero during the interval of maximum current and vice-versa. Hence the total efficiency is $\eta_t = 100\%$ whereas the practical efficiency for the fundamental power follows from a simple Fourier analysis of the current and voltage wave shapes as $\eta = (8/\pi^2)\,\eta_t = 81\%$ [5.33]. The dynamic $I(V)$ characteristic required from a voltage-controlled two-terminal device to yield the optimum oscillation waveforms of Fig. 5.9a is sketched in Fig. 5.9b. Both the

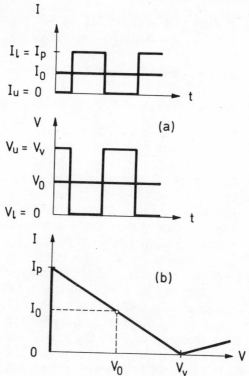

Fig. 5.9 (a) Ideal current and voltage waveforms for 100% power conversion efficiency: $I_0 = I_p/2$, $V_0 = V_v/2$. After Penfield [5.33].
(b) Dynamic $I(V)$ characteristic of an ideal two-terminal device, necessary for obtaining the waveforms of (a).

valley current I_v and the voltage V_p at the current peak have to approach zero. As in devices made from TE material these quantities are finite, as in any practical n.d.c. device, the 'square-wave' efficiency can never reach the ideal value of 81%.*

The conceptually simplest operational mode of TE devices is an LSA oscillation with homogeneous r.f.-field distribution. In this case the dynamic $I(V)$ characteristic is proportional to the static $v(E)$ characteristic of the TE material if the frequency considered remains sufficiently below the intrinsic cut-off frequency f_{TE}. If space-charge-control requirements are disregarded, the maximum obtainable efficiency can be calculated in complete analogy with tunnel diodes or other devices showing stable negative conductance [5.34, 5.35]. The values obtained in this way constitute only rough estimates but may serve to provide a comparison for different TE materials and to elucidate the influence of the device temperature. Considering a square-wave field swing between the field values E_1 and E_u the efficiency for the fundamental power is easily calculated by applying Fourier analysis. Independently of the detailed shape of the $v(E)$ characteristic one obtains

$$P_0 = \frac{1}{4}\, enLA(v_1 + v_u)\,(E_1 + E_u) \tag{5.40a}$$

and

$$P_1 = \frac{8}{\pi^2}\, enLA(v_1 - v_u)\,(E_1 - E_u) \tag{5.40b}$$

and thus according to eqns (5.38) and (5.39) for the 'square-wave' efficiency

$$\eta = \frac{8}{\pi^2}\, \eta_t = \frac{8}{\pi^2}\, \frac{v_1 - v_u}{v_1 + v_u}\, \frac{E_u - E_1}{E_u + E_1} \tag{5.41}$$

with the bias value being $\qquad E_{h0} = (E_1 + E_u)/2.$ $\qquad\qquad$ (5.42)

For a coarse estimate of efficiency it may suffice to restrict the r.f.-field swing to the negative differential-mobility region of the $v(E)$ characteristic, i.e. to set $E_1 = E_p$ and $E_u = E_v$ [4.77]. At 300 K one has in GaAs $E_v \gg E_p$ and $v_p/v_v \approx 2\cdot5$ (see Fig. 2.8) thus yielding $\eta \approx 35\%$. Corresponding values for InP are (based on Fig. 5.10; cf. p. 152 ff., however) $E_v/E_p \approx 2\cdot0$ and $v_p/v_v \approx 5\cdot8$ yielding $\eta \approx 19\%$; for $In_{0\cdot2}Ga_{0\cdot8}Sb$ they are $E_v/E_p \approx 2\cdot9$ and $v_p/v_v \approx 3\cdot8$ yielding $\eta \approx 23\%$ [5.36a, b].

An increase in efficiency may be possible if the r.f. field is allowed to swing into the positive differential-mobility regions [5.35]. This applies particularly to materials where the positive differential mobility near the velocity peak and/or valley is relatively low compared with the average value $(v_p\text{-}v_v)/(E_v\text{-}E_p)$ in the falling part of the $v(E)$ characteristic. In TE material this criterion is not generally fulfilled for a significant distance away from the velocity peak and, hence, it is more appropriate to assume $E_1 \approx E_p$. However, the velocity valley is usually very flat and it pays to have

---

*See p. 167 ff. for still more favourable waveforms yielding $\eta \to \eta_t$.

the r.f. field swing far beyond the valley point E_v. An elegant geometrical method can be used in such a case for determining the maximum efficiency [5.37]. The method is applicable for any chosen shape of the $v(E)$ characteristic. Equation (5.41) is plotted in a plane of v_u/v_1 against E_u/E_1 with efficiency as a constant parameter yielding a family of curves as shown in Fig. 5.10. Thus each point of the plane is characterized

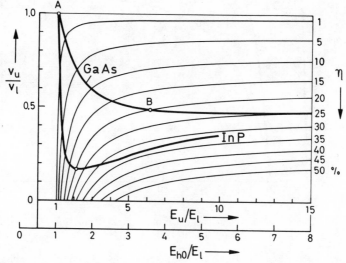

Fig. 5.10 Square-wave efficiency loci in the plane of v_u/v_1 against E_u/E_1 according to eqn (5.41). Superimposed are the theoretical $v(E)$ characteristics at 300 K for GaAs and InP with $E_1 = E_p$. Adapted from Hilsum [5.37].

by a definite efficiency value. Considering a sample that oscillates between the lower turning point of the field at (E_1, v_1) and some upper turning point B (E_u, v_u) the efficiency of the particular oscillation can be read off immediately from B. Consequently, an efficiency analysis can be made by appropriately superimposing the $v(E)$ characteristic upon the series of efficiency curves. Starting at A with the lower turning point (E_1, v_1) on the characteristic the intersections B give the upper field values necessary for obtaining the appropriate efficiency. The maximum efficiency follows by locating the efficiency curve that is tangential to the $v(E)$ characteristic. The necessary bias field for a particular oscillation then follows from eqn (5.42). In Fig. 5.10 the normalized bias field E_{ho}/E_1 is indicated on the abscissa.

As an example, Fig. 5.10 contains the theoretical room temperature $v(E)$ characteristic of GaAs (from Fig. 2.8) and compares it with that* of InP [5.36b]. For the superposition, $E_1 = E_p$ has been chosen for the reasons given above. The efficiency values thus obtained are plotted as a function of bias field in Fig. 5.11 which also shows the behaviour for $In_{0.2}Ga_{0.8}Sb$ [$v(E)$ after [5.37]]. As can be seen, the optimum efficiency for InP and $In_{0.2}Ga_{0.8}Sb$ is found to be higher than that obtained from the

*Cf. remarks on p. 152 ff.

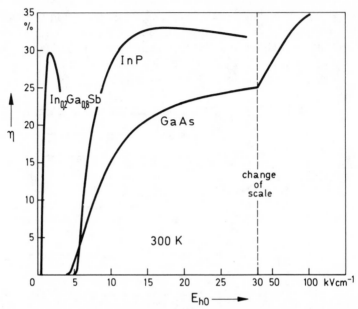

Fig. 5.11 Theoretical bias-field dependence of optimum efficiency for TE devices made from GaAs, InP and $In_{0.2}Ga_{0.8}Sb$.

rough estimate given above. Thus, the influence of the rising part of the characteristic beyond the valley becomes apparent. It is to be noted that the semiconductors InP and particularly $In_{0.2}Ga_{0.8}Sb$ exhibit the optimum efficiency at relatively low bias fields whereas for GaAs the corresponding maximum occurs only at very high bias fields. This is a disadvantage for GaAs since a higher power density is necessary to achieve good efficiency, leading to increased heat removal problems (see Section 6.3.2). Apart from these difficulties the ultimate limit of the bias field is given by dielectric breakdown due to carrier multiplication. On the other hand, the accuracy of the theoretical $v(E)$ curves of InP and $In_{0.2}Ga_{0.8}Sb$ used for calculating the above efficiencies is still questionable. So far, experimental results on the $v(E)$ characteristic of InP indicate a general shape very similar to that of GaAs [5.38].

Owing to the form of the *calculated* drift-velocity/field characteristics high efficiency values are predicted for the semiconductors InP and $In_xGa_{1-x}Sb$ as discussed above. This calculation [5.36a–c] is based on a 'three-level' electron transfer which is thought responsible for the favourable characteristic in these materials with a high peak-to-valley ratio for the drift-velocity [see normalized $v(E)$ characteristic for InP beyond peak field in Fig. 5.10].

A high peak-to-valley ratio results from a rapid electron transfer within a narrow electric-field range. Rapid electron transfer occurs if the central valley and the relevant set of satellite valleys are only weakly coupled. The degree of coupling is determined by the deformation potential Ξ which relates the deformation of the band structure caused by lattice vibration to the dilatational strain (see Section 2.2). At weak coupling electrons rapidly attain high energies and, consequently, an intense transfer to the satellite valleys is required to stabilize the distribution function. Similar as in the case of Ge [5.105] it is believed [5.36] that *weak* coupling ($\Xi \approx 1 \times 10^8$ eVcm^{-1}) exists in InP between the (000) central valley and the $\langle 111 \rangle$ satellite valleys which constitute the lowest satellite set, positioned about 0·4 eV above the central valley [6.33b] (see Table 6.1).

Because of the weak coupling between central valley and the $\langle 111 \rangle$ valleys, the next higher satellite valleys, which are found along the $\langle 100 \rangle$ directions at an energy of about 0·7 eV above the central minimum (Table 6.1, [6.33b]) become also important. Coupling between these and the central valley is thought to be *stronger* ($\Xi \approx 1 \times 10^9$ eVcm^{-1}). By transfer to these $\langle 100 \rangle$ valleys the remaining high-energy electrons of the central valley are able to dissipate energy; otherwise unwanted early impact ionization would set in. The strong coupling to the $\langle 100 \rangle$ valleys, together with strong coupling between the $\langle 111 \rangle$ and $\langle 100 \rangle$ satellite valleys themselves, however, causes a steady rise in drift velocity beyond the minimum point of the characteristic. This is a disadvantage since it leads to a drop in efficiency as the electric field increases (see Fig. 5.11).

The electron transfer from the central valley to the $\langle 111 \rangle$ satellite valleys is a relatively slow process because of the weak coupling. The electrons are not able to respond to rapid variations in electric field (see effect on cut-off frequency of InP devices, Fig. 5.7). Also the diffusion rate is markedly increased. It has been suggested [5.36a, d] that these effects tend to impede the formation of dipole domains. A definite experimental proof of the ability for domain inhibition is, however, still lacking, cf. [5.38a–c].

The high value of negative differential mobility predicted for three-level-transfer semiconductors should result in lower noise generation [e.g. eqn(5.74)]. Experiments with InP amplifiers [5.76] support such a contention. But as above, measurements of $v(E)$ curve in InP did not reveal a particularly steep slope [5.38]. The experimental curves are actually more consistent with calculations based on the simpler two-level transfer [6.33a]. So that question remains still open whether a three-level mechanism is indeed involved in these materials and/or whether the numerical values for the deformation potentials have been chosen correctly in the calculations which yielded the rapid fall in velocity with increasing electric field.*

5.2.3 Temperature Dependence

The temperature dependence of the efficiency is of considerable practical interest since self-heating of a TE device, particularly in C.W. operation (Section 6.3.1), causes the semiconductor chip to be at temperatures considerably higher than ambient. Hence, it is important to investigate the influence of the temperature on the $v(E)$ characteristic (Sections 2.3 and 6.3.2). Figure 5.12 shows square-wave efficiency curves for GaAs based on $v(E)$ characteristics that are calculated by the Monte-Carlo technique (Fig. 2.10). The numerical $v(E)$ curves have been approximated by the analytical formula (3.4), choosing the following parameters:

T/K	$\mu_1/\text{cm}^2 \text{ (Vs)}^{-1}$	$v_\text{v}/10^7 \text{ cm s}^{-1}$	$E_\text{a}/\text{kV cm}^{-1}$	v_p/v_v
300	8 000	0·85	4·0	2·45
500	4 300	0·65	4·2	1·91

*Recent work actually support the suspicion that two-level transfer mechanism is dominant also in InP; see Fawcett, W. and Herbert, D. C., *J. Phys. C (Solid State Phys.)* 7 (1974), p. 1641. Low InP noise measure experimental results could be related to a lower diffusion constant; see Hammer, C. and Vinter, B. *Electron. Lett.* 9 (1973), p. 9.

The efficiency drops with increasing temperature, e.g. for $E_{h0} = 30$ kV cm^{-1} from 30% at 300 K to 22% at 500 K. This drop results essentially from a decrease of the peak-to-valley ratio, v_p/v_v, from 2·45 to 1·91. A somewhat lower temperature dependence is obtained for InP [5.39].

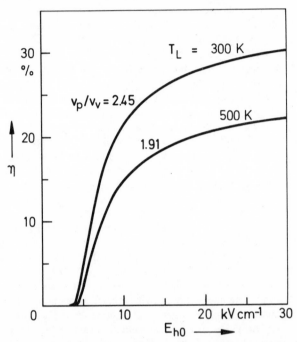

Fig. 5.12 Theoretical influence of temperature on the maximum square-wave efficiency of GaAs devices based on the Monte-Carlo $v(E)$ characteristics derived by Ruch and Fawcett using a Monte Carlo technique [2.59].

5.2.4 Bias and Circuit Dependence

The optimum square-wave efficiency as estimated in the previous sections is hardly achievable in practice. This is caused by the combined influence of non-ideal circuit load, space-charge control requirements or actual dipole-domain formation, and the deviation from the static $v(E)$ characteristic at higher frequencies. These different effects will now be discussed for a GaAs TE device at room temperature.

The published literature on efficiency calculation for the various modes in GaAs TE devices is rather extensive. A variety of different $v(E)$ relationships have been used; thus one has to be cautious when comparing calculated efficiency values. Here, we shall compare different modes only if the efficiencies are based on the same $v(E)$ characteristic. For ease in mathematical handling the $v(E)$ characteristic is approximated analytically throughout. Generally, the expression (3.4) is used. A further simplification by employing a straight-line approximation is made in special cases. For a unified analytical treatment of the various modes based on such a straight-line approximation see, for instance [1.30].

5.2.4.1 *Simple Asymptotic Models*

(a) *LSA Mode.* So far we have discussed efficiencies resulting from a *voltage square wave* for a homogeneous r.f. field distribution within the TE device, i.e. for the ideal LSA mode. If a simple high-Q parallel RCL circuit (Fig. 3.19) is employed there occurs only a single resonance where the circuit represents a real load resistance R_L. At any of the higher harmonic frequencies it constitutes a short circuit. Accordingly the voltage waveform is a pure *sine wave*. The current waveform follows from the $I(V)$ characteristic and the resulting 'sine-wave' efficiency is calculated employing Fourier analysis as in the square-wave case above. For the LSA mode the maximum sine-wave efficiency and the corresponding load resistance $R_L = -R_{ac}$ can be determined by assuming again uniform r.f. field distribution (cf. Section 4.2.1, particularly Fig. 4.1). The results based on the $v(E)$ relationship of eqn (3.4) with the 300 K parameters of Section 5.2.3 are shown in Fig. 5.13. If the space-charge control requirements are not imposed the dashed curves are obtained, otherwise the maximum efficiency and the corresponding load resistance become a function of the n_0/f ratio as indicated by the solid curves. In Fig. 5.13a the square-wave efficiency with $E_1 = E_p$ is included. The efficiency is reduced considerably for the sine-wave oscillation, particularly at high bias field E_{h0}. For the $v(E)$ relationship chosen ($v_p/v_v = 2\cdot45$) the optimum efficiency is about 18 % at a bias field near 17 kV cm^{-1} and an n_0/f ratio between 5 and 10 $\times$ 10^4 s cm^{-3}.

(b) *Dipole-Domain Modes.* Turning now to the dipole-domain modes we will evaluate both square-wave and sine-wave efficiencies. The $I(V)$ characteristic is changed from a scaled replica of the $v(E)$ curve to a curve of the form of Fig. 3.15 exhibiting a static and a dynamic branch. [Its normalized version in the velocity/field plane, Fig. 3.15a, was called $v(E_h)$ characteristic in Chapter 3]. For the efficiency evaluation we assume a large n_0L product, i.e. approximate the dynamic branch by a horizontal straight line corresponding to the saturation (valley) velocity v_v, and set $E_T = E_p$. As a consequence, domain formation and extinction times are small compared to the transit time and can be neglected.

In general, a voltage *square wave* does not induce an antiphase square wave in the current because of the dynamic domain properties resulting in a hysteresis between the two $I(V)$ branches. In the delayed *dipole-domain mode* an antiphase current square wave is obtained only if the oscillation period T is twice the transit time τ (domain transit angle $\theta = \pi$) and $E_1 \approx E_p$ is adjusted to a value slightly below threshold. Then, the domain is triggered when the mean field E_h moves from E_1 to E_u and hence the current is fixed on the dynamic branch. At the instant when E_h reverts from E_u to E_1 the domain exits at the anode and the current resumes the peak value. In the *quenched dipole-domain mode* an antiphase current square wave is always obtained if the voltage square-wave amplitude is large enough to sweep

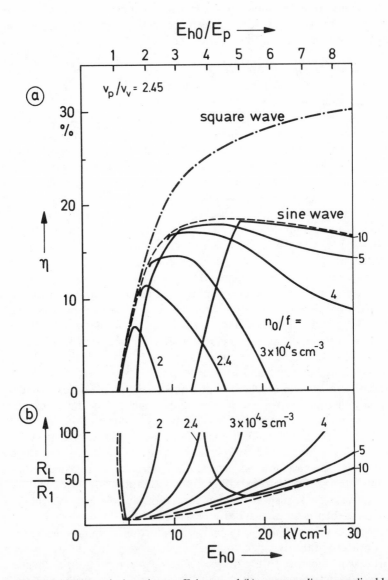

Fig. 5.13 (a) Theoretical maximum efficiency and (b) corresponding normalized load resistance for sine-wave voltage oscillations in the LSA mode of a GaAs device. Dashed curves are for condition without space-charge control. The calculation is based on the $v(E)$ relationship of eqn (3.24) with the 300 K parameters of Section 5.2.3 ($v_p/v_v = 2 \cdot 45$). After Copeland [4.9]. Maximum square-wave efficiency for $E_1 = E_p$ is also shown.

the sample field below the domain sustaining value during each cycle. However, since E_1 is then smaller than E_p the efficiency is considerably lower than in the delayed-dipole-domain case. The efficiency even tends to zero for $n_0 L \to \infty$ [dynamic $v(E_h)$ branch is a horizontal straight line!]. This draw-back can in principle be overcome by employing a modified voltage square wave [5.40]: one sets $E_1 \approx E_p$ for the square wave but superimposes a very narrow spike on the leading edge of each square pulse thereby ensuring that the domain is quenched during each cycle (Fig. 5.14). For negligible quenching times this spike can be considered as infinitely narrow, and the same efficiency values are obtained as in the delayed dipole mode

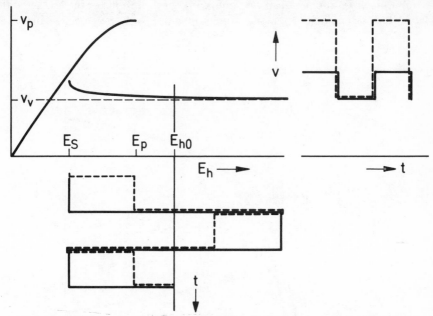

Fig. 5.14 Square-wave oscillation in quenched dipole-domain mode; solid curve: small efficiency; dashed curve modified voltage square wave leading to maximum efficiency.

with $T = 2\tau$. Figure 5.15a shows this efficiency as evaluated from eqns (5.41) and (5.42) with $E_1 = E_p$ and $v_u = v_v = v_p/2\cdot49$. It is somewhat higher than that of ideal LSA operation (with space-charge control disregarded) at low bias fields E_{h0} but approaches it at large E_{h0}. For the dipole-domain modes avalanche breakdown in the high domain fields limits the accessible range of E_{h0} more severely than for LSA operation (see Section 6.4, particularly Fig. 6.25). Hence LSA oscillators can be operated at higher bias levels leading to higher efficiencies in practice.

The theoretical efficiency evaluation in the dipole-domain modes is somewhat more involved when the voltage waveform is *sinusoidal*. The current waveforms have to be derived from the two-branch $v(E_h)$ characteristic taking the dynamic properties of the domain into account (cf. Section 3.5, particularly Figs 3.23 and 3.25). In the delayed dipole-domain mode the domain transit angle $\theta = 2\pi\tau/T$ strongly influences the current waveform and hence the efficiency which is calculated from the d.c. and fundamental components in the same way as for the LSA mode (cf. Section 4.2.1: the formulas presented there hold also for the dipole-domain modes if E is replaced by the average field E_h and $v(t)$ is the appropriate normalized current waveform of the type of Figs 3.23 and 3.25. For details see [5.41, 3.44, 3.46, 3.48]). Thus the maximum efficiency is correlated to a particular transit angle θ. (Figs 5.15 and 5.16). As in the LSA mode this dependence is absent in the quenched dipole-domain mode (Figs 5.18 and 5.19).

(c) *Comparison of the Modes.* A comparison of the results for the LSA, and the delayed dipole-domain (d.d.), modes is presented in Figs 5.15 to 5.17. Figure 5.15a

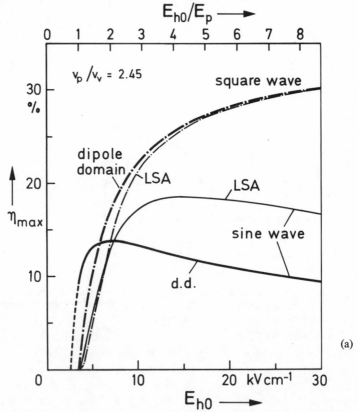

Fig. 5.15 Theoretical comparison between LSA and delayed dipole-domain (d.d.) mode for sine-wave voltage oscillations of GaAs device: (a) maximum efficiency, (b) corresponding normalized load resistance, and (c) output power as a function of bias field. $v(E)$ relationship as in Fig. 5.13 ($v_p/v_v = 2·45$); for d.d. curves $n_0L \to \infty$. The voltage square-wave case for $E_1 = E_p$ is indicated in (a) and (b). Sine-wave calculation after Heinle [5.96]. The LSA curves are identical to those of Fig. 5.13a without space-charge control.

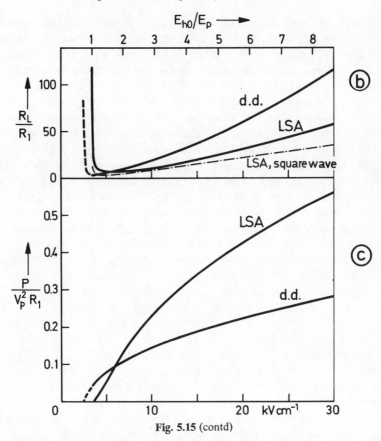

Fig. 5.15 (contd)

compares the maximum efficiency in sine-wave operation to the square-wave efficiency discussed above. The efficiency reduction is much stronger for the delayed-dipole-domain mode than for LSA. Interestingly, the optimum value for the delayed dipole-domain efficiency of 13·8% occurs at a relatively low bias field, viz. $E_{h0} \approx 2E_p$, compared with the optimum LSA efficiency of 18·6% which requires $E_{h0} \approx 4E_p$. This behaviour results from the strong current decrease when a domain forms above the peak field E_p. In Fig. 5.15a the square-wave efficiency becomes smaller than the maximum sine-wave efficiency at low bias fields E_{h0}. This is artificial, since the simplification $E_1 \approx E_p$ was made for estimating square-wave efficiencies. Such an approximation is justified only for high bias levels E_{h0} and leads to an underestimation of the efficiency at low E_{h0}.

Positive efficiencies are obtained in the delayed dipole-domain mode even below the threshold field E_p (dashed part of d.d. curves in Fig. 5.15). This is possible since a dipole domain already forms, resulting in a strong current decrease, if the r.f. voltage swing is just large enough to reach beyond E_p. Because of the hysteresis property between the dynamic and static branches of the $v(E_h)$ characteristic a net negative sample resistance averaged over one r.f. period may still result. This interesting property of the dipole-domain oscillations of a TE device in a resonator can be used to trigger a stable device into continuous oscillations by applying a short constant-voltage pulse [8.4] or a short r.f. pulse with suitable frequency and amplitude [8.11b] to the bias at pre-threshold (memory applications, see Section 8.4.2).

Figures 5.15b and c show the behaviour of the load resistance and the r.f. output power, respectively, corresponding to the maximum sine-wave efficiency of Fig. 5.15a. The ratio of LSA, to delayed-dipole-domain, output power approaches almost a factor of two at high bias levels demonstrating the superiority of LSA operation. It is to be noted that the load-resistance curves are of inverse character to the maximum-efficiency curves: large efficiencies are associated with small load resistances. This statement also holds when comparing the two different modes and is further substantiated by Fig. 5.16a which shows the efficiency as a function of load resistance at a fixed bias field of $E_{h0} = 10$ kV cm^{-1}. Here, roughly the same efficiencies are obtained at a given normalized load resistance R_L/R_1. However, the LSA mode can be more heavily loaded ($R_L/R_1 \geq 11.5$ as compared to $R_L/R_1 \geq 17.5$ for the d.d. curve), giving a further increase in efficiency and leading to a higher maximum value (17.2% as compared to 13.2% for the d.d. curve).

Whereas the LSA curve of Fig. 5.16a is frequency independent and is realizable as long as the space-charge control requirements are met, the d.d. curve has the normalized domain transit angle $\theta/2\pi = f/f_\tau$ as parameter. The d.d. curve connects points of highest efficiency for a *fixed* transit angle (cf. Fig. 5.19a). This results in the two-valued nature of the delayed dipole-domain efficiencies. For the selected bias value the delayed dipole-domain mode has a maximum efficiency of 13.2% at $f/f_\tau = 0.804$. The LSA curve is readily derived from Fig. 4.1 using the curve branches for $\Gamma < 1$ (net decay of space charge) and eliminating the r.f. amplitude $\hat{E}_h$ (note: $R_{ac} = -R_L$). The maximum efficiency coincides roughly with $\Gamma = 1$. The space-charge control requirement, $6h_p < n_0/f < 5h_N$ (see Section 4.2.3), limits the accessible part of the LSA efficiency curve for a given frequency or n_0/f value. This can be seen from Fig. 5.16b which shows the complete tuning range that allows space-charge control. For maximum efficiency this range narrows to a discrete frequency. At large load resistances (small efficiency) the tuning range approaches a band of roughly 4:1. In the case of the delayed dipole-domain mode the tuning range is limited by the frequency parameter of the efficiency curve as shown in Fig. 5.16c. The maximum tuning range at low efficiencies approaches a band of only 1.38:1. Figure 5.16 demonstrates that for wide tuning ranges a penalty of reduced efficiency has to be paid in both modes.

The asymptotic tuning ranges for $\eta \to 0$ ($R_L \to \infty$: dashed horizontal lines in Figs 5.16c and b) are plotted in Fig. 5.17 as a function of the bias field E_{h0}. Both η and R_L have, of course, to remain finite in practice, particularly in the delayed dipole-domain mode because transition to the quenched dipole-domain mode may occur above a critical R_L, thereby causing a gap in the tuning range (see Fig. 5.19b).

160

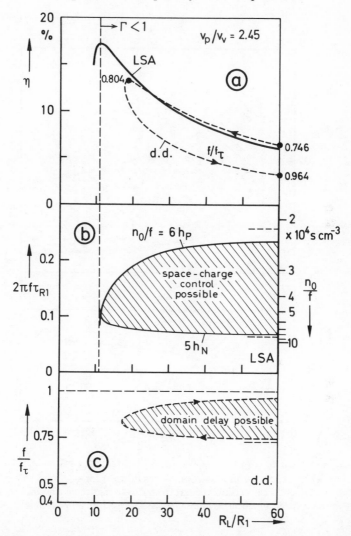

Fig. 5.16 Theoretical load dependence of sine-wave efficiency and tuning range in LSA and delayed dipole-domain (d.d.) mode exhibited by GaAs device at bias $E_{h0} = 10$ kV cm^{-1}. The LSA frequency is normalized to the positive dielectric relaxation frequency $(2\pi\tau_{R1})^{-1} = en_0\mu_1/(2\pi\varepsilon)$, the d.d. frequency to the transit-time frequency f_τ. After Heinle [5.96].

However, these accessible R_L values are sufficiently large to make the asymptotic frequency boundaries acceptable approximations for the maximum tuning range. The frequency which yields maximum efficiency is shown by the dashed curves in Fig. 5.17. For LSA operation it corresponds to $5h_N = 6\,h_P$ ($\Gamma = 1/e = 0.368$, see Section 4.2.3), a condition occurring close to the theoretical efficiency maximum

161

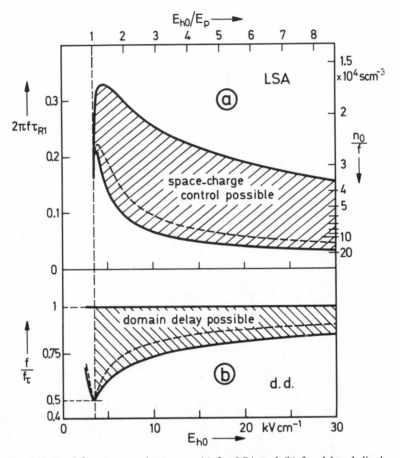

Fig. 5.17 Total frequency tuning ranges (a) for LSA and (b) for delayed dipole-domain (d.d.) operation as a function of bias field E_{h0} for sine-wave voltage oscillations of GaAs device. Dashed curve corresponds to maximum efficiency (Fig. 5.15a, sine wave). Frequencies normalized as in Fig. 5.16. After Heinle [5.96].

obtained if space-charge control is disregarded (see Fig. 5.16). Thus, the dashed curves in Fig. 5.17 define the frequency parameter along the sine-wave efficiency, load resistance, and output-power curves of Fig. 5.15 in each particular mode. Figure 5.17 indicates a general reduction of achievable frequency with bias in the LSA mode and an increase in the delayed dipole-domain mode. On the other hand, the tuning range increases for LSA and decreases for delayed dipole-domain operation. Note that in both modes the maximum efficiency occurs in the lower-frequency part of the tuning range.

In the quenched dipole-domain (q.d.) mode sine-wave efficiencies are even lower than in delayed dipole-domain operation as demonstrated in Fig. 5.18, because of

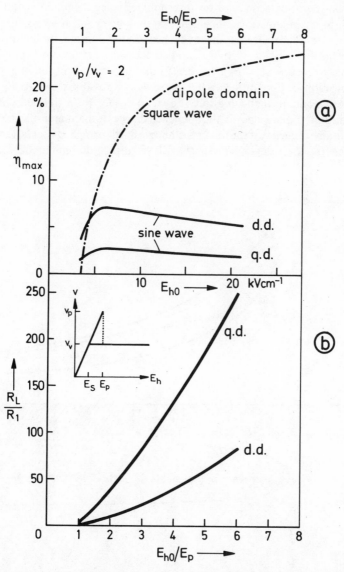

Fig. 5.18 Theoretical comparison of delayed (d.d.) and quenched (q.d.) dipole-domain modes (GaAs): (a) maximum efficiency, (b) corresponding normalized load resistance as a function of bias field for sine-wave voltage oscillations. Inset shows $v(E_h)$ characteristic approximated by straight lines with $v_p/v_v = 2$. Adapted from Carroll [1.31]. Maximum square-wave efficiency of the dipole-domain modes is shown for comparison.

163

the larger voltage swings needed for domain quenching ($E_1 < E_s$). This is a result of the strong current decrease below threshold reducing the fundamental r.f. current amplitude by an amount not compensated by the somewhat larger voltage amplitude.

The smaller efficiencies of the quenched dipole-domain oscillation are again caused by the higher values of the load resistance necessary for synchronizing domain quenching (to ensure $E_1 < E_s$). As seen from Fig. 5.19 (corresponding to Fig. 5.16) synchronization of domain delay (i.e. $E_1 < E_p$) is possible down to $R_L/R_1 = 19.5$ (full curve), domain quenching only down to $R_L/R_1 = 73$ (dashed curve). As in LSA operation, the efficiency in the quenched dipole-domain mode is independent of the oscillation frequency if the load resistance is fixed (provided oscillation is possible at all). The frequency dependence in the delayed dipole-domain mode is indicated in

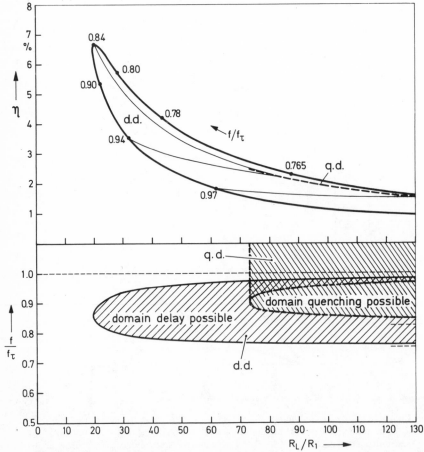

Fig. 5.19 Theoretical load dependence (a) of efficiency and (b) of frequency tuning range, in the domain modes of GaAs device at a bias $E_{h0} = 2.8\ E_p \approx 10\ \text{kV cm}^{-1}$ for sine-wave oscillation. The $v(E_h)$ characteristic is the same as in Fig. 5.18. After Pollmann [5.42].

Fig. 5.19a by a few representative efficiency curves with constant domain transit angle $\theta/2\pi = f/f_r$. Their upper end points for which E_1 becomes equal to E_p define the envelope already presented in Fig. 5.16a, their lower end points lie on the (dashed) q.d. curve. Thus a smooth transition between the two dipole-domain modes takes place at the latter points (they correspond to the case where quenching occurs at the instant of domain exit at the anode in the limit $E_1 = E_S$).

The curves of Figs 5.18 and 5.19 have been prepared by using a $v(E_h)$ characteristic somewhat different from the one on which the previous figures have been based. Here, the static branch of Fig. 3.15a was approximated by a straight line and the peak-to-valley ratio was chosen to be $v_p/v_v = 2$ (inset of Fig. 5.18b). The influence of the changed $v(E_h)$ characteristic becomes apparent by comparing the d.d. curves to the corresponding ones of Figs 5.15 and 5.16. The reduction of the sine-wave efficiency results from both the reduced peak-to-valley-ratio and the more rapid velocity decrease for $E_h < E_p$ on the straight line as compared to the rounded curve of Fig. 3.15a. The square-wave efficiency of the dipole-domain modes is also indicated in Fig. 5.18a. Its reduction as compared to the corresponding curve of Fig. 5.15a is caused by the smaller peak-to-valley ratio only (because of the assumption $E_1 = E_p$).

The quenched *multiple* dipole-domain mode can be treated by employing a similar device model [3.59]. Simultaneous build-up and quenching of a series of domains needs less time than the corresponding transient processes for a single domain; the neglect of these transient times is, thus, justified up to higher operating frequencies. An important difference to the quenched single dipole-domain mode results from the smaller voltage swings needed for quenching multiple domains since the domain sustaining voltage $E_S L$ becomes larger approaching almost the threshold value [3.11]. This increase in E_S can easily be understood from the steady-state domain characteristic of the excess potential $\Phi_{D\infty}$ against the outside field $(E_{1\infty})$ (see Fig. 3.13). For several domains in series this characteristic is stretched along the Φ_D axis by a factor that is roughly equal to the number of domains. Since the slope of the 'device line' remains unchanged, a higher E_S value results when locating its tangential position to the stretched $\Phi_{D\infty}$ $(E_{1\infty})$ characteristic. This increase of E_S leads to higher efficiencies. When approximating $E_S \approx E_p$ the efficiencies of the quenched multiple dipole-domain mode approach the values obtained for delayed dipole-domain operation if the dynamic $v(E_h)$ branch is still simplified by a horizontal straight line at the valley velocity v_v [3.59].

Figure 5.20 provides a summary of the frequency tuning and loading behaviour of the different modes at a constant bias field. The representation follows directly from plots of the type of Figs 5.16 and 5.19 and contains also hybrid-mode operation (Section 3.5.4). The calculations [3.54] are based on a straight-line approximation of the $v(E)$ characteristic with a peak-to-valley ratio of $v_p/v_v = 2.49$. The $n_0 L$ product of the device is chosen to be $2 \times 10^{12}\,\mathrm{cm^{-2}}$. Using this particular value a conversion of the n_0/f scale into a f/f_r scale can be performed, thus allowing a comparison of all modes on the same frequency scale (cf. Fig. 5.1). It demonstrates very clearly the wide tuning ranges and high efficiencies achievable in the LSA and hybrid modes.

The hybrid mode appears to fill the gap between quenched dipole-domain and LSA operation rather smoothly. In the device model used for the hybrid mode a time constant τ_D is introduced for domain formation [3.53]. When $E_h(t) > E_p$ the sample field is assumed to be homogeneous within the time τ_D (LSA regime). At the end of this interval an abrupt transition to a mature dipole domain is assumed, i.e. the device current drops to the dynamic $v(E_h)$ branch and stays there until spontaneous quenching occurs when $E_h(t) = E_S$ (dipole-domain regime). Thus the model yields

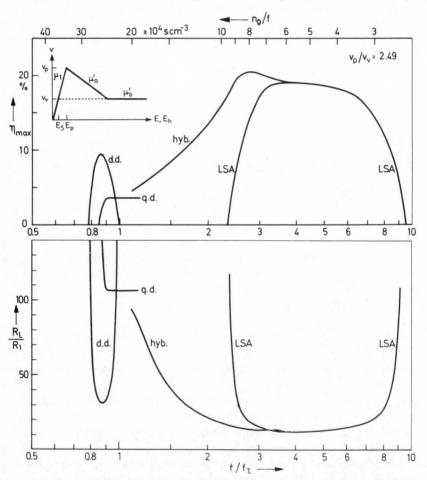

Fig. 5.20 Theoretical frequency tuning and loading behaviour of a GaAs TE device with $n_0 L = 2 \times 10^{12}$ cm^{-2} at a bias field of $E_{h0} = 4\, E_p \approx 14$ kV cm^{-1} for sine-wave voltage oscillation: (a) maximum efficiency, (b) corresponding normalized load resistance as a function of normalized frequency f/f_τ or $n/_0 f$ ratio. Based on the $v(E)$ and $v(E_h)$ characteristics of the inset. ($\mu_1 = 7\,000$ cm^2 (Vs)$^{-1}$, $\mu'_a = -1\,500$ cm^2 (Vs)$^{-1}$, $\mu'_b = 0$, $E_p = 3 \cdot 2$ kV cm^{-1}, $v_p/v_v = 2 \cdot 49$). Hybrid and LSA mode adapted from Huang and MacKenzie [3.53]. Dipole-domain modes after Frey [5.30].

a smooth transition to the genuine LSA case for a small enough oscillation period $T < \tau_\mathrm{D}$.

(d) *Influence of Higher Harmonics.* As outlined above, simple sinusoidal voltage waveforms lead to considerably lower efficiencies than the ideal square waveform. Hence, it is advantageous to employ a circuit that approximates the square waveform in the voltage as closely as possible. The Fourier components of a square wave contain only odd harmonics. If the current, as well as the voltage, also has a square waveform then the ratio of the voltage to the current amplitude in each individual harmonic remains unchanged. Thus the load resistance R_L has to be the same at all odd harmonics and be zero at the even ones. One may simply verify that for the odd harmonics it is required

$$R_\mathrm{L} = \frac{V_\mathrm{u} - V_1}{I_1 - I_\mathrm{u}} = \frac{\mu_1(E_\mathrm{u} - E_1)}{v_1 - v_\mathrm{u}} R_1 \tag{5.43}$$

or, if E_u is replaced by the bias field E_h0 according to eqn (5.42),

$$\frac{R_\mathrm{L}}{R_1} = \frac{2\mu_1(E_\mathrm{h0} - E_1)}{v_1 - v_\mathrm{u}} \tag{5.44}$$

A plot of R_L/R_1 for LSA operation with $E_1 = E_\mathrm{p}$ was included in Fig. 5.15b. A relatively simple resonant circuit that fulfils the above requirements of establishing a square wave in the voltage is a frequency-independent resistance R_L shunted by a transmission line which is a quarter wavelength long at the fundamental frequency and is shorted at its end, see p. 457 ff., [5.43].

Another circuit that allows an approach to voltage square-wave operation fairly well is an inductance in series with a high-Q parallel resonant circuit [5.43]. For obtaining a substantial increase in efficiency it is sufficient, however, to employ a circuit that supports, in addition to the fundamental, only the third harmonic frequency [5.34, 5.45]. For the LSA mode this is demonstrated by the dashed curve ($R_{\mathrm{L}3} > 0$) in Fig. 5.21a. Up to about 13 kV cm^{-1} the efficiency approaches the square-wave value (calculated with $E_1 = 0.95\ E_\mathrm{p}$) quite well, but falls short at higher bias fields. As indicated by the dashed curves of Fig. 5.21b the load condition $R_{\mathrm{L}1} \approx R_{\mathrm{L}3}$ that is suggested from the square-wave case is fulfilled only in the lower bias range. The square-wave load resistance, eqn (5.44) with $E_1 = 0.95\ E_\mathrm{p}$, is plotted for comparison.

The addition of only the second harmonic to the pure sine wave in the voltage waveform ($R_{\mathrm{L}2} > 0$, dotted curves in Fig. 5.21) leads to a still higher increase in efficiency in the low bias range. There, it approaches the *total* r.f. efficiency of a voltage square wave $\eta_\mathrm{t} = (\pi^2/8)\eta$, i.e. almost *all* r.f. power is conveyed into the fundamental. An efficiency peak occurs close to $E_\mathrm{h0} = 11$ kV cm^{-1} where $R_{\mathrm{L}2}$ tends to infinity. The corresponding waveforms are shown in Fig. 5.22: the voltage follows almost a half sinusoid, and the current approaches a square wave. These waveforms lead to the highest fundamental efficiency theoretically possible [5.43]. Since the half sinusoid contains only even higher harmonics and the square wave only

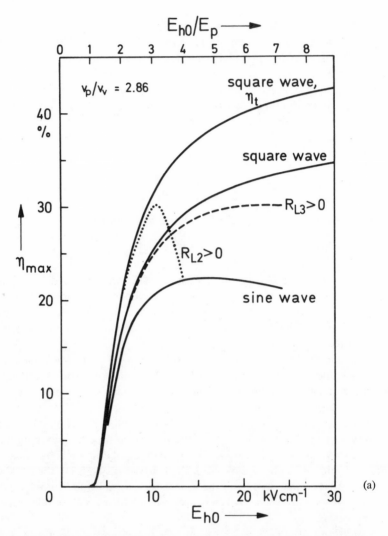

Fig. 5.21 Theoretical increase of maximum efficiency shown by GaAs device in the LSA mode by adding harmonics to the sine wave of the voltage: $R_{L2} > 0$ denotes only second harmonic, $R_{L3} > 0$ only third harmonic added; (a) maximum efficiency, (b) corresponding normalized load resistances and (c) frequency tuning ranges at maximum efficiency as a function of bias field. Based on the $v(E)$ relation of eqn (3.4) with $\mu_1 = 9\,000$ cm² (Vs)$^{-1}$, $v_v = 8 \times 10^6$ cm s^{-1}, $E_a = 4$ kV cm^{-1} ($v_p/v_v = 2\cdot86$). Square-wave efficiencies for $E_1 = 3\cdot2$ kV cm$^{-1} = 0\cdot95\,E_p$. Adapted from Copeland [5.45].

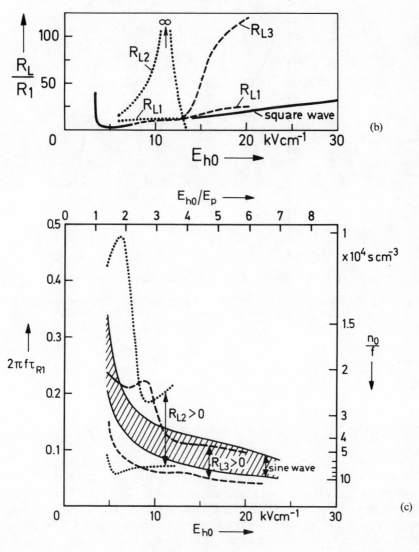

Fig. 5.21 (contd)

odd ones, all r.f. power is concentrated in the fundamental: the load resistance is infinite for even harmonics and zero for the higher odd harmonics. Fourier analysis shows that in such an idealized case the efficiency and load resistance at the fundamental are

$$\eta = \frac{v_1 - v_u}{v_1 + v_u} \frac{E_u - E_1}{E_u + (\pi - 1)E_1} \tag{5.45}$$

169

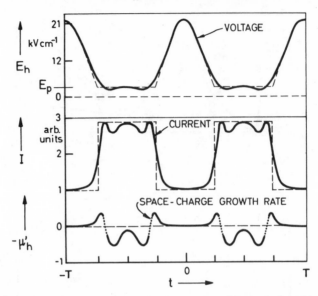

Fig. 5.22 Waveforms for high-efficiency LSA operation of GaAs device with second-harmonic voltage component at bias $E_{h0} = 11 \cdot 0$ kV cm^{-1}. After Copeland [5.45].

and

$$R_{L} = \frac{\pi}{4} \frac{\mu_1(E_u - E_1)}{v_1 - v_u} R_1, \tag{5.46}$$

respectively, and the bias field is expressed as

$$E_{h0} = [(\pi - 1)E_1 + E_u]/\pi. \tag{5.47}$$

The postulated square-wave shape of the current with the half-sinusoidal voltage is realizable only in the delayed dipole-domain mode ($v_u = v_v$) if $\theta = \pi$ or, approximately, in LSA operation if the field swings far enough into the velocity saturation region $v_u \approx v_v$. Additionally, the condition $E_1 = E_p$ is required. Then both, η and R_L are larger by a factor of $\pi^2/8 = 1 \cdot 23$ than the corresponding values of the voltage square wave, eqns (5.41), (5.42) and (5.44), at the same bias field E_{h0}. Thus, η becomes just equal to the *total* r.f. conversion efficiency η_t of the voltage square wave. The required circuit properties, however, are difficult to achieve in practice.

Incorporation of higher harmonics into the voltage waveform changes the frequency tuning behaviour of a particular mode considerably. As an example, Fig. 5.21c shows tuning ranges which allow space-charge control and maximum efficiency to be obtained in the LSA mode. (The sine-wave range corresponds to the dashed curve of Fig. 5.17a. There, the range has contracted to a single curve because of the different parameters of the $v(E)$ relationships used.) In particular by adding a second

170

harmonic the tuning range widens up. A similar behaviour is shown in Fig. 5.23 for the delayed dipole-domain mode at a given bias field. Maximum efficiency and corresponding load resistances are represented in the form of Fig. 5.20 with (solid) and without (dashed) a second-harmonic voltage component. The second-harmonic voltage content shifts the lower tuning limit downwards, this extension causing also an increased optimum efficiency. The addition of more harmonics leads to a further improvement of the optimum efficiency [5.46b].

(e) *Role of* n_0L *Product*. So far only highly-idealized device models have been employed in discussing the load influence on conversion efficiency and r.f. output

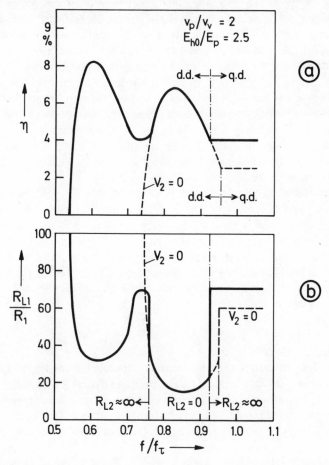

Fig. 5.23 Extension in frequency-tuning range and increase in maximum efficiency for delayed (d.d.) and quenched (q.d.) dipole-domain mode of GaAs device at bias $E_{h0} = 2 \cdot 5\, E_p \approx 8 \cdot 7\, \text{kV cm}^{-1}$: (a) maximum efficiency, (b) corresponding normalized load resistance against frequency. $V_2 = 0$, sine-wave voltage oscillation. Calculation based on $v(E_h)$ relation of Fig. 5.18. After Frey [5.46a].

171

power for the different modes. These ideal models can be considered as asymptotic models for homogeneously-doped devices with large n_0L product. In this case the space charge drifts only a negligibly short distance in devices operated in the LSA mode and space-charge transient times are small in the domain modes.

As we have seen in Fig. 5.1 the different modes tend to converge towards smaller n_0L products, becoming indistinguishable near the critical n_0L value. Here, we have another extreme situation where the space charge drifts virtually over the whole sample length and space-charge transient times are of the same order of magnitude as the oscillation period, and the essential features in the oscillating device are cyclic growth and decay of space charges in the entire bulk of the device rather than homogeneous electric field or steady-state space-charge drift. The space-charge wave model of Section 4.3, which leads to stable negative conductance for devices with subcritical n_0L product (Section 4.4), can be used as the appropriate asymptotic model to describe the behaviour of the device-load combination for oscillating devices with small n_0L product. However, the small-signal analysis presented in Sections 4.3 and 4.4, has to be adjusted to include the non-linear effects that limit the oscillation amplitude and hence determine the r.f. output power and conversion efficiency [5.47]. Frequency tuning is based on the frequency behaviour of the (small-signal) device impedance, thus the condition $E_1 < E_p$ necessary for domain or space-charge control becomes irrelevant. Tuning of oscillations is, then, possible even if E_1 remains above E_p.

The space-charge-wave model can be adjusted also for calculating saturation output power and efficiency of the *amplifying modes* [5.30]. Efficiencies of the reflection-type amplifier approach the corresponding oscillator efficiencies since the r.f. field distribution is similar in both cases. On the other hand, the efficiency of the unilateral travelling-wave amplifier (Section 7.2.2) is reduced considerably, particularly at high gain. This is a consequence of the travelling r.f. field which is allowed to reach the non-linear range only at the anode, whereas the d.c. power is distributed more or less evenly throughout the entire semiconductor volume, cf. [5.77, 7.162].

5.2.4.2 *Complex Models*

For closing the gap between the two extreme approaches for large and small n_0L product more sophisticated models have been developed particularly for dipole-domain-mode operation. Such complex models have the disadvantage, however, of yielding only a limited amount of information. In particular the optimization with respect to efficiency or to other parameters is tedious and slow or may even become impossible.

In the simplified models for domain delay and quenching discussed above, the reactive properties of the dipole domain are lumped into a single capacitance in parallel to an element representing the non-linear $I(V)$ behaviour of the device. This capacitance can then be regarded as part of the circuit lowering the actual oscillation frequency. An improved approach considers the dipole domain as a separate entity leading to an equivalent circuit for the device of the form presented in Fig. 3.21. This automatically introduces a time constant $\tau_D = R_1 C_D$ for domain growth and decay. It becomes possible, for example, to consider a series resonant circuit as the load which establishes as sine wave of the device *current*. Efficiencies are higher than those obtained for the sine-wave *voltages* discussed above [1.31, 3.38b]. Because of the capacitive nature of the sample reactance it may suffice to provide a series LR circuit as a load for inducing an approximate sine wave of the current [3.54], cf. also [5.107].

In the delayed dipole-domain mode a new domain usually forms at the cathode while the old one is still being discharged at the anode. This effect can be included in the analysis by adding the appropriate equivalent circuit for the second domain in series to the single-domain circuit of Fig. 3.21 [3.54].

It should be pointed out that the domain equivalent circuit was actually derived for a *mature* domain in transit and thus can describe the domain *transient* processes only approximately, cf. [3.66]. For higher accuracy the domain formation and quenching phases can be described by separate time constants which determine exponential time functions for the device current or domain excess potential that are unaffected by the change in terminal voltage during the particular phase [5.48]. Models that neglect the reactive domain properties in the mature domain phase and take into account these transient time constants only have also been proposed [1.30, 3.51, 5.49, 5.50]. A semi-empirical model based on an experimental dynamic $I(V)$ characteristic of the device is very useful for estimating the expected conversion efficiency of actual oscillators [5.51].

The incorporation of dipole-domain time constants yields the expected decrease in efficiency with increasing frequency in the quenched dipole-domain mode, leading eventually to an upper frequency limit as estimated in Section 5.1.2. Another result is an upper bias voltage limit for quenched dipole-domain operation [5.48]. Above this critical value the r.f. voltage swing necessary for domain quenching becomes so large that, at the maximum terminal device voltage, the domain would become larger than the device length. This causes the device conductance to become positive. Thus only hybrid or LSA operation remains possible. In practice this bias limit might even be reduced because of avalanche breakdown setting in before the domain has grown to the device length (cf. Figs 3.11 and 3.12).

Improved efficiency calculations of the LSA mode take into account the dynamics of an accumulation-layer domain [4.10, 4.16]. Thus, space-charge control requirements, as treated independently in the homogeneous-field model (Section 4.2.3), are obtained with better accuracy. Introducing an empirical space-charge quenching time [5.97] greatly simplifies the analysis.

All the models mentioned above rely on certain *a-priori* assumptions about the domain that can be justified only by a more rigorous analysis of the fundamental dynamic equations. Only such an analysis which treats the entire semiconductor bulk as a single entity in which the electrons obey the fundamental dynamic laws in time

173

and space, is able to characterize device performance with respect to *all* possible modes. Copeland [3.10] extended the early computer study of McCumber and Chynoweth [3.4] to investigate a TE device connected to a high-Q parallel resonant load circuit (sinusoidal voltage). Some of his results are shown in Fig. 5.24. Both

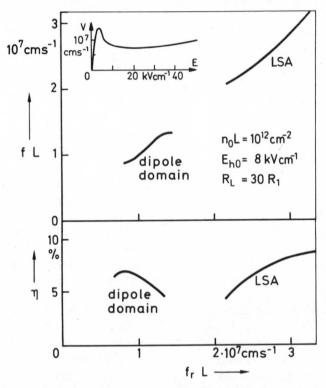

Fig. 5.24 Oscillation frequency f and efficiency η as a function of the resonant frequency f_r of the external circuit in dipole-domain and LSA-mode GaAs device. Obtained from rigorous computer analysis for sinusoidal voltage waveform. After Copeland [3.10]. (We thank J. A. Copeland for providing us with the original figures.)

dipole-domain and LSA mode operation were obtained (the dipole-domain mode carries symptoms of quenched operation, but was not further specified; it is possible that the n_0L product of 10^{12} cm^{-2} is already too small for a clear distinction). The efficiency values of the figure can be compared only qualitatively with the efficiencies shown in Fig. 5.20 which is a similar plot for the simplified device models, since quite different $v(E)$ characteristics were assumed in the two cases—such rigorous computer models were also applied to more complex circuits which generate non-sinusoidal voltage waveforms across the device [4.13, 5.98–5.102].

174

A more sophisticated analysis takes account of random doping fluctuations in the sample [3.11]. Then virtually all possible oscillation modes including quenching of multiple dipole domains, can be simulated. When reducing the amount of the doping fluctuation, the efficiency of the latter mode increases and eventually changes smoothly into LSA operation, cf. [6.76]. Figure 5.25 shows an example of such behaviour. The higher doping fluctuations require lower sub-threshold voltages to be reached for space-charge or domain quenching

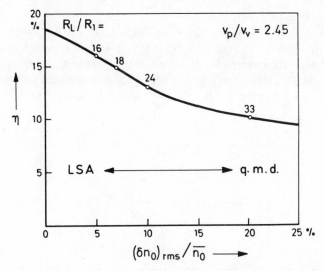

Fig. 5.25 Theoretical maximum efficiency and corresponding normalized load resistance during transition from LSA to quenched multiple-dipole-domain (q.m.d.) operation versus normalized r.m.s. value of doping fluctuations in the GaAs sample for sinusoidal voltage waveform. $v(E)$ characteristic assumed is same as in Fig. 5.13. Sample data: $n_0 = 6 \times 10^{14}$ cm^{-3}, $L = 70$ μm, $E_{h0} = 2\cdot 8$ $E_p \approx 10$ kV cm^{-1}, $f = 15$ GHz. After Thim [3.11].

during the r.f. cycle. The increased voltage swing increases the required load resistance R_L and hence decreases the efficiency. This is in general agreement with experimental observations [5.52]. Similar effects are expected from other doping inhomogeneities [4.14] and from carrier-density fluctuations induced by noise [6.77]. Single dipole-domain, rather than multiple dipole-domain, operation was obtained in computer simulations of samples with random doping fluctuations only when, additionally, a major doping inhomogeneity was present, or, for frequencies close to the transit-time frequency if the sample voltage never dropped below threshold. Samples with an n_0L product below the critical value form accumulation layers instead of dipole domains but may still yield high efficiencies if quenching is achieved (e.g. 10% as compared to 13% in the quenched multiple-dipole-domain mode, [3.11]).

For such 'subcritically'-doped samples the efficiency improves if the cathode configuration is changed from an ideal ohmic contact to a contact with constant carrier injection: computer simulation studies of a homogeneously-doped sample [5.53] showed a remarkable increase in efficiency from 1·7 to 16·5%. Efficiency improvement by other device boundary effects, like geometrical shaping [5.54] or dielectric surface loading [4.73] are also possible.

175

Of practical interest is the effect on efficiency of non-uniform doping profiles since these are often present in epitaxially-grown GaAs material (Section 6.1). Investigations show that normally a drop in efficiency results [5.69], but an increase can be expected in special cases [5.54, 5.69b]. A suitable doping gradient can with advantage be used to compensate a mobility gradient caused by a temperature profile in C.W. operation, thereby increasing the LSA efficiency by several percent [4.14].

5.2.4.3 *Experimental Verification*
In an experimental check on the theoretical findings, a pulsed bias voltage has to be applied to the device to avoid complications caused by self-heating of the semi-conductor chip. Experimentally obtained d.c./r.f. power conversion efficiencies in pulsed GaAs TE oscillators are in remarkably good agreement with those expected from the simplified theoretical device models described earlier. Figure 5.26 shows the

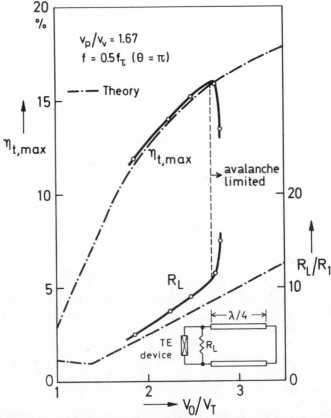

Fig. 5.26 Experimental and theoretical maximum value of total d.c./r.f. conversion efficiency η_t and corresponding normalized load resistance R_L/R_1 as a function of bias voltage for delayed dipole-domain operation of GaAs device in the circuit of the inset. Device parameters: $R_1 = 20\,\Omega$, $V_T = 58.5$ V, $v_v/v_p = 0.6$. Adapted from Kino and Kuru [5.43a].

total efficiency η_t and the corresponding load resistance determined for a sample oscillating under optimum load conditions in the delayed dipole-domain mode at a domain transit angle $\theta = \pi$. The load circuit (inset) established a square wave for the r.f. voltage [5.43]. The theoretical curves were derived from eqns (5.41), (5.42) and (5.44) by optimizing η_t for a straight-line approximation of the $v(E_h)$ characteristic (inset to Fig. 5.18). The peak-to-valley ratio was determined experimentally from the current oscillations with constant voltage drive ($R_L = 0$). Beyond a critical bias voltage of about $V_0 = 2\cdot 7\ V_T$ the experimental efficiency drops sharply. This can be explained by avalanche breakdown in the domain since the upper turning point of the terminal voltage, V_u, becomes too high [5.43], cf. p. 248.

In another experiment with dipole-domain-mode oscillations the efficiency at the fundamental frequency of a practical sample reached 21 % with a lumped-element resonator [5.55, 6.88a]. The bias-voltage dependence is shown in Fig. 5.27. For

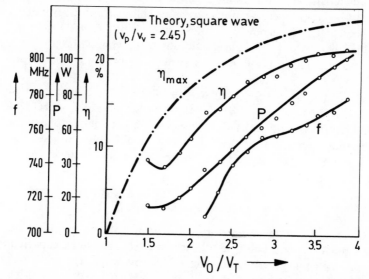

Fig. 5.27 Experimental maximum efficiency, r.f. output power, and oscillation frequency as a function of bias voltage for GaAs device oscillating in dipole-domain mode. After Heeks *et al.* [6.88a]. Dash-dotted curve is maximum square-wave efficiency of dipole-domain modes from Fig. 5.15a.

comparison the theoretical maximum square-wave efficiency from Fig. 5.15a is indicated ($v_p/v_v = 2\cdot45$). The behaviour agrees qualitatively very well, thus the somewhat lower experimental efficiency seems to be attributable mainly to an actually lower peak-to-valley ratio. A similar qualitative agreement was obtained for the bias dependence of the load resistance [5.55]. The highest efficiency of a pulsed dipole-domain-mode oscillator measured so far is 32·2 % at 1·09 GHz [5.56], whereas in LSA operation values of 14·6 % at 1·75 GHz [5.57] and 27·5 % at 3 GHz [5.95] have been achieved. However, these figures should not be used for a final judgement between the different modes. They only demonstrate that experimental efficiencies *can* approach the theoretical values. Efficiencies as they are *generally* obtained in practice are below 10% since both the device parameters and the circuit are far from optimum. Additionally, heating effects in C.W. operation (cf. Section 6.3) reduce efficiency (Section 5.2.3). Improvements in device technology (Chapter 6) and in circuit design (Section 7.1.1) should make higher efficiencies more common.

5.2.5 Frequency Dependence

As discussed in Section 5.1.3, the finite electron-scattering relaxation times lead to an upper frequency limit of the negative differential mobility. In the large-signal case of an oscillator these scattering times cause, at microwave frequencies, time-dependent deviations from the static $v(E)$ characteristic and, hence, from the corresponding $I(V)$ characteristic used for each particular mode in the above discussion. This results in a lowering of the r.f.-power conversion efficiency, which eventually drops to zero at the cut-off frequency. The time-dependent $v(E)$ relationship is generally termed the 'dynamic' $v(E)$ characteristic not to be confused with the 'dynamic' dipole-domain branch of the $I(V)$ characteristic introduced in Section 3.4.4.2 (Fig. 3.15).

We now consider an LSA oscillator in which space-charge accumulation is neglected. In a highly-simplified manner the small-signal results of Section 5.1.3 may be generalized for large-signal sinusoidal-voltage operation. Employing the single-relaxation-time approximation for the microwave differential mobility, eqn (5.29), we define the a.c. microwave admittance of the oscillating device by an average mobility (cf. Fig. 5.4a):

$$Y_m/G = (\overline{\mu'_m}/\overline{\mu'}) \approx \overline{\mu}/\overline{\mu'} + \frac{1 - (\overline{\mu}/\overline{\mu'})}{1 + j\omega\tau_{eff}} \qquad (5.48)$$

where G is the a.c. conductance for $\tau_{eff} = 0$. Thus the finite relaxation time τ_{eff}, which is a function of the bias field E_{h0}, has the effect of adding a capacitance $C_{eff} = \mathrm{Im}\, Y_m/\omega$ to the geometrical capacitance C_h of the sample. The real part of the a.c. admittance, on the other hand, defines the load resistance $R_{L,m} = -1/\mathrm{Re}\, Y_m$ that sustains a particular microwave oscillation. For the average static mobilities we set approximately $\overline{\mu'}/\mu_1 \approx -R_1/R_L$ and $\overline{\mu} \approx v_{h0}/E_{h0}$, thus

$$-(\overline{\mu}/\overline{\mu'}) \approx \frac{R_L}{R_1} \frac{v_{h0}}{\mu_1 E_{h0}}. \qquad (5.49)$$

Here, $R_L = -1/G$ is the load resistance for $\tau_{eff} = 0$, the case treated in the previous sections. The static-mobility ratio $\overline{\mu}/\overline{\mu'}$ is a function both of the bias field E_{h0} and of the r.f. amplitude E_h through R_L. One readily obtains from eqn (5.48) the frequency dependence of $R_{L,m}$ and C_{eff} for a given ration $\overline{\mu}/\overline{\mu'}$:

$$R_{L,m} = R_L \frac{1 + (\omega\tau_{eff})^2}{1 + (\overline{\mu}/\overline{\mu'})(\omega\tau_{eff})^2} \qquad (5.50a)$$

178

$$C_{\text{eff}} = \frac{\tau_{\text{eff}}}{R_{\text{L}}} \frac{1 - \overline{\mu/\mu'}}{1 + (\omega\tau_{\text{eff}})^2}. \tag{5.50b}$$

The microwave load resistance $R_{\text{L,m}}$ increases, whereas the effective capacitance C_{eff} decreases, with increasing frequency ω. The simplicity of the model justifies the assumption that the d.c. power is unchanged by the fact that τ_{eff} is finite. For a sinusoidal voltage waveform the r.f. power is $P = (1/2)\hat{E}_{\text{h}}^2/R_{\text{L,m}}$ and hence we may conclude that the microwave efficiency η_{m} is inversely proportional to $R_{\text{L,m}}$ for a given r.f. amplitude $\hat{E}_{\text{h}}$. With eqn (5.50a) this yields directly

$$\eta_{\text{m}} = \eta \frac{R_{\text{L}}}{R_{\text{L,m}}} = \eta \frac{1 + (\overline{\mu/\mu'})(\omega\tau_{\text{eff}})^2}{1 + (\omega\tau_{\text{eff}})^2} \tag{5.51}$$

where η is the efficiency at the load resistance R_{L} for $\tau_{\text{eff}} = 0$. Thus the results of the homogeneous-field LSA model described in Section 5.2.4 are readily adjusted for microwave frequencies by using eqns (5.49)–(5.51). An example for the efficiency calculated from eqn (5.51) is shown in Fig. 5.28 (solid line). The efficiency drops sharply at higher frequencies reaching zero (equivalent to $R_{\text{L,m}} = \infty$) at the intrinsic cut-off frequency

$$f_{\text{TE}} = \frac{1}{2\pi\tau_{\text{eff}}\sqrt{(-\overline{\mu/\mu'})}} \tag{5.52}$$

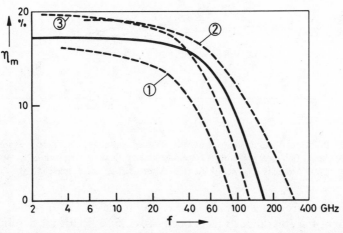

Fig. 5.28 Efficiency decrease for GaAs devices at higher microwave frequencies caused by finite electron relaxation times at bias $E_{\text{ho}} = 10 \text{ kV cm}^{-1}$. Solid line: eqn (5.51) with $\tau_{\text{eff}} \approx \tau^{(1)} = 0.8$ ps (from Fig. 5.5c), $\eta = 17.2\%$ ($R_{\text{L}} = 11 R_1$ or $\hat{E}_{\text{h}} = 7.8 \text{ kV cm}^{-1}$ from Figs 5.15 and 4.1), and $\overline{\mu/\mu'} = -1.43$ [from Figs 4.2 and 5.15b according to eqn (5.49)]. Dashed lines: (1) after Butcher and Hearn [5.58]; (2) after Ohmi *et al.* [5.60]; (3) after Curtice and Purcell [4.15] (all dashed curves for r.f. amplitude $\hat{E}_{\text{h}} = 8 \text{ kV cm}^{-1}$).

which is 166 GHz for the numerical values of Fig. 5.28. The analytical formula is compared with curves (dashed) obtained (1) from rigorous calculations based on the time-dependent Boltzmann equation using displaced-Maxwellian distribution functions for the electrons in each valley [5.58], cf. [5.59]; (2) from a two-carrier model with finite relaxation time between the two types of carrier [5.60]; (3) from a simplified two-valley model with finite energy relaxation time [4.15]. The analytical formula (5.51) appears to be a rather good approximation; for ensuring optimum agreement one even could consider the quantities $\bar{\mu}/\bar{\mu}'$ and τ_{eff} as adjustable parameters. A qualitatively similar drop in efficiency has been obtained for InP in calculations making use of the Monte-Carlo technique [5.92].

Figure 5.29 shows some dynamic $v(E)$ characteristics at microwave frequencies and the corresponding drift-velocity (current) waveforms for a sinusoidal r.f. terminal

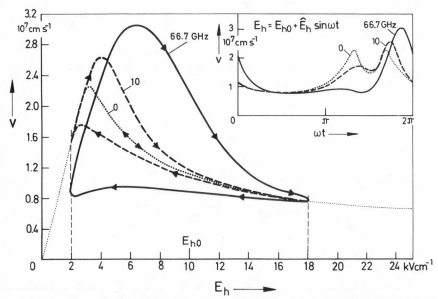

Fig. 5.29 Several dynamic GaAs drift-velocity/field characteristics and corresponding drift-velocity (current) waveforms for sinusoidal r.f. field with amplitude $\hat{E}_h = 8$ kV cm^{-1} at bias $E_{h0} = 10$ kV cm^{-1}. After Curtice and Purcell [4.15].

field as obtained from the simplified two-valley model [4.15]. A similar behaviour was found for the other models mentioned above [5.58–5.60] and, additionally, for the iterative method of calculating the electron distribution function [5.16]. The phase shift of the current waveforms with increasing frequency is clearly seen. This shift is a consequence of the capacitive contribution of electron relaxation and causes the simultaneous reduction of the negative sample conductance.

The finite scattering times of the electrons also influence the domain modes, in a qualitatively similar manner. Detailed calculations have, to the authors' knowledge, not yet been performed. It has been shown, however, that this factor influences the steady-state shape of the travelling domains [5.61].

5.2.6 Power-Impedance Product

It has become customary to compare the r.f.-power/frequency limitation of various types of microwave solid-state device in terms of their r.f.-power/impedance product at the maximum operating frequency f_{max}, the latter considered to be determined by the geometrical dimensions of a particular device chip. This product serves as a good figure-of-merit for the principal r.f. power handling capabilities as a function of f_{max} since it depends solely on basic material parameters. It does not, however, include an assessment of device performance with regard to d.c./r.f. power conversion efficiency and heat dissipation. Calculations of the power-impedance product have been carried out for bipolar transistors [5.62, 5.63], varactors [5.63, 5.64], tunnel diodes [5.63], and IMPATT diodes [5.63], as well as for TE devices [3.10, 5.65, 5.66].

Generally, for sinusoidal r.f. voltages across the active semiconductor, the power-impedance product of a negative-resistance device is given simply by

$$PR_L = (1/2)\hat{V}^2. \tag{5.53}$$

Here, the reactive properties of the device are considered to be part of the circuit and, hence, the modulus of the device 'impedance' is equal to the load resistance, $|R| = R_L$. According to eqn (5.53) it is the limitation imposed on the r.f. voltage amplitude $\hat{V}$ that determines the maximum of PR_L. Because of its principal bulk nature one has $\hat{V} = \hat{E}_h L$ for a TE device. The maximum device length L is directly related to the maximum operating frequency f_{max}. The terminal electric-field amplitude $\hat{E}_h$, on the other hand, is ultimately limited either by the existence of a finite valley field E_v in the $v(E)$ characteristic or by dielectric breakdown occurring in the region of maximum field within the device. The latter mechanism causes the limitation in TE devices made from GaAs.

For the delayed dipole-domain mode, one has to stipulate $\hat{E}_h \leq E_{h0} - E_S$ to avoid domain quenching, and f_{max} is, as discussed in Section 5.1, equal to the transit-time frequency, eqn (5.1). Hence eqn (5.53) becomes [5.65]

$$PR_L = (E_{h0} - E_S)^2 v_D^2 / 2 f_{max}^2. \tag{5.54}$$

In the case of those modes which are not transit-time limited, i.e. quenched (single and multiple) dipole-domain, hybrid and LSA oscillations, PR_L remains a function of the $n_0 L$ product if the appropriate eqns (5.6), (5.7), (5.9) and (5.11) are used to express L by the maximum operating frequency. Allowing $n_0 L \to \infty$ another length limiting mechanism sets in as a result of the finite propagating velocity c of electromagnetic waves within the device material: to ensure phase coherent oscillation of the homogeneous part of the sample field one has to impose a restriction* $L \lesssim \lambda/4$

---

*This restriction implies that the device is to be operated in a fundamental-mode closed resonator. For operation in an open plane-wave resonator it should be possible to increase this dimension considerably [4.6a].

where $\lambda = c/f_{max}$ [5.66]. For reasonable efficiency one requires $\hat{E}_h \lesssim E_{h0}$. Hence, from eqn (5.53) follows

$$PR_L = E_{h0}^2 c^2 / 32 f_{max}^2. \tag{5.55}$$

At a given bias field $E_{h0} \gg E_s$, the power-impedance product of the modes which are not transit-time limited is thus larger by a factor of $(c/4\bar{v}_D)^2$ than its corresponding value for transit-time operation. For the frequencies of interest (1–100 GHz) the influence of the conductive properties of the semiconductor on electromagnetic wave propagation is negligible [5.66] yielding $c = c_0/\sqrt{(\varepsilon/\varepsilon_0)}$, where c_0 is the free-space velocity of light. Thus one obtains for GaAs the factor $(c/4\bar{v}_D)^2 \approx 5 \times 10^4$.

As already mentioned, the maximum tolerable bias field E_{h0} in GaAs is determined by dielectric breakdown at the maximum terminal field $E_h = E_{h0} + \hat{E}_h \approx 2E_{h0}$. Its magnitude depends largely on whether or not a high-field domain forms during an oscillating cycle and on the nature of such a domain. We consider here only the two extreme cases of a triangular steady-state domain in the delayed dipole-domain mode and an entirely homogeneous field in the LSA mode. The latter case readily yields $E_{h0} \lesssim E_{bd}/2$ where E_{bd} is the dielectric breakdown field. In the former case one needs a relationship between the terminal field and the maximum domain field $E_{2\infty}$, as given in Fig. 3.11. This figure demonstrates that the condition $E_{2\infty} \lesssim E_{bd}$ severely reduces the maximum possible bias field $E_{h0} \approx E_h/2$, particularly at higher n_0L products (see Fig. 5.26 for experimental evidence!). In Section 6.4 breakdown phenomena occurring in the domain will be discussed in more detail (Fig. 6.25).

The breakdown field E_{bd} is a slightly falling function of device length L in the LSA mode, or of domain width b_∞ in the delayed dipole-domain mode [5.66]. Since b_∞ itself is related to L (Fig. 3.12) and L is to be expressed by the maximum operating frequency f_{max}, the decrease in power-impedance product with f_{max} becomes somewhat less pronounced than indicated explicitly by eqns (5.54) and (5.55). If we consider only a limited frequency range, e.g. between about 1 and 100 GHz, this complication can be neglected and E_{bd} can be approximated by a constant material parameter which for GaAs is $E_{bd} \approx 180$ kV cm^{-1} (cf. Section 6.4). Using this value one obtains a maximum bias field $E_{h0} \approx 90$ kV cm^{-1} for the homogeneous LSA mode and, assuming a device with $n_0L \approx 2 \times 10^{12}$ cm^{-2}, a field $E_{h0} \approx 25$ kV cm^{-1} for the delayed dipole-domain mode. The corresponding plots of the power-impedance product, PR_L, against the maximum operating frequency, f_{max}, are given in Fig. 5.30a. They are compared with similar plots for some silicon p-n junction devices. Whereas the transit-time limited TE oscillator, as a consequence of the high domain fields, is somewhat inferior to the silicon devices, the LSA oscillator which is not limited by transit time by far exceeds their performance. Results similar to those for the LSA oscillator follow for an arrangement consisting of a direct series connection of several TE devices with subcritical n_0L product [4.41]. Such a configuration can be operated both as oscillator and amplifier.

The absolute r.f. power level available from a device follows from the plots of Fig. 5.30a if the optimum (i.e. minimum possible) load resistance R_L is known. As

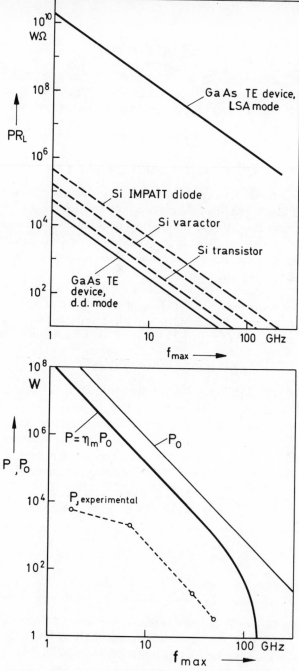

Fig. 5.30 (a) Theoretical power-impedance product as a function of the maximum operating frequency f_{max} for the delayed dipole-domain (d.d.) mode and the LSA mode of a TE GaAs device being dielectric-breakdown limited, in comparison with the performance of various Si pn junction devices (transistor is bipolar type). Curves for Si devices taken from [5.63], with a breakdown field of 200 kV cm^{-1}, a carrier saturation velocity of 6×10^6 cm^{-1}, and an avalanche parameter of $m = 6$ (R_L is interpreted as modulus of device impedance). (b) Theoretical frequency behaviour of the achievable r.f. power P in an LSA GaAs device with its area limited by skin, and standing wave, effects; parameters chosen: $\sigma_1^{-1} = 0.5\,\Omega$ cm, $E_{h0} = 25$ kV cm^{-1}, $R_L/R_1 = 50$. Experimental values from [5.57, 5.67, 5.68].

we have seen in the previous section, this resistance becomes infinite and hence $P \to 0$ at the cut-off frequency of the TE effect. To show this vanishing of the active properties in the case of a GaAs LSA oscillator we examine the maximum r.f. power in the form

$$P(f) = \eta_m(f)P_0$$

with

$$P_0 = J_0 E_{h0} LA \approx 0.75 \, E_p E_{h0} \sigma_1 LA. \tag{5.56}$$

Here, $\eta_m(f)$ is the frequency-dependent d.c./r.f. conversion efficiency as approximated by eqn (5.51), σ_1 the low-field conductivity and LA the total volume of the active semiconductor bulk. In addition to the limitation on L discussed before, i.e. $L \lesssim \lambda/4$, we now have to introduce appropriate limitations for the current-carrying area $A = wd$. One dimension, say d, is limited by field variations inside the crystal due to skin effect and standing waves [6.70b, 6.76]; the other, w, is again assumed to be limited by the $\lambda/4$ criterion, i.e. $w \lesssim \lambda/4$. If the field variations within the crystal are required to be restricted to 10% [5.66], an LSA device of $\sigma_1^{-1} = 0.5 \, \Omega\text{cm}$ and $R_L/R_1 = 50$ is limited to $d \leq 1.3 \, \text{cm GHz}/f$. With these parameters and by choosing, conservatively, a bias field of $E_{h0} = 25 \, \text{kV cm}^{-1}$, we obtain numerically from eqn (5.56)

$$P_0(f_{max}) = 7.7 \times 10^8 \, \text{W GHz}^3/f^3{}_{max}.$$

This function is plotted in Fig. 5.30b together with $P(f_{max}) = \eta_m(f_{max}) \times P_0(f_{max})$ where η_m is taken from eqn (5.51) with $\eta = 18\%$, $\bar{\mu}/\bar{\mu}' = -2.1$ and $\tau_{eff} = 0.8 \, \text{ps}$. It should be pointed out that another practical area limitation becomes important at low frequencies. The necessary load resistance R_L may fall below values that can be realized, since $R_1 = 1/\sigma_1 d = 0.38 \, \Omega f/\text{GHz}$. This problem arises for frequencies below 1 GHz. Figure 5.30b also shows the following experimental values:

P/W	$\eta_m/\%$	f/GHz	Ref.
6 000	14·6	1·75	[5·57]
2 100	5	7	[5·57]
20	7·5	30	[5·67]
3·5	0·5	50	[5·68]

In C.W. or high-duty-cycle pulsed operation, limitations caused by heat removal problems (Section 6.3) become dominant, particularly at lower frequencies. This severely decreases the obtainable r.f. power levels (Section 7.1.2.1).

5.3 NOISE PERFORMANCE

5.3.1 General Remarks
A basic noise contribution in TE amplifiers and oscillators originates from statistical drift-velocity fluctuations of the charge carriers caused by the various scattering

processes. In an (ohmic) resistance R_Ω at thermal equilibrium these carrier velocity fluctuations give rise to a noise voltage u (thermal Nyquist, or Johnson noise) which obeys the relation, cf. [5.71]

$$(\overline{u^2}/\Delta f) = S_u = 4k_B T_0 R_\Omega. \tag{5.57a}$$

Here, $\overline{u^2}$ is the mean square of the open-circuit voltage fluctuations across the resistance R_Ω at a temperature T_0, and Δf is the bandwidth for which $\overline{u^2}$ is determined. The quantity S_u denotes the spectral noise density and corresponds to an average power per unit frequency. An expression equivalent to eqn (5.57a) is

$$(\overline{i^2}/\Delta f) = S_i = 4k_B T_0 G_\Omega \tag{5.57b}$$

where $\overline{i^2}$ is the mean square of the short circuit current fluctuations in the conductance $G_\Omega = 1/R_\Omega$ whereas S_i denotes the corresponding spectral density of the mean-square current. The highest noise power *available* from the resistance R_Ω follows directly from the matched circuit condition as

$$P_N = \frac{\overline{u^2}}{4R_\Omega} = \frac{\overline{i^2}}{4G_\Omega} = k_B T_0 \Delta f. \tag{5.58}$$

The temperature T_0 can be expressed by the diffusion constant D and the mobility μ of the fluctuating carriers using Einstein's relation, eqn (2.8),

$$T_0 = \frac{eD}{k_B \mu}. \tag{5.59}$$

Our discussion on (thermal) noise in active TE semiconductors devices that are operating in a steady state (but not, of course, in thermal equilibrium) will be based on the following assumptions:

(1) A quasi-equilibrium condition exists that allows us to define a 'noise temperature' T_N.

(2) The noise temperature T_N is determined by an average diffusion constant D and the differential microwave mobility μ'_m of the average drift velocity of the carriers in the different valleys in analogy to eqn (5.59) cf. [5.76a], viz.

$$T_N = \frac{eD}{k_B |\mathrm{Re}\mu'_m|}. \tag{5.60}$$

At frequencies $f \ll f_{TE}$ one simply has $\mathrm{Re}\mu'_m = \mu'$. Besides the contributions from the different valleys as manifested in eqn (2.9), D contains also components from diffusion processes associated with carrier scattering between the different valleys ('straggling diffusion', [5.70, 5.72], cf. also [5.90]). For frequencies approaching the reciprocal of the scattering time constant a generalized frequency-dependent diffusion constant $D(\omega)$ results.

(3) The Nyquist noise formula (5.57) or (5.58) can be applied for the noise

temperature T_N, replacing R_Ω by the absolute value of the real part of the dynamic impedance $Z(\omega)$ of the semiconductor volume considered. This leads to

$$S_u = 4k_B T_N |\mathrm{Re} Z| \tag{5.61a}$$

$$S_i = 4k_B T_N |\mathrm{Re} Y| \tag{5.61b}$$

$$P_N = k_B T_N \Delta f. \tag{5.62}$$

The generalized thermal noise introduced in this way will be called 'diffusion noise' [5.71b].

(4) An additional noise contribution in a negative-mobility semiconductor results because of diffusion noise amplified by the mechanism of space-charge-wave growth.

The above concepts can be justified by a rigorous study of the noise processes in bulk semiconductors under non-equilibrium conditions using the concept of the 'impedance field' in a semiconductor medium [5.70, 5.72–5.74].

5.3.2 Amplifier Noise

We start with a short review of general noise relationships cf. [5.71]. The (spot) noise figure F of an amplifier, as a linear two-port device connected between signal source and load, is defined as the ratio of the noise output power of the actual noisy amplifier to the noise output power obtained in the hypothetical case of the amplifier assumed to be noiseless, i.e.

$$F = \frac{P_{N,2}}{P_{NS,2}} = 1 + \frac{P_{NA,2}}{P_{NS,2}} = 1 + \frac{P_{NA,2}}{g P_{NS,1}} \tag{5.63}$$

where 'output' is characterized by the subscript 2, 'input' by the subscript 1. The quantity $P_{N,2} = P_{NS,2} + P_{NA,2}$ represents the total noise output power consisting of a contribution from the signal source at standard temperature, $P_{NS,2}$, and from the amplifier itself, $P_{NA,2}$. The power gain of the amplifier is denoted by g. In addition to the noise figure F, the noise measure

$$M = \frac{F-1}{1-1/g} = \frac{P_{NA,2}}{(g-1)P_{NS,1}} \tag{5.64}$$

is a widely-used figure-of-merit.

We now specify eqns (5.63) and (5.64) for a reflection-type amplifier (see Fig. 5.31a) which is characterized by identical values of load impedance Z_L and source impedance Z_S as embodied by the appropriate circulator port. Noting that $g = rr^\star$, where $r = (Z - Z_S)/(Z + Z_S)$ is the complex reflection coefficient and the star indicates the complex-conjugate quantity (cf. Section 7.2.1), one obtains by using the formulae of Fig. 5.31a

$$F = 1 + \frac{\overline{u_A^2}}{\overline{u_S^2}} \frac{4|Z_S|^2}{|Z - Z_S|^2}$$

186

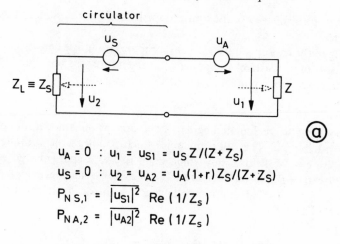

$$u_A = 0 \; : \; u_1 = u_{S1} = u_S Z/(Z+Z_S)$$
$$u_S = 0 \; : \; u_2 = u_{A2} = u_A(1+r)Z_S/(Z+Z_S)$$
$$P_{NS,1} = \overline{|u_{S1}|^2} \; \text{Re} \, (1/Z_s)$$
$$P_{NA,2} = \overline{|u_{A2}|^2} \; \text{Re} \, (1/Z_s)$$

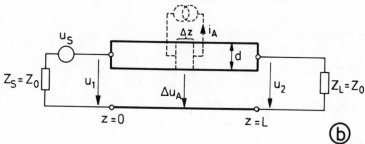

$$i_A = 0 \; : \; u_1 = u_{S1} = u_S/2$$
$$u_S = 0 \; : \; u_2 = u_{A2}$$
$$P_{NS,1} = \overline{|u_{S1}|^2} \; \text{Re} \, (1/Z_0)$$
$$P_{NA,2} = \overline{|u_{A2}|^2} \; \text{Re} \, (1/Z_0)$$

Fig. 5.31 Noise equivalent circuit of a TE amplifier and basic relations for noise voltages: (a) reflection-type amplifier (dotted arrows indicate direction of wave propagation), (b) travelling-wave transmission amplifier.

or

$$M = \frac{\overline{u_A^2}}{\overline{u_S^2}} \; \frac{2|Z_S|^2}{-ZZ_S^* - Z^* Z_S} \cdot$$

For the high-gain condition $g \to \infty$, i.e. $Z_S \approx -Z$, this simplifies to

$$M = F - 1 = \frac{\overline{u_A^2}}{\overline{u_S^2}} = \frac{S_{uA}}{S_{uS}} \tag{5.65}$$

where

$$S_{uS} = 4k_B T_0 \text{Re} Z_S = 4k_B T_0 |\text{Re} Z| \tag{5.66}$$

187

is the spectral noise density of the source impedance Z_S at the standard temperature $T_0 = 290$ K.

Referring to the concepts outlined above, the spectral noise density of a stable TE device consists of a 'resistive' component S_{uR} given by eqns (5.61a) and (5.60) and a 'space-charge-wave' component S_{uw}, i.e.

$$S_{uA} = S_{uR} + S_{uw}. \tag{5.67}$$

The component S_{uw} is calculated here in a one-dimensional model for *homogeneous* d.c. field distribution. The diffusion noise fluctuations at the cathode boundary cause a noise electric field $E_N(\omega)$ which gives rise to a space-charge wave of the form $-\delta E_N \exp(-jkz)$ as outlined in Section 4.3.2 for sinusoidal fields (cf. eqn (4.18) with $\hat{E}_a = -\hat{E}_b = \delta E_N$]. The noise voltage across the device terminals is obtained by integration as

$$-u_w = \delta E_N \int_0^L \exp(-jkz) = \frac{j\delta E_N}{k} [\exp(-jkL) - 1]. \tag{5.68}$$

Thus the mean-square value becomes

$$\overline{|u_w|^2} = \frac{\overline{\delta E_N^2}}{|k|^2} (1 + \exp 2\alpha L - 2 \exp \alpha L \cdot \cos \beta_e L). \tag{5.69}$$

The mean-square noise field $\overline{\delta E_N^2}$ follows from the available diffusion noise power, eqn (5.62). The electrostatic energy density of the wave at the cathode is $\varepsilon\sigma \overline{E_N}/2$ leading to the power

$$P_N = \frac{1}{2} \varepsilon \overline{\delta E_N^2} v_{h0} A \tag{5.70}$$

carried by the wave with the group velocity v_{h0}. Equating expression (5.70) to expression (5.62) one obtains

$$\frac{\overline{\delta E_N^2}}{\Delta f} = \frac{2k_B T_N}{v_{h0} A} \tag{5.71}$$

and hence

$$S_{uw} = \frac{\overline{|u_w|^2}}{\Delta f} = \frac{2k_B T_N}{|k|^2 \varepsilon v_{h0} A} (1 + \exp 2\alpha L - 2 \exp \alpha L \cdot \cos \beta_e L). \tag{5.72}$$

Since S_{uR} is given by eqn (5.61a) the noise measure M becomes according to eqns (5.65), (5.66), (5.67) and (5.72)

$$M = \frac{T_{Neq}}{T_0} \tag{5.73a}$$

with

$$T_{Neq} = T_N \left(1 + \frac{R_w}{|\text{Re}Z|}\right) \tag{5.73b}$$

where we have introduced the abbreviation

$$R_W = \frac{1 + \exp 2\alpha L - 2 \exp \alpha L \cdot \cos \beta_e L}{2|k|^2 \varepsilon v_{h0} A}.$$ (5.73c)

The real part of the r.f. device impedance, ReZ, follows from eqn (4.48) and the wave propagation constant k from equation (4.21) if the influence of diffusion on space-charge-wave growth is negligible [5.72, 5.73]. For small space-charge-wave gain $\alpha L < 1$ (small $n_0 L$ product) one has $R_W \ll |$Re$Z|$ in the frequency range of substantial *negative* device resistance ReZ (where $\cos \beta_e L \approx 1$) and hence $M = T_N/T_0$ or, with eqn (5.60),

$$M = \frac{T_N}{T_0} = \frac{eD}{k_B T_0 |\mu'|}$$ (5.74)

if the considered frequencies are sufficiently below the intrinsic cut-off frequency f_{TE}. Thus a TE material with small diffusion constant D and large negative differential mobility μ' is required in order to obtain a low noise measure M. For GaAs at $T_0 = 300$ K, exhibiting a maximum $|\mu'_{h0}|$ of 2 000 cm^2 (Vs)$^{-1}$ (at $E_{h0} \approx 6$ kV cm^{-1}) and $D_{h0} \approx 400$ cm^2 s^{-1} one finds $M \approx 9$ dB. Figure 5.32 shows the noise measure M

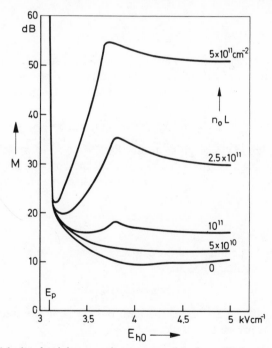

Fig. 5.32 Calculated minimum noise measure M of a reflexion-type TE GaAs amplifier as a function of bias field E_{h0} for different $n_0 L$ products, based on a $v(E)$ characteristic with a peak field of $E_p = 3 \cdot 1$ kV cm^{-1}. After Maloberti and Svelto [5.73].

at somewhat lower bias field values E_{ho} for different n_0L products as evaluated from the rigorous eqn (5.73), with the transit angle $\theta = \beta_e L$ adjusted to yield a minimum value of M [5.73]. In this evaluation $v(E)$ and $D(E)$ were chosen according to the measured GaAs curves (1) of Figs 2.9 and 2.10*. It is seen that M strongly increases with increasing n_0L product at the higher bias fields. Lowest noise figures F measured in a TE reflection-type amplifier are 15 dB for GaAs [5.75] and 7·5 dB for InP [5.76a]. For *non-uniform* d.c.-field distribution the noise figure theoretically increases, since the maximum of $|\mu'|$ can be established only locally within the semiconductor [5.72, 5.73]. This agrees with experimental findings [5.76b].

Turning now to the unilateral TE travelling-wave amplifier of the transmission type (see Fig. 5.31b), an internal noise measure M_i is calculated by using the concept of characteristic impedance Z_0 of space-charge waves in a guiding structure [5.74]. The impedance Z_0 is defined as the ratio of an equivalent 'waveguide voltage' to the carrier current. With the power gain $g = \exp 2\alpha L$ (cf. Sections 4.3 and 7.2.2) and the formulas of Fig. 5.31b we obtain from eqn (5.64)

$$\left. \begin{aligned} M_i &= \frac{4\overline{|u_{A2}|^2}}{|u_S|^2 \,(\exp 2\alpha L - 1)} \\[2mm] &= \frac{4S_{uA2}}{S_{uS} \,(\exp 2\alpha L - 1)} \end{aligned} \right\} \tag{5.75}$$

where S_{uS} is again given by eqn (5.66) with $Z_S = Z_0$. The spectral density S_{uA2} is calculated by summing over the noise contributions from the uncorrelated diffusion noise sources i_A in small segments Δz of the amplifier. One has according to eqn (5.61b)

$$S_{iA} = \overline{i_A^2}/\Delta f = 4k_B T_N |\sigma_z'| wd/\Delta z \tag{5.76}$$

if the frequencies are low enough to ensure $\text{Re}\,\sigma_m' \approx \sigma'$. The noise current i_A, as a carrier current, propagates through a segment Δz in form of a space-charge wave, i.e. $i_A \propto \exp(-jk_z z)$. The net noise contribution that emanates from the segment is therefore

$$\Delta i_A = \frac{\partial i_A}{\partial z} \Delta z = -jk_z i_A \Delta z \tag{5.77}$$

and, according to the definition of the characteristic impedance Z_0, the corresponding equivalent noise voltage becomes

$$\Delta u_A = Z_0 \Delta i_A. \tag{5.78}$$

When this noise voltage arrives at the terminal $z = L$ it has travelled a distance $L - z$, hence it has grown to

$$\Delta u_{A2} = \Delta u_A \exp[jk_z(z - L)]. \tag{5.79}$$

---

*Calculated noise measures based on theoretical diffusion constants are considerably lower; see Kallback, B., *Electron. Lett.* **9** (1973), p. 11.

Because the noise voltages from each segment are uncorrelated we have to sum up the mean-square values, i.e.

$$\overline{|u_{A2}|^2} = \sum_{\Delta z} \overline{|\Delta u_{A2}|^2} \tag{5.80}$$

Combining eqns (5.76) to (5.80) and replacing the summation by an integration from $z = 0$ to $z = L$ one eventually arrives at

$$S_{uA2} = \frac{\overline{|u_{A2}|^2}}{\Delta f} = 2|k_z|^2 |Z_0|^2 k_B T_N |\sigma_z'| wd \frac{\exp 2\alpha L - 1}{\alpha} \tag{5.81}$$

where the wave-vector component k_z in eqn (5.79) has been expressed by $k_z = \beta + j\alpha$. By inserting the derived S_{uA2} and the quantity S_{uS} of eqn (5.66) into eqn (5.75) we obtain for the internal noise measure

$$M_i = \frac{2|k_z|^2 |Z_0|^2 |\sigma'| wd}{\alpha \mathrm{Re} Z_0} \frac{T_N}{T_0}. \tag{5.82}$$

This equation simplifies considerably for an amplifier that consists of a thin TE-active layer of GaAs grown on a passive semi-insulating substrate. In this case one has $\beta \approx \beta_e \gg \alpha \approx \beta_e \alpha_{Rz} d$ [cf. Section 4.3.3, particularly the paragraph below eqn (4.44a): note $\varepsilon_I = \varepsilon_{II} = \varepsilon$ and $a \approx d$ since the permittivity of air above the epitaxial layer is negligibly small compared with that of GaAs]. Additionally, the characteristic impedance Z_0 is real [5.74], viz. $Z_0 = 1/(\omega \varepsilon w)$. Keeping in mind the definitions (4.20) and (4.30) for β_e and α_{Rz}, eqn (5.82) simplifies to [5.74]:

$$M_i = 2\frac{T_N}{T_0} = 2\frac{eD}{k_B T_0 |\mu'|}, \tag{5.83}$$

where T_N has been expressed by eqn (5.60) with $\mathrm{Re}\mu_m' \approx \mu'$. This is just twice as high as the noise measure of the high-gain reflexion amplifier with small $n_0 L$ product, eqn (5.74). For a GaAs device with homogeneous d.c. field distribution at $E_{h0} = 6$ kV cm^{-1} one finds for instance $M_i \approx 12$ dB. The *external* noise measure M is expected to be larger due to unavoidable coupling losses at the r.f. input port of the amplifier. This reduces the output noise of the source impedance simultaneously with the r.f. signal, but not the noise generated by the amplifier. Experimental noise figures are still scarce, being of the order of 30 dB for GaAs devices [5.77, 5.78].

5.3.3 Oscillator Noise

As pointed out at the beginning of Section 5.1.2.1 only resonator-controlled TE oscillators exhibit noise levels sufficiently low for usual microwave applications. Noise fluctuations exert their influence in such oscillators by frequency (FM) and amplitude modulation (AM) of the generated r.f. signal ('carrier'). Since the amplitude of a negative-resistance oscillator is relatively stable because of the strong non-linearities of the negative resistance, FM noise is generally more severe. One distinguishes between base-band noise originating from impedance fluctuations at low frequencies and r.f. noise caused by voltage or current fluctuations at frequencies in the vicinity of the carrier frequency, e.g. [5.79]. In the former case the modulation of the carrier is caused by a mixing process, in the latter case by phase-locking action, e.g. [1.31]. The inherent properties of the combination of semiconductor sample and circuit give rise to a correlation between AM and FM noise: the non-linearity of the sample reactance converts AM to FM noise, whereas the frequency dependence of the load resistance, to the contrary, converts FM to AM noise, e.g. [5.79].

We confine ourselves here mainly to a discussion of r.f. noise using the classical formulas for negative-resistance oscillators derived by Edson [5.80]. If the available noise power in a given frequency band Δf_m is denoted by P_N, the noise power that is frequency-modulated upon the carrier in a single side band reads, cf. eqn (7.5),

$$P_{N, FM} = \frac{1}{2Q_L^2} \left(\frac{f}{f_m} \right)^2 P_N \tag{5.84}$$

where f is the carrier frequency, f_m the separation of the r.f. noise frequency from the carrier ('modulation frequency') and Q_L the loaded quality factor of the circuit with the device susceptance included. Thus, for low FM-noise performance high-Q circuits are required. For the root-mean-square frequency deviation, cf. [5.81],

$$(\delta f)_{rms} = f_m \sqrt{\frac{2P_{N, FM}}{P}} \tag{5.85}$$

follows, with eqn (5.84),

$$(\delta f)_{rms} = \frac{f}{Q_L} \sqrt{\frac{P_N}{P}} \tag{5.86}$$

where P is the power at the carrier frequency f. The analogous equation for amplitude-modulation noise reads

$$P_{N, AM} = \frac{2P_N}{s^2 + 4Q_L^2(f_m/f)^2}. \tag{5.87}$$

The quantity s denotes here a saturation parameter (usually of the order of 10) resulting from the non-linearities of the negative device resistance in the steady-state oscillating condition. It is responsible for the smallness of $P_{N, AM}$ as compared with $P_{N, FM}$.

Considering only diffusion noise, P_N follows from eqn (5.62), where T_N has to be calculated from a large-signal analogue to eqn (5.60) [5.82]. For this purpose we use eqn (5.65) for the noise measure of a reflection amplifier of high gain g, the equation being valid also for the negative-resistance oscillator (since $g = \infty$). In analogy to eqn (5.73a) we replace T_N by an equivalent temperature T_{Neg} since similar to the amplifier the oscillating TE semiconductor is in general not a homogeneous body with respect to the r.f. field distribution. Thus we have

$$P_N = k_B T_{Neq} \Delta f_m = k_B T_0 M \Delta f_m. \tag{5.88}$$

Considering the device susceptance as part of the load circuit, the resonance condition requires $Z_s = Z_L = R_L$ (cf. Fig. 3.22) and eqn (5.65) for $g = \infty$ becomes

$$M = \frac{S_{uA}}{S_{uS}} = \frac{S_{uA}}{4T_0 k_B R_L}. \tag{5.89}$$

For an LSA oscillator* (cf. equivalent circuit in Fig. 4.6) we neglect the field deformation caused by the accumulation layer and, hence, obtain

$$S_{uA} \approx 4k_B T_{Nh} |R_{ac}|$$

or, because $|R_{ac}| \approx R_L$,

$$M \approx T_{Nh}/T_0. \tag{5.90a}$$

Expressing T_{Nh} in analogy to eqn (5.60) we find

$$M = \frac{\overline{D_h}}{D_1} \frac{R_L}{R_1}, \tag{5.90b}$$

where $|\mathrm{Re}\ \mu'_m|/\mu_1$ has been replaced by its large-signal analogue $R_1/|R_{ac}| \approx R_1/R_L$, μ_1 by the low-field diffusion constant D_1 according to eqn (2.8) or (5.59), and D by some average value of the diffusion constant during one oscillation cycle, $\overline{D_h}$.

For a dipole-domain oscillator (cf. equivalent circuit in Fig. 3.21b) the capacitances C_h and C_D of low-field region and domain, respectively, are neglected and, hence, one obtains

$$S_{uA} = 4k_B(T_L R_1 + T_{ND}|R_D|). \tag{5.91}$$

Here, R_D is the dynamic domain resistance in the oscillating state. Thus eqn (5.89) becomes

$$M = \frac{R_1}{R_L} + \frac{\overline{D_D}}{D_1} \frac{|R_D|^2}{R_1 R_L}, \tag{5.92}$$

with T_{ND} being estimated from eqn (5.60) in a similar manner as for T_{Nh} above. D_D is an average diffusion constant within the domain during one r.f. cycle. Since generally one has $R_1 \ll R_L \approx |R_D|$ (cf. Figs 5.15 and 5.18) eqn (5.92) simplifies to

$$M \approx \frac{\overline{D_D}}{D_1} \frac{R_L}{R_1}. \tag{5.93}$$

---

*For a different, more rigorous analysis of FM noise in LSA oscillators see [5.88].

In this approximation the domain capacitance C_D can be considered again as part of the load circuit influencing only the resonant frequency. Thus a similar expression is obtained as eqn (5.90) valid for the LSA case. To minimize the noise measure, the normalized load resistance R_L/R_1 should be made as small as possible. This is the same condition as for maximum efficiency (Fig. 5.16a). Since an LSA oscillator can be 'loaded-down' more strongly (lower R_1/R_L) this type of oscillator is expected to have a somewhat lower noise measure M than a dipole-domain oscillator. Considering the latter for $R_L/R_1 = 20$ and assuming $\overline{D_D} \approx 2D_1$ [cf. curve (1) in Fig. 2.10] eqn (5.93) predicts a 'noise measure' ($g \to \infty$) of $M = 40 = 16$ dB.

As an experimental example we discuss a measured FM-noise spectrum of a Gunn domain oscillator (Fig. 5.33). Diffusion noise, at a fixed impedance, has a

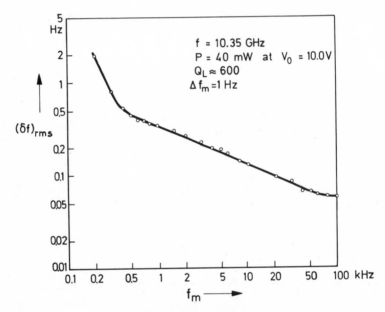

Fig. 5.33 Measured FM-noise spectrum of a GaAs CW dipole-domain oscillator. Noise data measured in 100 Hz and 1 kHz bandwidth and converted to $\Delta f_m = 1$ Hz bandwidth. After Omori [5.87].

frequency-independent spectral density [eqn (5.61)] and one speaks therefore of 'white' noise. Experimentally, white noise is observed in the FM-noise spectrum only at relatively large modulation frequencies f_m (Fig. 5.33). The values for M as calculated from the measured $(\delta f)_{rms}$ by using eqns (5.86) and (5.88) fall into a range that is consistent with eqn (5.93), [5.82]. From Fig. 5.33 one finds, for instance, a constant $(\delta f)_{rms} = 0.064$ Hz for $f_m > 60$ kHz. With the numerical oscillator parameters quoted in the figure this leads to $M \approx 21$ dB. This is somewhat higher than predicted by the simplified theory. Additional white noise is expected in a dipole-domain oscillator from time jitter in domain nucleation with the resonator exerting a filtering action [5.89, 8.84].

At lower modulation frequencies the experimental FM noise behaves as $(\delta f)_{rms} \propto 1/\sqrt{f_m}$, and assuming the validity of eqns (5.86), (5.88) and (5.89), the noise power density has to obey $S_{uA} \propto 1/f_m$ ('flicker' noise spectrum). It can be explained by base-band noise up-converted in a mixing process [5.79]. Correlation measurements between FM noise of a dipole-domain oscillator and low-frequency current noise of the non-oscillating device below threshold confirm this explanation [5.83–5.85]. Flicker noise may be the result of electron trapping with relaxation times distributed within a certain time interval. As a result of this, carrier-concentration fluctuations occur and these induce fluctuations in the dipole-domain velocity and hence in the device impedance [5.84]. The device contacts [5.85] as well as the device surface [5.86] have been revealed as additional sources for flicker noise. Trap states with a defined time constant influence the current noise spectrum by characteristically enhancing the noise below the trapping time constant [6.132]*. The relatively strong increase of the FM noise in the example of Fig. 5.33 at $f_m < 400$ Hz may be the result of such a trap state having a long time constant of about 3 ms. However, it could be also a contribution of the power supply [5.87].

An anomalously high current noise is observed in an extremely limited range of applied d.c. electric field near the TE threshold due to plasma instabilities [5.91]. This might be also a source of additional noise in TE oscillators in which the r.f. voltage swings below threshold.

REFERENCES

5.1 Copeland, J. A., Characterization of bulk negative resistance diode behaviour. *IEEE Trans. Electron Dev.* **ED–14** (1967), p. 461.

5.2 Bosch, B. G., Transferred-electron oscillators and amplifiers. *Proc. 7th Int. Conference MOGA, Hamburg, 1968, Nachrichtentechn. Fachber.* **35** (1968), p. 388.

5.3a Suematsu, Y., Nishimura, Y., Yamazawa, M. and Ikegami, T., A classification of behaviour of bulk negative mobility devices derived from a small-signal theory. *Proc. IEEE* **56** (1968), p. 1736.

5.3b Suematsu, Y., Nishimura, Y., Yamazawa, M. and Ikegami, T., Small-signal charac-teristics of Gunn-effect devices at frequencies higher than the transit-time frequency. *Electron. Commun. Japan* **52**–C (1969), p. 118.

5.4a Ridley, B. K., The inhibition of negative resistance dipole waves and domains in n-GaAs. *IEEE Trans. Electron Dev.* **ED–13** (1966), p. 41.

5.4b Sasaki, A. and Takagi, T., Conditions for space-charge wave growth and differential negative resistance in "two-valley" semiconductors. *Proc. IEEE* **55** (1967), p. 732.

5.5 Das, P. and Staecker, P. W., Upper frequency limit for Gunn oscillator imposed by carrier energy-relaxation time. *Electron. Lett.* **2** (1966), p. 258; also *Proc. 6th Int. Conference MOGA, Cambridge, 1966,* p. 274 (IEE Conf. Publication no. 27).

5.6a Vlaardingerbroek, M. T., Kuypers, W. and Acket, G. A., Energy relaxation time of hot electrons in GaAs. *Phys. Lett.* **28A**, (1968), p. 155.

---

*For details on noise caused by multi-level traps see De Cacqueray, A., Blasquez, G. and Graffeuil, J., *Solid State Electronics* **16** (1973), p. 853.

5.6b   Vlaardingerbroek, M. T., Boers, P. M. and Acket, G. A., High-frequency conductivity and energy relaxation of hot electrons in GaAs. *Philips Res. Repts* **24** (1969), p. 379.

5.7   Rees, H. D., Hot electron effects at microwave frequencies in GaAs. *Solid-State Commun.* **7** (1969), p. 267.

5.8   Zimmerl, O., Relative influence of relaxation times on the frequency behaviour of GaAs. *Phys. Stat. Sol. (A)* **1** (1970), p. 29.

5.9   Ohmi, T., A limitation on frequency of Gunn effect due to the intervalley scattering time. *Proc. IEEE* **55** (1967), p. 1739.

5.10   Grasl, L. M., Hillbrand, H. A. and Zimmerl, O. F., Comments on "A limitation on frequency of Gunn effect due to the intervalley scattering time". *Proc. IEEE* **56** (1969), p. 245.

5.11   Das, P. and Bharat, R., Hot electron relaxation times in two-valley semiconductors and their effect on bulk microwave oscillators. *Appl. Phys. Lett.* **11** (1967), p. 386.

5.12a Grasl, L. M. and Zimmerl, O. F., Small signal high-frequency conductivity of GaAs. *Arch. Elektr. Übertragung* **23** (1969), p. 165.

5.12b Kaneda, A. and Abe, M., Frequency dependence of microwave mobility of hot electrons in n-type GaAs. *J. Japan. Soc. Appl. Phys.* **40**, Supplement (1971), p. 125 (Proc. 2nd Conference on Solid State Devices, Tokyo, 1970).

5.13   Heinle, W., Small-signal relaxation times for the lattice scattering in n-type GaAs. *J. Appl. Phys.* **41** (1970), p. 2160.

5.14   Heinle, W., Displaced-Maxwellian transport calculation for InP and InAs$_{0.2}$P$_{0.8}$. *Phys. Lett.* **35A** (1971), p. 365.

5.15   Zimmerl, O. F., Frequency limitations of 3-level oscillators. *Electron. Lett.* **6** (1970), p. 728.

5.16   Rees, H. D., Time response of the high field electron distribution function in GaAs. *IBM J. Res. Dev.* **13** (1969), p. 537.

5.17   Zimmerl, O. F. and Grasl, L. M., The drifted Maxwellian approach to the Gunn effect in GaAs. *Arch. Elektr. Übertragung* **24** (1970), p. 100.

5.18   Grasl, L. M. and Zimmerl, O. F., Frequency behaviour of space-charge-wave amplifiers in GaAs. *Phys. Stat. Sol. (A)* **2** (1970), p. 391.

5.19   Fisher, R. E., Generation of subnanosecond pulses with bulk GaAs. *Proc. IEEE* **55** (1967), p. 2189.

5.20   Flemming, P. L., Self-modulation of pulsed GaAs oscillators. *Proc. IEEE* **54** (1966), p. 799.

5.21a Tsai, W. C. and Rosenbaum, F. J., Bias circuit oscillations in Gunn devices. *IEEE Trans. Electron Dev.* **ED–16** (1969), p. 196.

5.21b Abkevich, I. I., Self-modulation of microwave oscillations generated in Gunn diodes. In: *Physics of p-n Junctions and Semiconductor Devices* (Eds: S. M. Ryvkin and Yu. V. Shmartsev), p. 255 (New York, London, Consultants Bureau, 1971).

5.22a Hobson, G. S. and Martin, B., External negative differential (END) conductance of Gunn oscillators. *Proc. European Microwave Conference, London, 1969*, p. 219 (IEEE Conf. Publication no. 58).

5.22b Spiwak, R. R., Frequency conversion and amplification with an LSA diode oscillator. *IEEE Trans. Electron Dev.* **ED–15** (1968), p. 614.

5.22c Majborn, B., On the possibility of millimeter wave amplification with an X-band LSA oscillator. *Proc. European Microwave Conference, London, 1969*, p. 227 (IEEE Conf. Publication no. 58).

5.23a Thim, H. W., Linear microwave amplification with Gunn oscillators. *IEEE Trans. Electron Dev.* **ED–14** (1967), p. 517.

5.23b Abdel 'Fatakh, Kh. A. and Rzhevkin, K. S., Amplification of microwave oscillations within GaAs in the presence of domain generation (in Russian). *Radiotekh. i Electron.* **15** (1970). English translation: *Sov. Radio Eng.* and *Electron Phys.* **15** (1970), p. 1056.

5.24   Bosch, B. G. and Engelmann, R., Proposal for a Gunn-domain amplifier with increased power amplification and cut-off frequency (in German). *Arch. Elektr. Übertragung* **22** (1968), p. 411.

5.25a   German Federal Patent No. 1.541.703 (inventor: B. G. Bosch; priority: Dec. 8, 1966).

5.25b   Aitchison, C. S., Possible Gunn effect parametric amplifier. *Electron. Lett.* **4** (1968), p. 15.

5.25c   Bhattacharya, T. K., Dynamic elastance of a Gunn diode. *Electron. Lett.* **5** (1969), p. 408.

5.26a   Sasaki, A., Self-pumped parametric amplification and oscillation of Gunn-effect diodes. *Proc. IEEE* **59** (1971), p. 89.

5.26b   Sasaki, A., Transient behaviour and characteristics of the high-field domain in Gunn-effect diodes. *The Radio and Electronic Engineer* **39** (1970), p. 81.

5.26c   Sasaki, A., Parametric change in admittance of Gunn-effect diodes (in Japanese). Rept. 69–25 (1969) of Professional Group on Electron Devices of Jap. Inst. Electronics and Comm. Engrs.

5.27   Carroll, J. E., Resonant-circuit operation of Gunn diodes: a self-pumped parametric oscillator. *Electron. Lett.* **2** (1966), p. 215.

5.28a   Aitchison, C. S., Corbey, C. D. and Newton, B. H., Self-pumped Gunn-effect parametric amplifier. *Electron. Lett.* **5** (1969), p. 36.

5.28b   Borodovskii, P. A. and Buldygin, A. F., Amplification of microwave oscillations in a Gunn diode in which a travelling domain is periodically excited and annihilated by an external signal (in Russian). *Fiz. Tekh. Poluprov.* **5** (1971), p. 247. English translation: *Soviet Physics-Semiconductors* **5** (1971), p. 211.

5.29   Kuno, H. J., Self-pumped parametric amplification with GaAs transferred electron devices. *Electron. Lett.* **5** (1969), p. 232.

5.30   Frey, W., AEG-Telefunken, Research Institute, Ulm, private communication (1971).

5.31   Robson, P. N. and Hashizume, N., Resistive parametric mixing in GaAs oscillators at microwave frequencies. *Electron. Lett.* **6** (1970), p. 120.

5.32   Rees, H. D., Free-electron parametric effects in GaAs Gunn oscillators. *Electron. Lett.* **6** (1970), p. 11.

5.33   Penfield, Jr., P., Proposed high-efficiency diode oscillator. *Electron. Lett.* **5** (1969), p. 387.

5.34   Sterzer, F., Power output and efficiency of voltage controlled negative resistance oscillators. *IEEE Trans. Electron Dev.* **ED–14** (1967), p. 718.

5.35   Carroll, J. E., Criteria for the assessment of transferred electron materials. *Electron. Lett.* **6** (1970), p. 393.

5.36a   Hilsum, C. and Rees, H. D., Three-level oscillator: a new form of transferred electron device. *Electron. Lett.* (1970), p. 277.

5.36b   Rees, H. D. and Hilsum, C., Three-level transferred-electron effects in InP. *Electron. Lett.* **7** (1971), p. 437.

5.36c   Hilsum, C. and Rees, H. D., A detailed analysis of three-level electron transfer. *Proc. 10th Int. Conference Physics of Semiconductors, Cambridge, Mass., USA, Aug. 1970*, p. 45 (Washington, D.C., US Atomic Energy Commission, 1970).

5.36d   Hilsum, C., Paper presented at 1971 Cornell University Conference on High Frequency Generation and Amplification, Ithaca, N.Y.

5.37   Hilsum, C., Simple graphical method for calculating the efficiency of bulk negative-resistance microwave oscillators. *Electron. Lett.* **6** (1970), p. 448.

5.38a   Boers, P. M., Measurement of the velocity/field characteristic of indium phosphide. *Electron. Lett.* **7** (1971), p. 625.

5.38b   'T Lam, H. and Acket, G. A., Comparison of the microwave velocity/field characteristic of n-type InP and n-type GaAs. *Electron. Lett.* **7** (1971), p. 722.

5.38c   Boers, P. M., Gunn effect and related phenomena in InP. Paper presented at European Semiconductor Device Research Conference, Munich, March 16–19, 1971.

197

5.38d Nielsen, L. D., Microwave measurement of electron drift velocity in indium phosphide for electric fields up to 50 kV/cm. *Phys. Lett.* **38A** (1972), p. 221.

5.38e Glover, G. H., Microwave measurement of the velocity-field characteristic of n-type InP. *Appl. Phys. Lett.* **20** (1972), p. 224.

5.39 Geraghty, S. and Gibbons, G., Power-impedance calculations for InP and GaAs bulk-negative-resistance oscillators. *Electron. Lett.* **6** (1970), p. 583.

5.40 U.S. Patent No. 3,486,132 (inventor: S. P. Yu, priority: March 20, 1968).

5.41 Warner, F. L., Extension of the Gunn-effect theory given by Robson and Mahrous. *Electron. Lett.* **2** (1966), p. 260.

5.42 Pollmann, H., Calculations on the load behaviour of Gunn oscillators (in German). AEG-Telefunken, Research Institute, Ulm, Technical Report FI-U Nr. 49/71 (1971).

5.43a Kino, G. S. and Kuru, I., High-efficiency operation of a Gunn oscillator in the domain mode. *IEEE Trans. Electron Dev.* **ED–16** (1969), p. 735.

5.43b Kino, G. S., Correction to "High-efficiency operation of a Gunn oscillator in the domain mode". *IEEE Trans. Electron Dev.* **ED–19** (1972), p. 290.

5.44 Day, G. F., Dow, D. G., Mosher, C. H. and Vane, A. B., The achievement of high efficiency in Gunn effect devices. *Proc. Int. Symp. on GaAs, Reading, 1966*, p. 189 (Institute of Physics and Physical Society, London, 1967).

5.45 Copeland, J. A., LSA oscillator waveforms for high efficiency. *Proc. IEEE* **57** (1969), p. 1666.

5.46a Frey, W., Influence of a second harmonic voltage component on the operation of a Gunn oscillator. *Proc. 8th Int. Conference MOGA, Amsterdam, 1970*, p. 2/38 (Kluwer, Deventer, 1970).

5.46b Popov, V. I., Calculation method for multi-resonant circuit Gunn self-oscillator. *Proc. 1971 Eur. Microwave Conference, Stockholm, Sweden*, Vol 1, p. A3/4:1 (Stockholm, The Royal Swedish Academy of Engineering Sciences, 1971).

5.47 Suematsu, Y. and Nishimura, Y., Impedance diagram of the Gunn diode. *Proc. IEEE* **54** (1966), p. 1095.

5.48a Khandelwal, D. D. and Curtice, W. R., A study of the single-frequency quenched-domain mode Gunn-effect oscillator. *IEEE Trans. Microwave Theory and Techniques* **MTT–18** (1970), p. 178.

5.48b Curtice, W. R. and Khandelwal, D. D., Multifrequency operation of a quenched-domain mode Gunn-effect device. *Proc. IEEE* **59** (1971), p. 416.

5.49a Ikoma, T. and Yanai, H., Effect of external circuit on Gunn oscillation. *IEEE J. Solid-State Circuits* **SC–2** (1967), p. 108.

5.49b Khandelwal, D. D., Curtice, W. R., Ikoma, T. and Yanai, H., Comment on "Effect of external circuit on Gunn oscillation". *IEEE J. Solid-State Circuits* **SC–4** (1969), p. 51.

5.50a Dizhur, D. P., Levinstein, M. E. and Shur, M. S., Computer calculations of the efficiency of the Gunn generator. *Electron. Lett.* **4** (1968), p. 444.

5.50b Dizhur, D. P., Levinstein, M. E., Nasledov, D. N. and Shur, M. S., Computer model of the Gunn diode. *Proc. 7th Int. Conference MOGA, Hamburg, 1968, Nachrichtentechn. Fachber.* **35** (1968), p. 436.

5.50c Levinstein, M. E., Pushkaroeva, L. S. and Shur, M. S., Influence of the second harmonic of a resonator on the parameters of a Gunn generator for transit and hybrid modes. *Electron. Lett.* **8** (1972), p. 31.

5.51 Chen, W. T. and Dalman, G. C., Dynamic i–v characteristics of a quenched mode oscillating Gunn diode. *Proc. IEEE* **58** (1970), p. 503.

5.52a Acket, G. A., The influence of doping fluctuations on limited space charge accumulation in n-type GaAs. *Phys. Lett.* **27A** (1968), p. 293.

5.52b Acket, G. A., The effect of inhomogeneities on limited space charge accumulation (LSA) in n-type gallium arsenide. *Proc. 7th Int. Conference MOGA, Hamburg, 1968, Nachrichtentechn. Fachber.* **35** (1968), p. 419.

5.53   Yu, S. P., Tantraporn, W. and Young, J. D., Transit-time negative conductance in GaAs bulk-effect diodes. *IEEE Trans. Electron. Dev.* **ED–18** (1971), p. 88.

5.54   Ladbrooke, P. H. and Carroll, J. E., Simple theory of improving the efficiency of Gunn oscillators. *Electron. Lett.* **4** (1968), p. 83.

5.55   Heeks, J. S., Au, H. K. and Woode, A. D., An experimental study of high peak power Gunn effect oscillators. *Proc. 7th Int. Conference MOGA, Hamburg, 1968, Nachrichten-techn. Fachber.* **35** (1968), p. 423.

5.56   Reynolds, J. F., Berson, B. E. and Enstrom, R. E., High-efficiency transferred electron oscillators. *Proc. IEEE* **57** (1969), p. 1692.

5.57   Jeppsson, B. and Jeppesen, P., A high power LSA relaxation oscillator. *Proc. IEEE* **57** (1969), p. 1218.

5.58   Butcher, P. N. and Hearn, C. J., Theoretical efficiency of the LSA mode for GaAs at frequencies above 10 GHz. *Electron. Lett.* **4** (1968), p. 459.

5.59   Nakamura, M., Frequency limit of LSA mode oscillations estimated from the hot electron problems in the high frequency electric field. *Japanese Journal Appl. Phys.* **8** (1969), p. 910.

5.60a Ohmi, T., Murayama, K. and Kanbe, H., Effect of the intervalley scattering time on LSA oscillations. *Proc. IEEE* **56** (1968), p.747.

5.60b Ohmi, T., Murayama, K. and Kanbe, H., Effect of intervalley scattering time on two-valley semiconductor oscillators. *Electron. Commun. Japan* **52–C** (1969), p. 145.

5.61a Székely, V. and Tarnay, K., Intervalley scattering model of the Gunn domain. *Electron. Lett.* **4** (1968), p. 592.

5.61b Székely, V. and Tarnay, K., Some aspects of the theory of Gunn effect. *Proc. of the 4th Colloquium on Microwave Communication*, Vol. V, (Akadémiai Kiado, Budapest, 1970) p. ME, SM–6/1.

5.62   Johnson, E. O., Physical limitations on frequency and power parameters of transistors. *RCA Rev.* **26** (1965), p. 163.

5.63   DeLoach, B. C., Recent advances in solid state microwave generators. In: L. Young (Ed.), *Advances in Microwaves*, vol. 2, p. 43, (Academic Press, New York, 1967).

5.64   Kodali, V. P., Fundamental power/frequency limitations of microwave semiconductor devices. *Electron. Lett.* **4** (1968), p. 311.

5.65   Sasaki, A., Frequency dependencies of power and efficiency of transit-time oscillations in two-valley semiconductors. *Proc. IEEE* **56** (1968), p. 1757.

5.66   Bosch, B. G., Power-frequency limitations of transferred-electron oscillators. *Arch. Elektr. Übertragung* **23** (1969), p. 377.

5.67   Eastman, L. F., Generation of high power microwave pulses using gallium arsenide. *Proc. Eur. Microwave Conference, London, 1969*, p. 208 (IEE Conf. Publication no. 58).

5.68   Camp, W. O. and Kennedy, W. K., Pulse millimeter power using the LSA mode. *Proc. IEEE* **56** (1968), p. 1105.

5.69a Hasegawa, F. and Suga, M., Effects of doping profile on the conversion efficiency of a Gunn diode. *IEEE Trans. Electron Dev.* **ED–19** (1972), p. 26.

5.69b Mircea, A. E., Computer optimized design of pulsed Gunn oscillators. *IEEE Trans. Electron Dev.* **ED–19** (1972), p. 21.

5.70   Shockley, W., Copeland, J. A. and James, R. P., The impedance field method of noise calculation in active semiconductor devices. In: *Quantum Theory of Atoms, Molecules and the Solid State*, p. 537 (Academic Press, New York, 1966).

5.71a King, R., *Electrical Noise* (London, Chapman and Hall Ltd., 1966).

5.71b Mumford, W. W. and Scheibe, E. H., *Noise Performance Factors in Communication Systems* (Dedham, Mass., Horizon House, Microwave Inc., 1968).

5.71c Van der Ziel, A., *Noise: Sources, Characterization, Measurement* (Englewood Cliffs, N.J., Prentice Hall, Inc., 1970).

5.72   Thim, H. W., Noise reduction in bulk negative-resistance amplifiers. *Electron. Lett.* **7** (1971), p. 106.

5.73   Maloberti, F. and Svelto, V., Noise spectrum and noise figure of a device with bulk negative differential mobility. *Alta Frequenza* **15** (1971), p. 667.

5.74   Kozdon, P. and Robson, P. N., Two port amplifiers using the transferred electron effect in GaAs. *Proc. 8th Int. Conference MOGA, Amsterdam, 1970*, p. 16/7 (Deventer, Kluwer, 1970).

5.75   Perlman, B. S., Upadhyayula, C. L. and Marx, R. E., Wide-band reflexion-type transferred electron amplifiers. *IEEE Trans. Microwave Theory and Techniques* **MTT–18** (1970), p. 911.

5.76a   Baskaran, S. and Robson, P. N., Noise performance of InP reflection amplifiers in Q band. *Electron. Lett.* **8** (1972), p. 137.

5.76b   Baskaran, S. and Robson, P. N., Gain and noise figure of GaAs transferred-electron amplifiers at 34 GHz. *Electron. Lett.* **8** (1972), p. 109.

5.77   Frey, W., Engelmann, R. W. H. and Bosch, B. G., Unilateral travelling-wave amplification in gallium arsenide at microwave frequencies. *Arch. Electron. Übertragungstechn.* **25** (1971), p. 1.

5.78a   Dean, R. H., Dreeben, A. B., Kaminski, J. F. and Triano, A., Travelling-wave amplifier using thin epitaxial GaAs layer. *Electron. Lett.* **6** (1970), p. 775.

5.78b   Dean, R. H. and Matarese, R. J., The GaAs travelling-wave amplifier as a new kind of microwave transistor. *Proc. IEEE* **60** (1972), p. 1486.

5.79   Hashiguchi, S. and Okoshi, T., Determination of equivalent circuit parameters describing noise from a Gunn oscillator. *IEEE Trans. on Microwave Theory and Techniques* **MTT–19** (1971), p. 686.

5.80   Edson, W. A., Noise in oscillators. *Proc. IRE* **48** (1960), p. 1454.

5.81   Ondria, J. G., A microwave system for measurements of AM and FM noise spectra. *IEEE Trans. Microwave Theory and Techniques* **MTT–16** (1968), p. 767.

5.82a   Harth, W., Equivalent noise temperature of bulk negative resistance devices. *Arch. Elektron. Übertragungstechn.* **26** (1972), p. 149.

5.82b   Ataman, A. and Harth, W., Intrinsic FM noise of Gunn oscillators. *IEEE Trans. Electron Dev.* **ED–20** (1973), p. 12.

5.83a   Faulkner, E. A. and Meade, M. L., Flicker noise in Gunn diodes. *Electron. Lett.* **4** (1968), p. 226.

5.83b   Meade, M. L., Relationship between F.M. noise and current noise in a cavity-controlled Gunn effect oscillator. *The Radio and Electronic Engineer* **41** (1971), p. 126.

5.83c   DeCacqueray, A., Blasquez, G. and Graffeuil, J., Experimental study of the correlation of F.M. and L.F. noise in Gunn oscillators. *Electron. Lett.* **8** (1972), p. 217.

5.84a   Matsuno, K., FM noise in a Gunn effect oscillator. *IEEE Trans. Electron Dev.* **ED–16** (1969), p. 1025.

5.84b   Matsuno, K., Low frequency current fluctuations in a GaAs Gunn diode. *Appl. Phys. Lett.* **12** (1968), p. 404.

5.85   Ataman, A., Herbst, H. and Harth, W., The influence of different contact materials on the noise performance of Gunn elements. *Arch. Elektron. Übertragungstechn.* **25** (1971), p. 396.

5.86a   Kuhn, P., Noise in Gunn oscillators depending on surface of Gunn diode. *Electron. Lett.* **6** (1970), p. 845.

5.86b   Kuhn, P., Modulation noise in Gunn oscillators (in German). *Nachrichtentechn. Zeitschrift* **25** (1972), p. 17.

5.87   Omori, M., Varian Associates, Palo Alto, Calif., private communication (1969).

5.88   Matsuno, K., Calculation of LSA oscillator noise. *Proc. IEEE* **56** (1968), p. 75.

5.89   Hobson, G. S., Source of f.m. noise in cavity controlled oscillators. *Electron. Lett.* **3** (1967), p. 63.

5.90   Qhmi, T. and Hasuo, S., Dynamic properties of electrons in two-valley semiconductors.

Proc. 10th Int. Conference Physics of Semiconductors, Cambridge, Mass., 1970, p. 60 (Washington, D.C., Atomic Energy Commission, 1970).

5.91 Matsuno, K., Critical fluctuations in GaAs in a d.c. electric field. *Phys. Lett.* **31A** (1970), p. 335.

5.92 Hillbrand, H. A., Dynamical study of the transferred-electron effect by a Monte-Carlo technique. *Electron. Lett.* **7** (1971), p. 495.

5.93 Copeland, J. A., GaAs bulk oscillators stir millimeter waves. *Electronics*, June 12, 1967 p. 91.

5.94a Taylor, B. C. and Colliver, D. J., Indium phosphide microwave oscillators. *IEEE Trans. Electron Dev.* **ED–18** (1971), p. 835.

5.94b Gibbons, G. and White, P. M., InP pulsed and C.W. millimetre-wave oscillators. *Electron. Letts.* **7** (1971), p. 150.

5.95 Eastman, L. F., Multiaxis radial circuit for transferred electron devices. *Electron. Lett.* **8** (1972), p. 149.

5.96 Heinle, W., AEG-Telefunken, Research Institute, Ulm, private communication (1971).

5.97 Camp, W. O., Jr., High-efficiency GaAs transferred electron device operation and circuit design. *IEEE Trans. Electron Dev.* **ED–18** (1971), p. 1175.

5.98 Dow, D. G., The LSA mode, control of efficiency and starting. *Proc. Cornell Conference "High Frequency Generation and Amplification"*, *1967*, p. 109 (Ithaca, N.Y., School of Electrical Engineering, Cornell University, 1967).

5.99 De Groot, J. and Vlaardingerbroek, M. T., Some numerical results on modes of oscillation in a transferred-electron device. *Proc. 8th Int. Conference MOGA, Amsterdam, 1970*, p. 20/34 (Deventer, Kluwer, 1970).

5.100 Shaw, M. P., Solomon, P. R. and Grubin, H. L., Circuit controlled current instabilities in n-GaAs. *Appl. Phys. Lett.* **17** (1970), p. 535.

5.101 Jeppesen, P. and Jeppsson, B., LSA relaxation oscillator principles. *IEEE Trans. Electron Dev.* **ED–18** (1971), p. 439.

5.102 Jeppsson, B. and Jeppesen, P., LSA relaxation oscillations in a waveguide iris circuit. *IEEE Trans. Electron Dev.* **ED–18** (1971), p. 432.

5.103 Chilton, R. H. and Kennedy, W. K., Jr., Multiple frequency operation associated with the LSA mode. *Proc. IEEE* **56** (1968), p. 1124.

5.104 Bosch, R. and Thim, H. W., Time evolution of the electron distribution function in GaAs Gunn devices. *Verhandlungen der Deutschen Physikalischen Gesellschaft* **6** (1971), p. 279. (European Semiconductor Device Research Conference, Munich, March 1971.)

5.105 McLean, T. P., The absorption edge of semiconductors. In: Gibson, A. F. and Burgess, R. E. (Eds), *Progress in Semiconductors*, vol. 5, p. 85 (London, Heywood, 1960).

5.106 Grubin, H. L., Shaw, M. P. and Conwell, E. M., Current instabilities in n-InP. *Appl. Phys. Lett.* **18** (1971), p. 211.

5.107 Guéret, P., Some non-linear properties of a circuit with a Gunn diode. *Proc. 8th Int. Conference MOGA, Amsterdam, 1970*, p. 20/40 (Deventer, Kluwer, 1970).

6
Material and Device Technology

6.1 SEMICONDUCTING MATERIAL

6.1.1 General Remarks

At the end of Chapter 2 it was concluded that essentially three criteria must be fulfilled by a semiconductor band structure to obtain, for n-type conduction, an electron-transfer mechanism leading to negative differential mobility of the electrons. These requirements are met in a number of III-V and II-VI compound semiconductors with direct band gaps, allowing electron transfer from the central (000) valley to the $\langle 100 \rangle$, or in some cases possibly the $\langle 111 \rangle$, satellite valleys as described in Chapter 2 (cf. the calculated band structures in [2.7]). All these semiconductors crystallize in the zinc-blende lattice structure which is sketched in Fig. 6.1.

Under normal environmental conditions, i.e. room temperature and atmospheric pressure, the TE effect has been observed experimentally, as well as in GaAs, in the polar compounds InP [1.19, 6.1]; cf. also [5.14, 5.35, 6.33], ZnSe [6.2], CdTe [3.28, 6.3], and also in the alloy crystals $GaAs_xP_{x-1}$ [1.22, 6.4a], $Ga_xAl_{1-x}As$ [6.4b] and in $In_xGa_{1-x}Sb$ [6.5]*. Since these latter alloy systems can be obtained without any miscibility gap (arbitrary composition x), one may be in a position to 'tailor' [6.168] the conduction-band structure in such a way as to create the optimum properties for the bulk negative differential-mobility effect. Other alloy semiconductors that might be of interest in this respect, but have not yet been investigated experimentally, are $InAs_xP_{1-x}$ [6.6a], $In_xGa_{1-x}P$, and $GaSb_xAs_{1-x}$.

On the other hand, the conduction-band structure of a semiconducting crystal can be affected by changing its temperature and/or by applying pressure. In this way, the TE effect has been obtained in InAs under the influence of hydrostatic pressure or uniaxial stress along a $\langle 111 \rangle$ crystal direction in excess of 14 kb [6.7], and in InSb at 77 K by increasing the hydrostatic pressure to more than 10 kb [6.8, 6.9].

---

*Some of these materials possibly operate in a 'three-level' electron transfer mechanism [5.36 a–c] rather than in the more basic transfer mechanism described in Chapter 2 which constitutes a two-level system (two sets of conduction-band valleys involved). A brief outline of the three-level effect was provided at the end of Section 5.2.2.

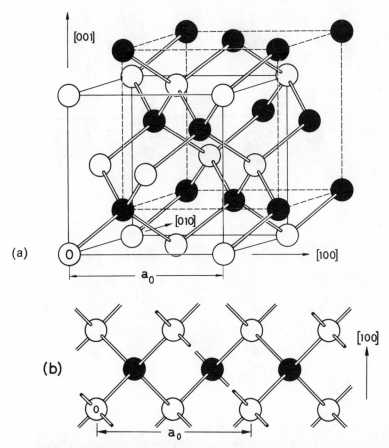

Fig. 6.1 The zinc-blende lattice, with a_0 denoting the lattice constant. In the case of the semiconductors considered, white balls: group-III or group-II atoms (e.g. Ga), black balls: group-V or group-VI atoms (e.g. As). (a) Representation as two penetrating face-centred cubic sub-lattices, Crystal bonds (double bars) shown within respective cubes only. (b)–(d) Schematic arrays of atoms along indicated directions: (b) and (c) projected onto (100) plane; (c) explains the property of easy cleavage of {110} planes; (d) projected onto (121) plane, shows the polarity of the [111] axis. In the special case of the *diamond* lattice all atoms are of the same kind (e.g. the group-IV atoms Ge). After [6.18 and 6.130].

In both cases the relatively high-lying ⟨111⟩ satellite valleys are lowered by the stress leading to a separation from the central valley smaller than the band gap.

A diversity of carrier transfer mechanisms causing bulk n.d.c. has been found under special environmental conditions in the elemental semiconductor Ge which crystallizes in the diamond lattice (being zinc-blende lattice with *identical* atoms, see Fig. 6.1) and possesses an 'indirect' band gap (i.e. satellite valleys of conduction band are lower than central valley). Both n- and p-type Ge have been shown to exhibit n.d.c. (review in [6.10]). In

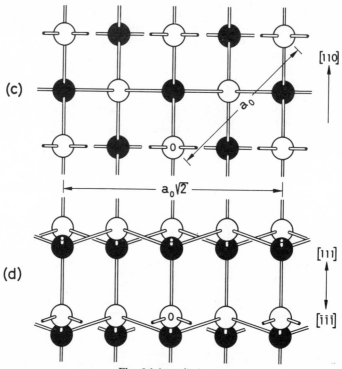

Fig. 6.1 (contd)

n-type Ge at temperatures below about 120 K electron transfer from the lowest ⟨111⟩ valleys to ⟨100⟩ ones seems to be responsible for current instabilities in ⟨100⟩ or ⟨110⟩ crystal directions [6.11]; see, however, [6.12]. At very low temperatures (<30 K) the strongly non-spherical symmetry of the ⟨111⟩ valleys causes bulk negative differential mobility in the ⟨111⟩ directions. This is because of an electron transfer within the four degenerate ⟨111⟩ valleys themselves despite the fact that no energy separation is actually present. (Due to a difference in heating rates electrons are effectively transferred from the three light mass valleys along the [Ī11], [1Ī1] and [11Ī] directions to the heavy-mass valley along the [111] direction which is parallel to current flow; [6.13]). Such a transfer also becomes effective at higher temperatures up to 300 K if uniaxial stress is applied perpendicular to the current direction to obtain a suitable energetic split of the ⟨111⟩ valleys. (Here, the most favourable configuration is a current flow in a [1Ž1] direction with stress in a [111] direction; [6.14, 6.15]). With a compression in the [111] or [100] direction n.d.c. has been observed even in *p-type* Ge at temperatures ≲ 160 K [6.16]. The energy degeneracy at the top of the valence band is lifted by the stress, so as to cause n.d.c. in the direction of stress by *hole* transfer from the light to the heavy-mass 'valley'. (Hole valleys are actually energy maxima in the valence band.)

In n-type InSb the TE effect has been observed at 77 K and atmospheric pressure even though the band gap is much smaller than the valley separation and hence impact ionization takes place at the same time [6.17a]. This indicates that mobility decrease by electron transfer is able to overcompensate carrier increase by an avalanche process. However, the effect is very weak and vanishes within 1 ns of the application of the field, since then impact ionization becomes dominant. (See also the critical remarks in [6.17b].)

Table 6.1 gives the properties of the above-mentioned semiconductors pertinent to the n.d.c. effect.

Experiments under high pressure and/or at low ambient temperatures have served mainly to clarify the basic properties of the electron transfer mechanism. For device application only material usable under normal environmental conditions is practical. The choice of material is further reduced since the peak field should be sufficiently small in order to avoid possible dielectric breakdown (or other carrier multiplication effects, e.g. ionization of deep traps), and the thermal conductivity should be high enough to allow high-average or C.W.-power performance without excessive heating. In ZnSe [6.1] and CdTe [6.30] the high Gunn-domain fields cause impact ionization and current runaway even slightly above threshold. This leaves GaAs, InP and possibly some of the alloy systems as material for practical devices.

InP shows a favourably high thermal conductivity but has received increased attention rather late because of difficulties in preparation. This problem seems to have been solved, and InP of a quality comparable to that of GaAs can be grown, at least in the research laboratory so far [6.166]. As a speculation, InP may eventually find even wider application than GaAs since it has been proposed to possess advantages with regard to oscillator efficiency (Section 5.2.2) and achieving domain-free operation [5.36, 6.1] as well as to noise behaviour [5.72, 5.76a], advantages which have also been predicted for the alloy $In_xGa_{1-x}Sb$.

Of the alloy semiconductors, $In_xGa_{1-x}Sb$ seems to be especially attractive owing to the low value of the threshold field and the potential advantages over GaAs as just mentioned. But here the material technology is complicated: only polycrystalline samples have been obtained up to the time of writing*. In addition, the thermal conductivity is extremely low, as is common for the alloy systems. On the other hand, since in oscillator applications the efficiency is expected to be at least as good as with GaAs devices (see Section 5.2.2; Fig. 5.11), the lower input density might still lead to an advantage of $In_xGa_{1-x}Sb$ over GaAs (see Section 6.3.2).

At present, commercially available TE devices are almost exclusively made from GaAs, due to its relatively low peak field E_p and a reasonably mature material technology that has already been advanced by the development of other devices such as tunnel diodes, varactor diodes, luminescent diodes, injection lasers and transistors (particularly field-effect transistors). Important physical properties of GaAs have been compiled in Table 6.2 where they are compared to those of InP and also those of the elemental semiconductors Ge and Si, the latter material being mostly used for making more conventional semiconductor devices, like transistors.

---

*Recently single crystals have been obtained: Michel, J. *et al.* and Miki, H. *et al.*, *5th Int. Symp. on GaAs and Related Comp.*, *Deauville (France)*, *1974*.

Table 6.1—Semiconductor material properties related to Gunn effect

Material	Composition	Current Direction	Temp. K	Thermal Conductivity κ $W(cmK)^{-1}$	Pressure kind§ p kb	Band Gap W_G eV	Valley Separation between	Valley Separation ΔW eV	Lower-Valley Mobility μ_1 $cm^2(Vs)^{-1}$	Mobility Ratio μ_1/μ_2	Gunn Effect observed	Measured Threshold Field E_T $kV\,cm^{-1}$	Calculated Peak Field E_p $kV\,cm^{-1}$	Calculated Critical Field§§ E_o $kV\,cm^{-1}$	Peak Velocity v_p $10^7\,cm\,s^{-1}$
n-AlAs			300	0·08[r]	normal	2·16[x]	⟨100⟩→(000)	0·74[x]			no				
n-GaP			300	0·77[r]	normal	2·24[a]	⟨100⟩→(000)	0·35[a]	300[b]		no				
n-GaAs		arbitr. arbitr. arbitr.	300 300 300 300	0·44[r]	normal, h 15, h 26, h 30	1·43[a] 1·57[c] 1·67[c] 1·71[c]	(000)→⟨100⟩	0·38[c] 0·24[l] 0·15[l] 0·12[l]	9 300[o]	43[d]–49[e]	yes; p≤26 kb	1·2–5·5[f,g,h] 1·9 (0·86)[m] 1·4 (0·64)[*]	4·0[l]	3·6[k]	2·2[j]
n-InP***		arbitr.	300	0·68[r]	normal	1·26[a]	(000)→⟨111⟩ (000)→⟨100⟩	0·40[ψ] 0·70[ψ]	4 600[b]	23[ψ]	yes	4·5–6·5[f]	5–11·5[n,π]	7·8[k]	2–3[n,π]
n-GaSb			300	0·33[r]	normal	0·70[a]	(000)→⟨111⟩	0·085[p]	4 000[b]	7·5[p]	no				
n-InAs		⟨111⟩ arbitr.	300 300 300	0·27[r]	normal, ⟨111⟩ {14, 16}, h 26	0·36[a]	(000)→⟨111⟩	1·28[n] 0·86χ	33 000[b]		yes; p≥14 kb	1·6[q] 1·45[q] ≈2χ		≈3·6[q]	
n-InSb		arbitr. arbitr. arbitr. arbitr. arbitr.	300 77 77 77	0·17[r]	normal, normal, h 4, h 8, h 12	0·18[a] 0·22[r] 0·28[r] 0·34[r] 0·40[r]	(000)→⟨111⟩	0·45[r] 0·41[r] 0·37[r] 0·33[r]	78 000[b] 600 000[r]		yes; T≤77 K, p≥8 kb for $E_T < E_{av}$	≈0·6[t]** ≈0·6[s]** 0·38[s] 0·24[s]			≈3[s]
n-ZnSe		arbitr.	300	≈0·2[u]	normal	2·6[ε]	(000)→⟨111⟩ ~0·51[v]		570[u]		yes	≈38[u]		28[u]	~1·5[u]
n-CdTe		arbitr.	300	0·075[v]	normal	1·5[ε]			1 100[v]		yes	11–15[v,k]		12·9[k]	~1·3[k]

Table 6.1—Semiconductor material properties related to Gunn effect (contd)

Material	Composition	Current Direction	Temp.	Thermal Conductivity κ	Pressure kind§	p	Band Gap W_G	Valley Separation between	ΔW	Lower-Valley Mobility μ_1	Mobility Ratio μ_1/μ_2	Gunn Effect observed	Measured Threshold Field E_T	Calculated Peak Field E_p	Calculated Critical Field§§ E_c	Peak Velocity v_p
n-GaAs$_x$P$_{1-x}$	x=0·33	arbitr.	300	0·22[τ]	normal		1·63[δ]	(000)→(100)	0·24[w]	≈5 000[y]		yes: x ⩾ 0·6	3·3 (0·86)[w]			
	x=0·66	arbitr.	300	0·18[τ]	normal		1·84[δ]		0·12[w]	≈3 000[y]			2·3 (0·59)†			
n-Ga$_x$Al$_{1-x}$As	x=0·90	arbitr.	300		normal		1·57[x]	(000)→(100)	0·25[x]	6 000[x]		yes: x ⩾ 0·77	2·8 (0·88)[x]			
	x=0·77	arbitr.	300		normal		1·76[x]		0·11[x]				2·0 (0·63)†			
n-In$_x$Ga$_{1-x}$Sb	x=0·7	arbitr.	300	0·048[γ]	normal		0·25[z]	(000)→(111)	0·40[α]	35 000[z]	~12[β]	yes: ? ≲ x < 0·7	≈0·6[z]			1·3[z]
	x=0·5	arbitr.	300	0·055[γ]	normal		0·36[z]		0·36[α]							
n-InAs$_x$P$_{1-x}$	x=0·2	arbitr.	300	0·19[τ]	normal		1·10[n]	(000)→(111)	0·95[n]					4·5–7·0[n]		2·3–3·2[n]
n-Ge		⟨100⟩ or ⟨110⟩	300	0·55[τ]	normal		0·665[ζ]	⟨111⟩→⟨100⟩	0·18[γ]	3 800[ζ]		yes: T ≲ 120 K	2·3[ϑ]			1·4
			77	3·5[τ]	normal		0·74[ζ]	⟨111⟩′→[111]	0[v]	36 500[ζ]			1·8[ϑ]			
			27	12·0[τ]	normal		0·745[ζ]	[111]→⟨111⟩′	0·082[λ]							
		[111]	4·2	6·7[τ]	[111]	8			0·153[λ]			yes:T ≲ 30K c.g.‡	0·002[v]			0·03–0·05[v]
		[1̄21]	300		[111]	15			0·082[λ]			p > 6·1 kb at 300 K	1·6[λ]			0·64
			27			8							0·43[λ]			≈2[λ]
p-Ge		⟨111⟩	140	1·4[τ]	⟨111⟩	15		(000)→(000)	0·060			yes: ††	1·16[λ]			
			140			19·5			0·078			T ≲ 160 K,	0·60[λ]			
			27	12·0[τ]		8·8			0·060			c.g.p>14kb	0·69[λ]			
			27			15			0·035			at 140K	0·20[λ]			
			4·2	6·7[τ]					0·060				0·008[ξ]			≈0·5[ξ]

§ h = hydrostatic; for uniaxial pressure crystal direction indicated.
§§ E_c is the critical dielectric breakdown field due to polar optical scattering, if the satellite valleys are neglected[σ].
** [n] normalized to threshold at atmosphere pressure in parentheses.
*** in avalanche region.
*** For more recent data on InP see James, L. W., J. Appl. Phys. 44 (1973), p. 2764.
† normalized to value at x = 1 in parentheses.
‡ other directions of current perpendicular to stress possible (weaker[ρ]).
†† also for current and stress along ⟨100⟩ (weaker)[λ].
⟨111⟩′ are the ⟨111⟩ valleys excluding the preferred [111] valley.

a [6.18]	b [6.19]	c [2.5b]	d [2.16]	e [2.16]	f [1.19]	g [6.20]	h [4.26]
q [2.29]	j [2.36]	k [3.28]	l [6.21]	m [1.21]	n [6.6, 5.36]	w [2.14]	p [6.22]
q [6.7]	r [6.8]	s [6.9]	t [6.17]	u [1.21]	v [6.3]	o [6.4a]	x [6.4b]
y [6.23]	z [6.7]	α [6.24]	β [6.25]	γ [6.26]	δ [6.3]	ε [6.28]	ζ [6.29]
η [6.11a]	ϑ [6.10a]	λ [6.10b]	ν [6.25]	ξ [6.16]	π [6.33a, 5.14]	ρ [6.34]	σ [6.35]
τ [6.110]	φ [6.15]	χ [6.7b]	ψ [6.33b]				

Table 6.2—Physical properties of pure GaAs and InP compared to those of pure Si and Ge at 300 K

	GaAs	InP	Ge	Si
Band gap W_G/eV				
at 0 K	1·52	1·42	0·746	1·165
300 K	1·43	1·26	0·665	1·12
Number ν of lowest equivalent C.B. valleys	1 at (000)	1 at (000)	4 in $\langle 111 \rangle$	6 in $\langle 100 \rangle$
Effective[+] mass $m_1^{\star(N)}/m_0$				
electrons	$\begin{cases} 0·072^a \\ 0·0665^h \end{cases}$	0·077	0·56	1·09
holes	$\approx 0·54$	$\approx 0·24$	0·29	0·55
Mobility μ_1/cm²(Vs)$^{-1}$				
electrons	9 300^b	4 500	3 800	1 450
holes	435	150	1 800	500
Intrinsic resistivity ρ/Ωcm	$3·7 \times 10^{8c}$	$\gtrsim 2 \times 10^{6d}$	47	$2·3 \times 10^5$
Dielectric constant $\varepsilon/\varepsilon_0$	$\begin{cases} 12·35 – 12·95^{h,1} \\ 11·6 \end{cases}$	10·9	16·0	11·8
Lattice constant a_0/Å	5·6535	5·8688	5·6575	5·4307
Density d/g cm^{-3}	5·316^d	4·787^d	5·327	2·329
Atom density/10^{22} cm^{-3}	4·43^d	3·96^d	4·45	5·00
Atomic weight^j	Ga: 69·72 As: 74·92	In: 114·82 P: 30·97	72·59	28·09
Coefficient of expansion $\alpha/10^{-6}$ K^{-1}	6·9^e	4·6^d	6·1	4·2
Thermal conductivity κ/W(cmK)$^{-1}$	0·44^f	0·68^f	0·55^f	1·45^f
Specific heat c_v/J(gK)$^{-1}$	0·36		0·31	0·76
Melting point T_M/°C	1 238	1 058	937^g	1 412^g
Vapour pressure at T_M p_v/b	0·99^d	21^d	$1·1 \times 10^{-9d}$	$7·5 \times 10^{-7d}$
Boiling point T_B/°C			2 827^g	3 145^g

Values for GaAs and InP taken from [6.18], for Ge and Si from [6.29], with the following exceptions: ^{a} [2.5], ^{b} [2.14], ^{c} [6.133], ^{d} [6.130], ^{e} [6.118a], ^{f} [6.110], ^{g} [6.134], ^{h} [6.136a], ^{1} [6.158], ^{j} [6.162].

[+]Combined-density-of-states effective mass in lowest valleys

$$\text{for electrons:} \quad m_1^{\star(N)} = \nu^{3/2}(m_L m_T^2)^{1/3}$$
$$\text{for holes:} \quad m_1^{\star(N)} = (m_h^{3/2} + m_1^{3/2})^{2/3}$$

where m_L and m_T denote the longitudinal and transverse electron mass, m_h and m_1 the heavy and light holes mass, respectively, and ν the number of equivalent valleys.

6.1.2 GaAs Crystal Growth

When preparing GaAs single crystals for use in TE devices, cf. [6.165], attention has to be focused on high purity and uniformity of the material. Otherwise a whole series of detrimental effects might occur as indicated below.

First of all, any existing impurities or native lattice defects constitute scattering centres for the carriers and hence degrade their mobility. This leads to a low peak-to-valley ratio of the $v(E)$ characteristic, i.e. according to Section 5.2 to a low d.c.-to-r.f. power conversion efficiency [4.16]. Secondly, impurity and defect levels are known to cause either high-field carrier trapping [3.32, 6.31] or the opposite effect, i.e. carrier multiplication by impact ionization [6.32a, 6.32b, 6.135]. In both cases additional undesired current changes occur and disturb the Gunn-effect performance. Thirdly, non-uniform doping of the GaAs crystal affects domain nucleation and propagation in the dipole modes and causes a fall in efficiency due to field distortion in the LSA oscillator mode [6.33]. Table 6.3 lists the doping impurities which for GaAs are important as donors, acceptors or deep traps. Their energy-level positions in the band gap are indicated in Table 6.4 together with levels that are not yet identified.

Table 6.3—Important impurity atoms for GaAs, indicated in the periodic chart of the elements (numbers are atomic radii in Å)

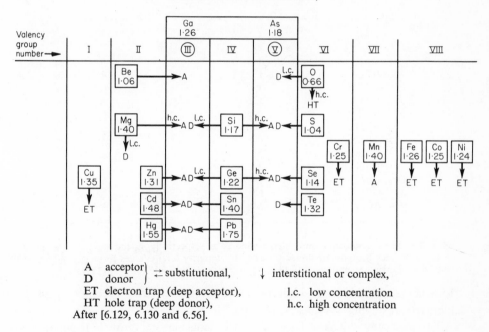

The net-doping concentration of GaAs material employed for TE devices lies usually in the range of $n_0 \approx 10^{14}$ to 10^{16} cm^{-3}. Material with $n_0 < 10^{14}$ cm^{-3} is rather difficult to prepare and is of possible interest only for amplifiers with sub-critical $n_0 L$ or $n_0 d$ product. On the other hand, material with $n_0 > 10^{16}$ cm^{-3} leads to such high fields in the domain (cf. e.g. Fig. 3.9) that dielectric breakdown may easily occur, particularly in long devices [6.36]. Thus it would be usable only for

Table 6.4—Impurity levels in GaAs at 300 K

$(E_c-E_I)/eV$	Impurity	A or D
0·002–0·005	Si, Ge, Sn, Se, Te	D
0·00586[d]	not identified	D
0·175[c]	not identified	A
0·30	not identified	A
0·47	not identified	A
0·55	not identified	A
0·62	not identified	A
0·67[e]	Cr	?
0·70	not identified	A

$(E_c-E_v)/2 = 0·715$, centre of band gap

$(E_I-E_v)/eV$		
0·64[b]	Cr	A
0·63	O	D
0·52	Fe	A
0·51	Cu	A
0·37	Fe	A
0·24	Cu	A
0·15	Cu	A
0·094	Mn	A
0·043[a]	Ge	A
0·026	Si	A
0·024	Zn	A
0·021	Cd	A
0·02	Cu	A
0·013	Mg	A

E_c and E_v is the conduction and valence band edge, respectively, E_I the impurity level position. A or D denotes acceptor or donor property of impurity level. Values from [6.129 and 6.130], with the following exceptions: [a] [6.131], [b] [6.128b], [c] [6.132], [d] [6.136], [e] [6.157].

LSA-mode operation. In addition, however, the electron mobility drops prohibitively at doping levels $n_0 > 10^{16}$ cm^{-3} [6.58].

Two basically different types of GaAs single-crystal growth techniques are commonly used for preparing device material, bulk growth and epitaxial growth. In the first case a relatively large single crystal is grown from the melt starting with a tiny seed crystal. The grown crystal is then cut along the desired crystal orientation into thin wafers which form the starting material for device work. In the second case a single-crystal wafer of bulk-grown material serves merely as a substrate for growing by deposition ('epitaxially') a thin layer of single-crystal GaAs suitable for device fabrication. Both techniques yield material exhibiting TE effect, but epitaxial crystals are generally of much higher quality. Epitaxial growth introduces additional flexibility by changing to substrate crystals other than GaAs if the crystal structures can be matched (hetero-epitaxy).

N-type *bulk* GaAs grown from the melt, without purposely introducing a dopant, by using the well-known Czochralski method (vertical growth) with rotating seed has room-temperature concentrations as low as $n_0 \approx 4 \times 10^{15}$ cm^{-3} with low-field mobilities of about $\mu_1 \approx 7\,000$ cm^2(Vs)$^{-1}$. Somewhat better results are obtained when using the Bridgman technique* (horizontal growth: 'boat-grown' crystals), namely $n_0 \approx 5 \times 10^{14}$ cm^{-3} and $\mu_1 \approx 8\,000$ cm^2(Vs)$^{-1}$ [6.37]. However, all bulk-grown material with a net-doping in the range suitable for TE devices shows strong carrier increase with increasing temperature due to a high concentration of deep-lying impurity levels (Fig. 6.4b). This property may lead to current-filament formation and destructive breakdown at C.W. operation [6.38], see also Section 6.3.2. In addition, the production techniques are not yet very reliable and the net-doping concentration turns out to be strongly inhomogeneous, particularly along the growth direction. Impurity control is extremely difficult owing to the high growth temperatures (melting point of GaAs: 1 238 °C) and the unavoidable contact of the melt with the crucible material which is usually quartz or graphite. Hence, today bulk-grown GaAs is scarcely used as TE device material, after having almost served exclusively in the early stages of device research. For more information on bulk-crystal growth see, for example, [6.130, 6.159].

N-type GaAs of high purity and acceptable quality is now available due to the perfection achieved in *epitaxial* crystal-growth processes. The desired GaAs layer is grown on the substrate wafers either from the vapour phase or from a gallium solution ('liquid phase'). Since the temperatures involved in epitaxial growth are only between 700 and 900 °C, less impurities are likely to penetrate into the growing crystal than with the melt technique.

Either highly-doped or semi-insulating GaAs substrates are employed which must be lapped and chemically polished at one side before use, in order to obtain a mirror-like surface. A high doping concentration of impurities like Sn, S or Te is provided in the substrate ($n_0 \approx 10^{18}$ to 10^{19} cm^{-3}; 'n$^+$-material') if it is to serve as one of the ohmic contacts to the active device. Semi-insulating material, on the other hand, with near-instrinsic carrier concentration ($n_0 \approx 10^7$ cm^{-3}; resistivity $\approx 10^8$ Ω cm; 'i-material') is relatively easily obtained by Cu, Fe, Ni, Co or Cr doping (with Cr giving the best results; [6.39]) because of the compensation effect of a deep-lying acceptor level close to the middle of the band gap [6.40], see Table 6.4. This material serves as an excellent support for devices of planar geometry suitable for integrated circuits [6.41]. Since semi-insulating GaAs is photoconductive [6.128] such devices must be protected from any ambient light.

In the epitaxial growth of GaAs from the *vapour phase*, transport of Ga and As

---

*Particularly with oxygen doping which, as a deep donor (Table 6.4), tends to compensate any acceptors without producing a high concentration of mobile electrons.

takes place in the form of certain volatile compounds with H_2 as a carrier gas; (as is used also in elemental form, being gaseous at the temperatures involved). From the various methods only two have proved to be suitable for the growth of high-purity layers: the best results have been obtained by the $AsCl_3$ method first described by Effer *et al.* [6.42, 6.43], whereas the AsH_3 method introduced by Tietjen and Amick [6.23] is more flexible and advantageous especially for the growth of mixed crystals.

The $AsCl_3$ process, which has been studied by a number of authors aiming to produce TE material, e.g. [6.44 to 6.49], uses an open-tube flow system as sketched in Fig. 6.2a. The horizontal furnace usually consists of two zones. Zone 1 contains the quartz boat with the Ga source at about 860 °C (melting point at 29·78 °C), whereas zone 2 establishes a temperature gradient and receives the GaAs substrate at a somewhat lower temperature of about 750 °C. Pd-purified H_2 is loaded with $AsCl_3$ by having it bubble through the $AsCl_3$ source (kept near room temperature) and then enters the furnace at the Ga-source side. The initial reaction taking place when the gas mixture heats up is

$$4 \, AsCl_3 + 6 \, H_2 \rightarrow 12 \, HCl + As_4. \tag{6.1}$$

While the HCl reacts with the Ga to form GaCl and $GaCl_3$, the arsenic vapour is completely absorbed by the Ga source until saturation occurs at 2·25 atom % As. Then a GaAs skin forms around the Ga source and elemental As starts to condense at the cold tube wall beyond the furnace. At this instant the substrate is inserted, with the polished surface upward, into the lower temperature zone, and epitaxial GaAs begins to grow thereon according to the reaction

$$6 \, GaCl + As_4 \rightarrow 4 \, GaAs \downarrow + 2 \, GaCl_3. \tag{6.2}$$

It may be advisable to raise the substrate temperature at first to about 850 °C by providing a short power increase to zone 2 of the furnace [6.49] or by pushing the substrate close to the Ga source [6.47], in order to vapour-etch a few microns off the substrate surface. Such a short etch can improve the substrate-epitaxial interface which has a tendency to become semi-insulating (formation of an interface i-layer). For achieving a high degree of abruptness in the transition from a highly doped n^+-substrate to the low-doped epitaxial n-layer, a prior coating of the rough back surface of the substance with an SiO_2 film ($\approx 5\,000$ Å thick) or high purity GaAs ($> 10 \, \mu$m thick) has proved to be valuable [6.125].

The purity of the source materials is of prime importance for satisfactory epitaxial growth. Presently Ga is commercially available pure to 99·9999 % and $AsCl_3$ to 99·999 %. The success of the $AsCl_3$ process in producing epitaxial layers of high quality is to a large extent due to this high purity of the starting materials.

Best results so far have been reported by Wolfe *et al.* [6.50a] with carrier concentrations n_0 at 300 K as low as $1·8 \times 10^{13}$ cm^{-3} and a total ionized impurity concentration of about 1×10^{14} cm^{-3}. The measured low-field mobilities μ_1 were around 8 000 cm² (Vs)$^{-1}$ at 300 K and as high as 210 000 cm² (Vs)$^{-1}$ at 77 K, with an absolute mobility peak of 340 000 cm² (Vs)$^{-1}$ occurring at 40 K. Thus the

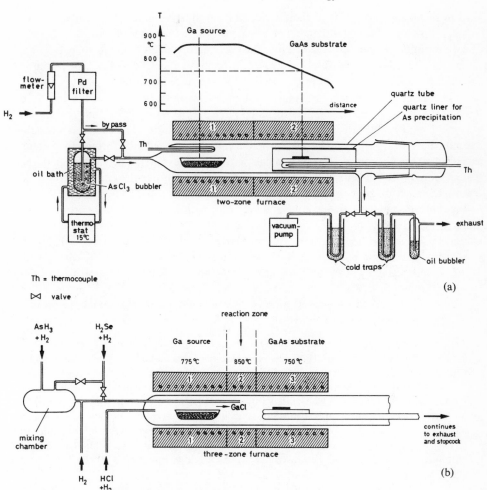

Fig. 6.2 Vapour-phase epitaxy. (a) $AsCl_3$ system with two-zone furnace and $AsCl_3$ bubbler. The furnace establishes the sketched temperature profile. After Gramann [6.49]. (b) AsH_3 system with three-zone furnace. The mixing chamber serves for intimate penetration of the n-type doping gas H_2Se with AsH_3 to achieve uniform doping; for rapid doping changes the chamber can be by-passed. After Berson *et al.* [6.51a].

77 K mobility has reached the theoretical estimate for pure lattice scattering (comprising polar, deformation-potential and piezoelectric scattering; cf. Section 2.2) of about 200 000 cm² (Vs)⁻¹ [6.45], (see also Section 6.1.3). Nevertheless the residual impurity content is still high. Mass-spectrographic analysis has revealed for instance O and C concentrations in the 10^{17} cm⁻³ range and Te as high as 9 × 10^{15} cm⁻³ [6.50b, c]. Even though Te and O are known to be donors (Table 6.4) and

C to be an acceptor most of their concentration must be in an electrically inactive state. To obtain layers with higher carrier densities, in the range of 10^{15} to 10^{16} cm^{-3} a donor impurity can be dissolved into the Ga source, like Sn or Ge [6.46, 6.49], or it can be introduced separately in gaseous form like S_2Cl_2 [6.41a]. Net-doping variations measured along the growth direction have been as low as $\pm 2.6\%$ [6.46].

The other successful vapour-phase system, based on AsH_3, is shown in Fig. 6.2b [6.51]: a three-zone furnace has to be employed, with zone 1 for the Ga source, zone 2 as a reaction region and zone 3 for the substrate. The AsH_3 is brought immediately into the centre zone 2 where it mixes with the GaCl formed by passing HCl over the Ga source. The system has two advantages: firstly, no saturation of the Ga source with As is necessary and, secondly, doping impurities can be introduced into the reaction zone in gaseous form very easily. Thus rapid changes of the carrier concentration in the growing layer and even production of p-n junctions are possible. However, the system seems to be rather susceptible to contaminations from AsH_3 and HCl since, without purposely doping, the lowest 300 K carrier concentrations n_0 obtained in the epitaxial layers were still as high as 3–8×10^{14} cm^{-3}, with 300 K mobilities μ_1 of up to 7 900 cm^2 (Vs)$^{-1}$ and 77 K mobilities of up to only 82 000 cm^2 (Vs)$^{-1}$ [6.51a].

A survey on growth and perfection of chemically-deposited epitaxial layers such as described above is found in [6.53].

A modification of the AsH_3 process, with the pure Ga source replaced by a gaseous organo-metallic compound [e.g. $Ga(CH_3)_3$] has been used to produce single-crystal GaAs films by hetero-epitaxy on insulating substrates. The substrates employed comprise sapphire (α-Al_2O_3), spinel ($MgAl_2O_4$) and beryllia (BeO), see [6.52]. Similarly as semi-insulating GaAs, such substrates are well suited for use in integrated circuits. BeO is especially attractive for high-power applications due to its high thermal conductivity κ [at 300 K one has $\kappa = 2.2$ W (cm K)$^{-1}$ in BeO, $\kappa = 0.33$–0.36 W (cm K)$^{-1}$ in sapphire according to [6.115]; compare with the κ-values of the semiconductors given in Table 6.1]. The low-field mobilities achieved so far in hetero-epitaxy GaAs are low, however [3 400 cm^2 (Vs)$^{-1}$ at 300 K with a carrier concentration of $n_0 \approx 1.3 \times 10^{17}$ cm^{-3}].

Crystal growth by *liquid-phase* epitaxy, as first described by Wolff et al. [6.54], is another process which has received much attention since the achievement of high purity GaAs by using the tilt-tube technique [6.55], see also [6.45] has been reported [2.14]. Basically, a solution of GaAs (single- or poly-crystalline ingots) in Ga, saturated at 850 °C, is brought into contact with a GaAs substrate and slowly cooled in an inert gas atmosphere (usually H_2) thereby inducing further crystalliza-tion. When the desired amount of GaAs has grown on the wafer surface, the solution is tilted off. As in the vapour-phase process, a short etch step may precede the growth, if necessary, by first raising the solution temperature slightly after the contact with the substrate has been made. Figure 6.3 shows a schematic representation of (a) the tilt-furnace system and (c) a typical temperature cycle.

Some modifications of the method of bringing solution and substrate into contact have been employed with variable success [6.56 to 6.58], such as the slide-frame

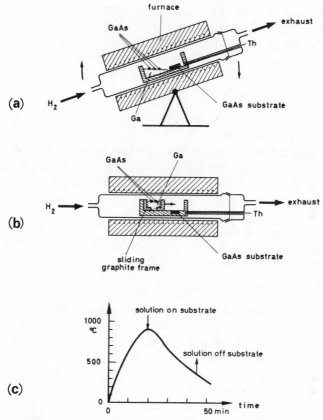

Fig. 6.3 Liquid-phase epitaxy. After [6.130]. (a) Tilt-tube techniques after Nelson [6.55]; (b) slide-frame technique after Beneking [6.57]; (c) typical temperature cycle during growth.

technique sketched in Fig. 6.3b. However, so far the simple tilt technique has yielded layers with the highest purity. Material properties have been achieved as good as those obtained with the $AsCl_3$ vapour transport, with low-field mobilities μ_1 at 77 K of up to 180 000 cm^2 (Vs)$^{-1}$ [6.58]. Net-doping concentrations n_0 at 300 K as low as $n_0 = 1.7 \times 10^{12}$ cm^{-3} have been reported [6.59] with mobilities μ_1 in the 8 000 cm^2 (Vs)$^{-1}$ range. There the mobility increased to 160 000 cm^2 (Vs)$^{-1}$ at 77 K and showed a peak of 250 000 cm^2 (Vs)$^{-1}$ at 50 K. It is interesting to note that the 300 K parameters of the single-crystal starting material dissolved into Ga were $n_0 = 5 \times 10^{15}$ cm^{-3} and $\mu_1 = 4\,500$ cm^2 (Vs)$^{-1}$. Hence, such a solution 're-growth' is an excellent way of purification. Higher carrier concentrations can be obtained by doping the Ga solution with Sn, Te or Se [6.56, 6.60 to 6.63]. Control of a net-doping range from 3×10^{14} to 5×10^{16} cm^{-3} has been achieved with fluctuations of only $\pm 5\%$ over a growth depth of 165 μm [6.63]. For high uniformity it seems to be more advantageous to grow at a constant temperature gradient between solution and substrate instead of cooling down the system [6.61, 6.62].

A definite advantage of the liquid-phase over the vapour-phase epitaxial growth process is its simplicity and the ability to grow relatively thick layers (>100 μm) in shorter times. For large-scale production purposes, however, the vapour-phase process appears to be particularly suitable.

6.1.3 GaAs Crystal Evaluation

After the material has been grown, it must be tested in regard to TE device application. Most important is the determination of carrier concentration n_0 and of the low-field mobility μ_1, is possible as a function of temperature.

Hall measurements using the standard Van-der-Pauw method on sheet-type samples with an otherwise arbitrary geometrical outline [6.64a, b] are most frequently used. (Epitaxial layers can be measured only on semi-insulating substrates; and n$^+$ substrate must be removed.) From the measurements the Hall constant R_H and the resistivity σ_1^{-1} are evaluated eventually yielding

$$\mu_1 \approx \sigma_1 R_H \quad \text{and} \quad n_0 \approx (eR_H)^{-1}. \tag{6.3}$$

The quantity $\sigma_1 R_H$ is frequently called the 'Hall mobility' since in special cases of electron scattering it may deviate from the drift mobility v/E because of different averaging procedures involved (see the general textbooks on semiconductor physics, like [6.29]). Examples for the temperature dependence of μ_1 and n_0 for high-quality GaAs epitaxial layers are shown in Fig. 6.4. In the upper temperature range the measured μ_1 curves are compared to theoretical curves based on lattice scattering only (polar, deformation-potential, and piezoelectric scattering). There, μ_1 increases with decreasing temperature since the lattice vibrations are reduced. At sufficiently low temperatures scattering at the ionized impurities becomes dominant. This decreases the mobility again, and thus a mobility peak is established. The peak, or the mobility value at some fixed temperature nearby (e.g. 77 K), serves as a good parameter for the purity of the material. In fact, the 77 K mobility could be correlated quantitatively to the total ionized-impurity density $N_D + N_A$ (N_D, ionized donor density; N_A, ionized acceptor density; $n_0 = N_d = N_D - N_A$, net-doping density) [6.137]. Ideally, for extrinsic material with the donor level lying shallow below the conduction band edge, n_0 should remain constant above about 50 K until the intrinsic carrier density (Fig. 6.4b, curve n_i) reaches the same order of magnitude. However a slight increase of n_0 usually occurs before that (Fig. 6.4b, curve 1) due to the contribution of deeper energy levels. In melt-grown material of low carrier concentration the deep levels even dominate the thermal behaviour (Fig. 6.4b, curve 2).

Another method for determining n_0 and μ_1 relies on measuring the longitudinal geometrical magnetoresistance [6.65]. It is especially suitable for obtaining data on finished TE devices of sandwich configuration (see Section 6.2.1). If the aspect ratio

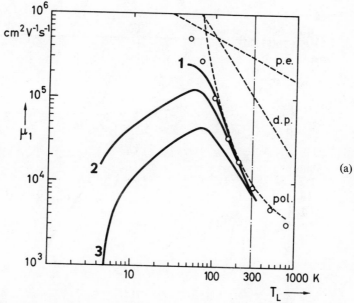

Fig. 6.4 Dependence of low-field electron mobility μ_1 and carrier density n_0 on lattice temperature T_L in GaAs. (a) $\mu_1(T_L)$; curves 1–3 measured for epitaxial material (curve 1: after Hicks and Manley [6.59], curves 2 and 3: after Bolger *et al.* [6.45]). Dashed are theoretical curves for polar (pol.), deformation potential (d.p.), and piezoelectric (p.e.) lattice scattering [6.45]; total lattice-scattering mobility μ_L (circles) is estimated using $\mu_L^{-1} = \mu_{pol.}^{-1} + \mu_{d.p.}^{-1} + \mu_{p.e.}^{-1}$
Material data:

curve	$n_0(300K)/cm^{-3}$	$(N_A + N_D)/cm^{-3}$
1	1.13×10^{13}	?
2	6.6×10^{14}	1.06×10^{15}
3	2.5×10^{15}	3.5×10^{15}

of the sample is $L/d < 0.39$, μ_1 follows, to an accuracy of better than 10%, from

$$\frac{\Delta R_M}{R_1} \approx (\mu_1 B_0)^2 \tag{6.4}$$

where ΔR_M is the resistance change by an applied d.c. magnetic induction field B_0. Similarly to relation (6.3), eqn (6.4) constitutes only an approximation for μ_1 because of the slightly altered averaging procedure for the mobility ('magneto-resistance mobility').

An optical method of determining n_0 makes use of the infra-red reflectance edge near the plasma frequency

$$\omega_{pl} = \left(\frac{n_0 e^2}{m^\star \varepsilon}\right)^{1/2} \tag{6.5}$$

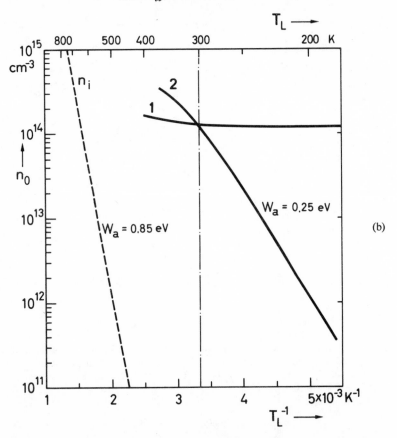

Fig. 6.4 (b) $n_0(T_L)$; curve 1 measured for epitaxial material (after Kang and Greene [2.14]), curve 2 for bulk material grown by Bridgman method showing carrier freeze-out (after Engelmann [6.143]). Curve n_i is theoretical intrinsic carrier density. W_a is an effective activation energy: $n \propto \exp(W_a/kT_L)$.

Material data:

curve	$\mu_1(300\ \text{K})\ \text{cm}^2(\text{Vs})^{-1}$	$(N_A + N_D)/\text{cm}^{-3}$
1	9 300	6.84×10^{14}
2	$\approx 5\ 000$	? (O-doped)

with the effective mass $m^\star$ as a known quantity. It is particularly useful for higher carrier densities n_0 but has also been applied for n_0 as low as $10^{15}\ \text{cm}^{-3}$ [6.66a].

All the measurement techniques described above give only an average carrier concentration n_0 in the layer or device. Additionally, however, it is important to know n_0 as a function of the crystal depth z ('net-doping profile'). For this purpose a metal layer of area A_b is evaporated onto the GaAs surface, thus forming a Schottky barrier. By measuring the barrier capacitance C_b as a function of applied voltage V_b one obtains, see for example [6.66b]

$$n_0(z) = \frac{C_b^3}{e \varepsilon A_b^2 (-dC_b/dV_b)} \tag{6.6a}$$

with

$$z = \varepsilon A_b / C_b. \tag{6.6b}$$

The measuring range of z is limited by the barrier breakdown voltage. This voltage increases with reduction in the net-doping density n_0 (e.g. 2 μm if $n_0 \approx 10^{16}$ cm^{-3}, and 80 μm if $n_0 \approx 10^{14}$ cm^{-3}). With higher density n_0 or thicker layers, therefore, the complete profile can be determined only in steps by successively etching the wafer to an appropriate depth [6.60].—A rather elegant technique for measuring the doping profile of semiconducting layers has been reported [6.67] which uses a modification of a method applied previously to pn-junction transistors [6.68]. It involves driving a Schottky barrier with a small constant r.f. current $\Delta I_t = \hat{I}_t \cos(\omega t)$ (e.g. a few hundred μA at 5 MHz). The depth of the depletion layer is varied by changing the d.c. voltage as in the above case of plotting $C_b(V_b)$. However, the inverse doping profile $n_0^{-1}(z)$ is now obtained by monitoring the r.f. voltage ΔV_b across the barrier at the fundamental frequency which is proportional to the depth z, and at the second harmonic which is proportional to n_0^{-1}:

$$\Delta V_b = \frac{\hat{I}_t z}{\omega \varepsilon A_b} \cos(\omega t) + \frac{\hat{I}_t^2}{4\omega^2 e \varepsilon A_b^2} \frac{1}{n_0} [\cos(2\omega t) + 1]. \tag{6.7}$$

A logarithmic converter and an X-Y recorder allow direct plotting of $\log n_0$ against z on a semilog graph paper. An example of such a plot for an epitaxial GaAs layer on a highly-doped substrate is shown in Fig. 6.5. The narrow lightly-doped layer at the interface is a common problem encountered in GaAs epitaxial layers; it is frequently missed by the $C_b(V_b)$ technique since the depletion layer may sweep across it when the voltage is increased by only 2%.—A method for determining the concentration of deep impurity levels by high-frequency capacitance measurements at low temperature (77 K) is described in [6.69].

By another technique for determining the net-doping profile, which is non-destructive (but has a relatively poor resolution) [6.70], resistance changes $\Delta R_{ph}(z) \propto \sigma_1^{-2}(z) . \Delta\sigma_{ph}$ as a result of photoconductivity $\Delta\sigma_{ph}$ are measured with an optical probe to a resolution of 100 μm. Assuming $\Delta\sigma_{ph}$ and the mobility μ_1 to be independent of z, $\Delta R_{ph}(z)$ becomes inversely proportional to n_0^2. The spatially constant $\Delta\sigma_{ph}$ has been confirmed, employing a quite involved probing technique, by Joule heating. A short pulse (length t_w) of high current density J generates a temperature profile

$$\Delta T(z) = J^2 \sigma_1^{-1}(z) t_w / c_v$$

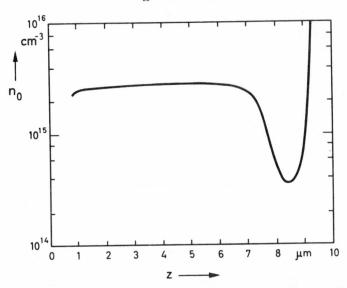

Fig. 6.5 Measured carrier density n_0 against depth z of an epitaxial n-type GaAs layer on an n$^+$-substrate. The lightly-doped boundary layer next to the substrate is a frequently-observed flaw in epitaxially-grown material. After Copeland [6.67a].

(c_v = specific heat capacity of GaAs). This temperature profile is measured by the shift of the optical absorption edge thus yielding the conductivity profile $\sigma_1(z)$ directly. As another check, $\sigma_1(z)$ was also obtained by probing with acousto-electric domains. The resolution of the optical probe could be improved down to 6 μm (see [6.70b]) thus making the technique useful for the shorter ($L < 100$ μm) GaAs devices generally used for microwave power generation.

For measuring the thickness of n-on-n$^+$ layers a non-destructive infra-red interference method can be used [6.71] but microscopic measurement after delineation by angle-lapping and staining, or cleaving and special etching, is also common, e.g. [6.58, 6.71a].

The determination of the velocity-field characteristic by microwave heating of the carriers [2.38, 2.39, 2.41, 2.50, 2.57, 2.58] has proved to be another important diagnostic tool for assessing material quality [2.43]. A rapid evaluation technique for GaAs epitaxial layers with respect to TE device quality was reported [6.72] which consists of measurements on finished TE devices and therefore is especially useful for production control. Several authors studied correlations of the material properties with device performance [6.31, 6.73–6.74]. A theoretical study of the influence of doping variations along the drift direction on the apparent threshold field and on dipole domain oscillations has been carried out in [6.75]. The detrimental effect of net-doping variations on the LSA mode (see Section 5.2) was investigated in [3.53, 6.76]; see also [6.77].

6.2 DEVICE PROCESSING

6.2.1 Device Configuration

Typical configurations of TE device chips are shown in Fig. 6.6. Type (a) is the simplest form: the whole chip consists of active material and is sandwiched between two metal layers forming the ohmic contacts. This type has been frequently constructed from bulk-grown material during the early stage of Gunn-effect research. Structures (b), (c) and (d) are made from epitaxial GaAs on a *highly-doped* substrate, the contacts again forming a sandwich [6.160]. At present they are the most common configurations used for commercial TE devices. In case (b) the top metal contact is confined to a small area in order to minimize surface breakdown and to reduce the active device area, e.g. [6.78]. Such a reduction can also be achieved by mesa etching as shown in types (c) and (d). Type (c) has been etched from the epitaxial side leaving the substrate as mechanical support, e.g. [6.79] and type (d), on the contrary, from the substrate side [6.80]. In the latter case the metal contact on the epitaxial layer is plated thickly enough to serve as support with excellent heat-sink properties (see Section 6.3).

The planar structures (e) and (f) on *semi-insulating* substrate [6.161] are particularly suitable for integrated circuits. Type (e) is obtained by selective and successive etching and epitaxial growth of n^+, n and n^+ layers with SiO_2 or Si_3N_4 films acting as masks for defining the desired geometry; thus this configuration still forms essentially a sandwich structure [6.41a, 6.81]. Type (f), however, in which the active layer is grown directly onto the semi-insulating substrate, is basically different since here the current in the active layer flows mainly in parallel to the contact surface (coplanar contact structure). The metal contacts are either applied directly to the surface of the active layer, e.g. [6.82] or on top of n^+ regions which are formed by selective etching and epitaxial growth, as in type (f) shown, thereby insuring a more uniform field at the contacts, e.g. [6.83]. For improving reliability (Section 6.4) the cross-sectional area of the active n layer is often increased near the anode contact as indicated in Fig. 8.19. Due to a high flexibility in contact and active-layer geometry (see e.g. [6.126, 6.127]) the sheet-type device (f) is highly suitable for special device applications, particularly in pulse generation and processing, see Chapter 8. There it may be required to incorporate an additional electrode for control purposes (see Figs 8.19, 8.20). The surface of planar structures is frequently protected by an insulating film, e.g. a SiO_2 layer sputtered onto the GaAs (surface passivation).

The GaAs chips are usually mounted with their lower side, as depicted in Fig. 6.6, on a heat sink which is part of the package. An exception is type (c) which is mounted frequently with the upper side ('up-side-down' mounting; [4.11]) because of a more favourable heat-flow condition. Good heat sinking is a necessity especially for C.W. operation (see Section 6.3). In the various sandwich structures (a) to (c) most of the heat generated by the d.c. power flows parallel to the electric current. In the coplanar contact structures, on the other hand, it flows perpendicular to the current. A photograph of a mounted chip of form (b), as obtained by a scanning electron microscope,

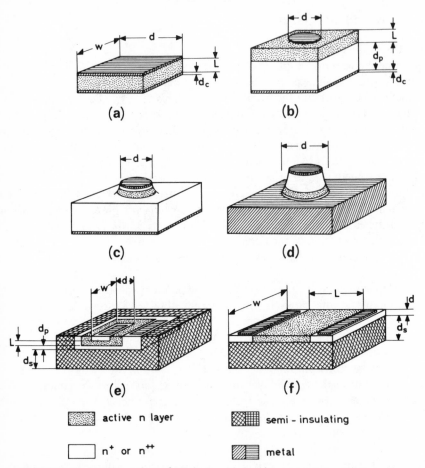

Fig. 6.6 Typical configurations of TE device chips (not to scale). L denotes the active-layer dimension in carrier drift direction; d/w, d_p, d_c, d_s the lateral dimension or the thickness of the active n layer, the passive n^+ or n^{2+} layer, the metal contact, and the semi-insulating substrate, respectively. Types (a) to (e) depict sandwich-contact structures, and (f) a coplanar-contact structure ('surface-oriented' or 'transverse' device).

In particular: (a) Bulk material sandwiched between two metal contacts (typically $d = w = 100\text{--}500\ \mu\text{m}$, $L = 20\text{--}300\ \mu\text{m}$, $d_c = 1\text{--}2\ \mu\text{m}$); (b) Epitaxial material on n^+ substrate, reduced (circular) metal top contact (typically $d = 50\text{--}150\ \mu\text{m}$, $L = 2\text{--}20\ \mu\text{m}$, $d_p = 50\text{--}100\ \mu\text{m}$, $d_c = 1\text{--}2\ \mu\text{m}$); (c) Same as (b) but with epitaxial n^{2+} contact, mesa-etched; (d) n^+–n epitaxial material mesa-etched from n^+ substrate side, electro-plated metal heat sink on n^{2+} contact side [6.80]; (e) Planar n^+-n-n^+ sandwich on a semi-insulating substrate fabricated by selective etching and epitaxial growth. The front face shows a cross-section (typically $d = 30\text{--}50\ \mu\text{m}$, $w = 100\text{--}150\ \mu\text{m}$, $L = 2\text{--}10\ \mu\text{m}$, $d_p = 10\ \mu\text{m}$, $d_s = 50\text{--}100\ \mu\text{m}$) [6.41a and 6.81]; (f) Coplanar contacts applied to epitaxial layer on a semi-insulating substrate (typically $L = 50\text{--}500\ \mu\text{m}$, $w = 100\text{--}1\,000\ \mu\text{m}$, $d = 2\text{--}20\ \mu\text{m}$, $d_s = 50\text{--}100\ \mu\text{m}$) [6.83].

is shown in Fig. 6.7. A top-view photograph of an element of the planar sandwich type (e), with an evaporated strip-line resonator directly connected to it, is given in Fig. 6.8 (monolithic integrated circuit). A coplanar-contact device of type (f) is shown in Fig. 6.9.

Fig. 6.7 Scanning electron micrograph of a GaAs chip of the type sketched in Fig. 6.6b mounted on a Au-plated Cu heat sink with ball-bonded top contact, for use in C.W. domain-mode oscillator operation. (By courtesy of R. Becker, AEG-Telefunken.)

With the sandwich-type devices (b), (c) and (d) the number of top contacts on an individual chip can easily be increased in order to raise the output power by enlarging the active device area. Such an arrangement has superior heat sinking properties as compared to a device with one large contact [4.11, 6.85, 6.86]. An array configuration can be fabricated also by parallelling separate chips as shown in Fig. 6.10 for type (b) devices. As a further possibility, increased output power has been achieved by direct series connection of devices. In this case it is of advantage to grow successively n and n$^+$ layers on a highly-doped substrate to form a multiple sandwich structure (Fig. 6.11, [6.86]). Better heat-sinking, however, is obtained by a series circuit of planar-type devices (e) or (f) with a parallel connection for the heat flow through the common semi-insulating substrate. The planar sheet-type device (f) may be of advantage also for high-power operation in the LSA mode, since the necessary large volume can easily be distributed to achieve optimum heat dissipation (parallel connection of several stripes).

Devices with contacts of different size [e.g. types (b) to (e)] usually exhibit an asymmetric current-voltage characteristic. For safe domain operation the smaller contact, close to which a somewhat higher electric field exists, has to be operated as the cathode (see Section 6.4).

The TE device chips are mostly mounted in packages similar to those used for

223

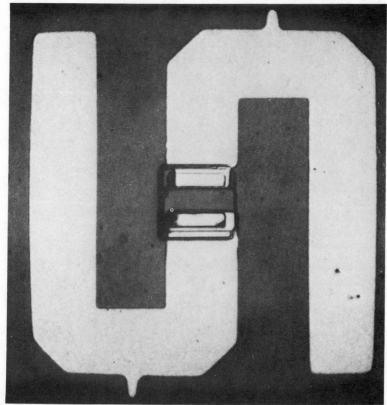

Fig. 6.8 Top view of a planar GaAs sandwich structure according to Fig. 6.6e with evaporated microstrip resonator forming a 'monolithic microwave integrated circuit'. (By courtesy of E. W. Mehal, Texas Instruments Inc.; [6.41a and 6.81].)

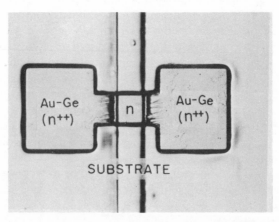

Fig. 6.9 Top view of a coplanar-contact GaAs device of the type shown in Fig. 6.6f. (By courtesy of K. Sekido, Nippon Electric Co. Ltd.; [6.83].)

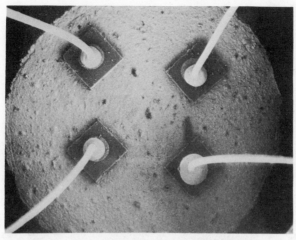

Fig. 6.10 Four sandwich-type GaAs chips mounted on a single heat sink. (By courtesy of R. Becker, AEG-Telefunken.)

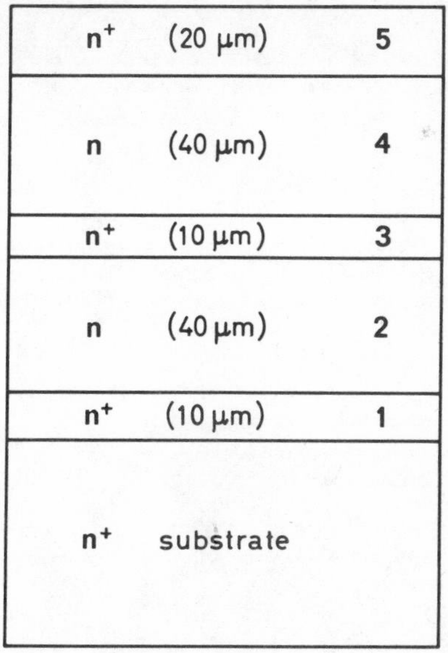

n^+	(20 μm)	**5**
n	(40 μm)	**4**
n^+	(10 μm)	**3**
n	(40 μm)	**2**
n^+	(10 μm)	**1**
n^+	substrate	

Fig. 6.11 Multi-layer epitaxial structure for series operation of TE devices. After Enstrom *et al.* [6.87].

other semiconductor microwave devices, like varactor diodes [3.41a–c]. Figure 6.12 shows some examples of such packages. A cross-sectional view of a package is given in Fig. 6.13. There, the top contact of the chip is connected by a bonded Au wire into a membrane which serves to reduce the detrimental series inductance.

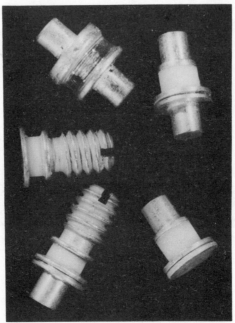

Fig. 6.12 Typical packages used for TE semiconductor devices. (By courtesy of H. Shah, AEG-Telefunken.)

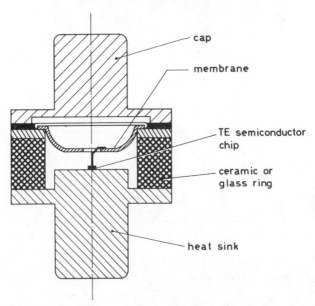

Fig. 6.13 Cross-sectional view of a device package. (By courtesy of H. Bendig, AEG-Telefunken.)

When TE devices are incorporated in integrated microwave circuits special mounting arrangements are preferred with and without individual encapsulation as shown in Figs 6.14 and 6.15.

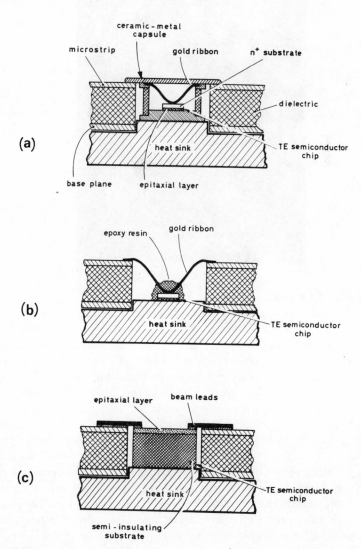

Fig. 6.14 Arrangement of a TE device in a microstrip circuit (cross-section, not to scale). After Heeks *et al.* [6.88a]. (a) Sandwich-type TE device encapsulated in a miniaturized pill package; (b) sandwich-type TE device mounted directly into the microstrip structure; protected by epoxy resin; (c) coplanar-contact type TE device with beam leads mounted directly into microstrip structure (electrically series-connected).

227

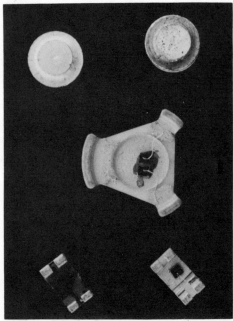

Fig. 6.15 Packaged TE devices for use in microstrip circuits. (H. Shah, AEG-Telefunken.)

6.2.2 Forming of Contacts

The contacts to the semiconductor chips should be essentially ohmic (see Section 3.3)*. Achieving satisfactory ohmic contacts on the relatively low-doped TE GaAs is not a straightforward task, however, and a certain attention has to be paid to this problem. We shall discuss a series of undesired effects that may occur during contact formation and indicate how they can be avoided to obtain acceptable results.

For TE-device work GaAs wafers oriented along {100} crystal planes (see first footnote in Section 2.1, p. 12) are preferably selected since the {100} planes, contrary to the {111} planes, do not show polarity effects (see Fig. 6.1b and d). Another advantage of the {100} planes is their rectangular intersection with the {110} cleavage planes (Fig. 6.1c), thus simplifying the final separation of the wafer into individual square chips. Wafers from bulk material usually are cut to a thickness of approximately 500 μm with a diamond saw. In the case of the epitaxial wafers the as-grown thickness of the active layer may range from less than 1 μm to about 100 μm. Before applying the contacts the wafers are lapped and eventually etch-polished to the thickness required for the devices (see e.g. [6.89]) and then degreased. The etch, e.g. dilute boiling NaOCl or sulphuric peroxide acid (3 H_2SO_4 : 1 H_2O_2 : 1 H_2O), has to remove at least the surface layer damaged by the mechanical processing (roughly to

---

*A doping-notch or surface-barrier type control region at the cathode is useful for a uniform d.c. field distribution in stable amplifier devices (p. 106) and for dipole nucleation in dipole domain operation (see Sect. 3.3.3, also pp. 231, 250 and Ch. 8.) Also here the specific contact resistance should be relatively low.

a depth of three or four times the grain size of the lapping compound). For degreasing, a sequence of hot TCE (trichlor ethylene), methanol and de-ionized water is used.

Basically three contacting techniques exist: (1) alloying of pre-forms; (2) evaporation of metal or alloy films through metal masks or over the whole wafer (possibly combined with wet plating) and subsequent heat treatment; and (3) epitaxial growth of a thin layer (usually 2–5 μm) of extremely high n-type doping (n^{2+} layer), either by the liquid- or by the vapour-phase process [the n^{2+} layer is then contacted by technique (2)]. The contacts must be barrier-free, should have mechanical strength (not too soft or too brittle) and thermal stability.

In technique (1) small spheres or disks of the contact metal or alloy such as pure Sn [6.89], Au-Ge eutectic or better a mixture of Au-Ge eutectic and Sn [6.90] are placed on the GaAs surface and heated beyond melting temperatures in a reducing atmosphere (H_2 or forming gas, i.e. a mixture of 80% N_2 and 20% H_2) with HCl acting as flux. Some GaAs is dissolved ('wetting' of GaAs) during this process in the melt and partly recrystallizes as an extremely thin n^{2+} layer during the subsequent cooling-down period in a way similar to liquid-phase epitaxy.

Better geometrical control is obtained with technique (2) when using metal masks. This method is to be preferred from the point of view of economy. A large variety of contact-metal compositions have been investigated, such as Ni-Sn, Ni-In, Au-Sn or Au-Sn-Cr(-Ni), Au-In, Au-Ge or Au-Ge-Ni, Ag-Sn and Ag-In-Ge, with a total film thickness of usually less than 1 μm [6.90–6.99]. A similar heating cycle as for technique (1) is used to form the ohmic contact. A flux, however, is generally not necessary. The heating cycle is critical in time and temperature, with an optimum maximum alloying temperature for obtaining minimum contact resistance as indicated in Fig. 6.16. The details of the cycle depend on the particular alloy composition and the net-doping concentration in the GaAs material [6.97]. Evaporating directly onto a heated GaAs wafer is also possible [6.103]. Pure Sn or Sn-based alloy contacts show good ohmic behaviour but have the disadvantage of relatively low melting temperatures (≈ 232 °C). Au-based alloy contacts, on the other hand, possess poor wettability to GaAs probably because of melting temperatures (e.g. 356 °C for Au-Ge eutectic) far below the Au-GaAs eutectic temperature of 450 °C [6.98]. Wetting and hence contact quality is greatly improved by adding Ni either as an overlay [6.93, 6.99a] or as a thin (≈ 500 Å) layer directly onto the GaAs surface [6.96]. In the latter case Ni-Cr or Cr is also suitable. However, the interface may still be not very planar as revealed by angle lapping. Ag-In-Ge alloys prove to have excellent wetting properties and yield satisfactory low specific contact resistances (resistance times contact area) ranging from below 10^{-4} Ω cm^2 on 0·1 Ω cm (or lower resistivity) GaAs to 10^{-3} Ω cm^2 on about 1 Ω cm GaAs. These contacts have the additional advantage of high melting temperatures (≈ 600 °C compared to a Ag-GaAs eutectic temperature of 650 °C) and are thus suitable for subsequent high-temperature device processing, like SiO_2 passivation, mounting, and bonding [6.94]. However, Ag-based alloy contacts are susceptible to ageing effects, even 'on the shelf', if not passivated e.g. by an Au overlay or a hermetically-sealed encapsula-

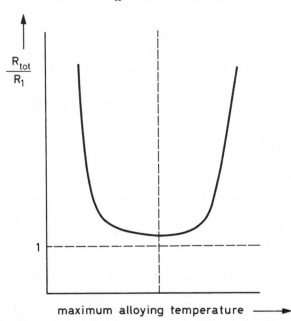

Fig. 6.16 Schematic plot of normalized total device resistance $R_{tot} = R_1 + R_c$ against maximum alloying temperature (after Paola and Knight [6.97a]); R_1 is the (low-field) bulk resistance, R_c the contact resistance.

tion [6.98]. For more details on ageing and the migration properties of Sn and In under bias conditions see Section 6.4.

Metallurgical contacts after technique (1) or (2) are mostly adequate for contacting bulk material. However, when contacting epitaxial TE GaAs, difficulties may arise due to the formation of a thin ($\approx 0{\cdot}1$–$2\ \mu$m) damaged crystal layer at the interface. This layer exhibits a higher resistivity, probably caused by an increased acceptor concentration [3.3], and seems to be present, at least over parts of the contacted area, even if it cannot be detected by the $C_b(V_b)$ measurement technique mentioned in Section 6.1.3 [6.99a]. At any rate its presence can be inferred from an increased dislocation density and is most likely responsible for a fairly high specific contact resistance obtained. Its effect on TE device performance is discussed in Section 6.4. The least disturbances were obtained with the Au-Ge-Ni contact. In bulk material the damaged interface layer has little effect because of the role of the deep-lying donors which bulk GaAs contains in a much higher concentration than epitaxial GaAs. The deep, still occupied, donors tend to compensate any acceptors introduced by the contacting process, without reducing the density of mobile carriers [6.99a]. On the other hand, the unoccupied deep energy levels cause the commonly-observed carrier freeze-out (see curve 2 of Fig. 6.4b). Such a model also explains the reduction in specific contact resistance generally obtained with increased net-doping density

n_0 in the GaAs, since the number of acceptors produced should be more or less independent of the doping, and hence the compensation becomes less effective for higher n_0. Temperature measurements yield for the contact resistance a dependence according to $\exp(W_a/kT_c)$ with a fairly large activation energy W_a ($W_a \approx 0{\cdot}1$ eV in the case of Ag-Sn contacts, [6.105]; T_c is the absolute temperature of contact).

Epitaxial contact formation after technique (3) [6.97 to 6.102], which is most widely used today, always yields good planar interfaces and, with special precautions, a high-resistivity interface layer can reliably be avoided. An epitaxial contact is automatically obtained when growing the active epitaxial n layer on an n^+ layer on an n^+ substrate. The second contact is produced in the form of an n^{2+} layer on top of the n layer by the same epitaxial techniques as described in Section 6.1.2 when a high concentration of dopants is introduced. In the case of the liquid-growth process, the highly-doped Ga solution may be replaced by a solution of GaAs in pure Sn. A consistent elimination of the formation of a detrimental interface layer could be achieved [6.99a] by etching about 5 μm off the n layer into the Ga solution whilst the furnace temperature was slightly raised just prior to growing the n^{2+} layer. A weak interface layer, however, seems to be advantageous for dipole-mode operation since it improves the domain nucleation properties [6.144b]; ('control region', see Section 3.3.3). For LSA devices, on the other hand, it should be avoided completely. An example for an interface layer has already been shown in Fig. 6.5.

It is fairly simple to contact the highly-doped n^{2+} GaAs by technique (2). Specific contact resistances turn out to be even lower than on n layers, for example about 10^{-5} Ω cm for $n_0 \approx 2 \times 10^{18}$ cm^{-3} [6.103, 6.104]. The undesired parasitic series resistance of the device [6.104] is thus very low and, in addition, it becomes less temperature-sensitive [6.105], (see also Section 6.3.2).

As a summary to this section, the authors can report, from their own experience, particularly satisfactory results obtained with an Au-Ge-Ni contact [6.93]. It can successfully be applied to bulk and epitaxial n-GaAs and, of course, also to epitaxially-grown n^{2+} contacting layers. Au-Ge-Ni contacts have proved to be satisfactory with regard to mechanical and thermal stability, low contact resistance, device efficiency, ageing, noise performance, and ease of soldering and bonding. The optimum metal-layer composition was found to be 500 Å Ni and 3 000 Å Au-Ge eutectic (12% per weight Ge) evaporated subsequently onto the GaAs surface. The alloying was carried out by heating the evaporated wafer to 420 °C for a period of 5 min.

6.2.3 Further Device Fabrication

The individual devices are obtained from the contacted wafer either by cutting with a diamond saw or by scribing and breaking. In the latter case a diamond scriber is led along intersection lines of the {100} wafer surface with perpendicular {110} cleavage planes. Breakage is then easily accomplished in an ultrasonic bath. Epitaxial wafers containing dot devices of the type (b) of Fig. 6.6 are scribed preferably from the backside in order to avoid mechanical damage to the active layer. For devices of the mesa type (c) or (d) of Fig. 6.6, mesa etching is made before separation, employing standard photolithographic techniques. A certain amount of overlap of the photo-resist mask beyond the metallized contact areas ensures a better geometrical control. With the reversed mesa form (d) of Fig. 6.6, an array of contact dots has to be made on the wafer substrate side, and the overall metal contact on the epitaxial side has to be plated-up with Au, Cu or Ag to about 50–100 μm [6.80]. The mesas are formed by completely etching through the semiconductor, thus leaving the plated metal sheet as support. The individual dice are cleaned (particularly degreased) and then they are mounted onto the, usually Au-plated, heat sink of the package (see Fig. 6.7). Soldering with Sn preforms is not very satisfactory due to a tendency of balling-up. Better results are obtained with Ge solder, especially if a thin (≈ 1 μm) Ge layer has been evaporated on top of the previously formed ohmic-contact area of the wafer prior to cutting it to dice [6.96a]. Lowest thermal heat resistance combined with good mechanical strength is obtained using a thermo-compression bonding process with ultrasonic activation. The top contact area (or areas) of the chip is connected to the package contact (or contacts) by bonding an Au wire or ribbon in-between (see Fig. 6.13). A good bond to the chip contact is ensured if the Au wire is ball-tipped (thermo-compression ball-bonding machine), see Fig. 6.7. Mesa devices of the type of Fig. 6.6c, widely-used today for microwave oscillators, are mounted frequently to the heat-sink 'up-side-down' (Fig. 6.17) in order to reduce the heat resistance. Instead of wire bonding, sometimes a simpler pressure contact is preferred [6.89]. In microwave integrated circuits the beam-lead technique [6.106], which can be applied particularly to planar devices (Fig. 6.6e and f), is advantageous.

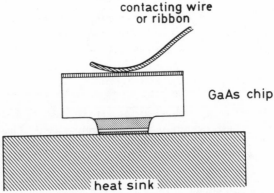

Fig. 6.17 'Up-side-down' mounted mesa chip ('flip chip'), [6.113], (not to scale).

6.3 THERMAL CONSIDERATIONS

6.3.1 Heat Sinking

The maximum average microwave power which a single TE device is able to generate is mainly limited by the highest temperature T_{to1} that can be tolerated within the device. In C.W. operation the d.c. power density converted to heat is a few 10^7 W cm^{-3} for applied electric fields of the order of the threshold field and for low-field resistivities of the active GaAs near 1 Ω cm. This high power density poses a limit to the dimensions of the sample in order to keep the temperature below T_{tol}, and as a consequence, effective heat sinking becomes of the utmost importance in TE-device design.

The exact value of T_{tol} is not yet definitely known for GaAs samples. At about 620 °C arsenic starts to evaporate because of its high vapour pressure, thus forming zones of pure metallic gallium [6.107]. Such a decomposition can be prevented by coating the GaAs with a layer of SiO$_2$. On the other hand, 1 Ω cm GaAs ($n_0 \approx 10^{15}$ cm^{-3}) becomes intrinsic at about 450 °C (cf. Fig. 4.6b); however, in the best material available the carrier concentration starts to increase at still lower temperatures probably because some deeper impurity levels are thermally activated (curve 1 of Fig. 6.4b). The most-widely used Au-based alloy contacts melt at about 350 °C. Ageing effects (Section 6.4), however, may occur even at somewhat lower temperatures. Hence, about 200 °C is considered to be a safe value for the highest tolerable temperature T_{tol}.

During domain operation at high d.c. drive, the anode contact of planar sheet devices may locally start to melt before a metallic breakdown channel develops [6.108]. It is proposed that hole injection and recombination at the anode is responsible for the additional heat generating mechanism there. Hence, the anode must be particularly well cooled; in sandwich-type devices the contact at the heat-sink side should always be biased with *anode* polarity for safe device performance (see also Section 6.4).

The maximum temperature T_{max} which is reached during operation in the active GaAs layer can be estimated from a simple heat-flow model (cf. [6.111 to 6.113]) in the following way. An 'up-side-down' mounted mesa structure (Fig. 6.17) is considered. Since the thermal resistance R_{th} on the top side is relatively large due to the fairly thick substrate in conjunction with the lack of a good heat sink, heat flow in this direction can be neglected. Similarly, any heat flow within the device chip in a lateral direction is negligibly small. Figure 6.18 is an enlarged view of the thermally significant part of Fig. 6.17. It is assumed that in the active region (a) of thickness L heat is generated uniformly, either in C.W. operation or in pulsed operation with sufficiently high duty cycle (ratio of on-to-off time) so as to ensure a steady heat flow determined by the averaged dissipated power. Between the active region (a) and the heat sink (h) we have a passive region (p) of thickness d_p if an epitaxial n^{2+} contact layer with negligible heat generation is employed, and, further, a metallic contact region (c) of thickness d_c connecting the device chip to the heat sink. Due to its thermal resistance, each region contributes a temperature rise ΔT above the ambient temperature T_0 thus establishing a temperature T_{max} at the top of the active region (a):

$$T_{max} = T_0 + \Delta T_h + \Delta T_c + \Delta T_p + \Delta T_a. \tag{6.8}$$

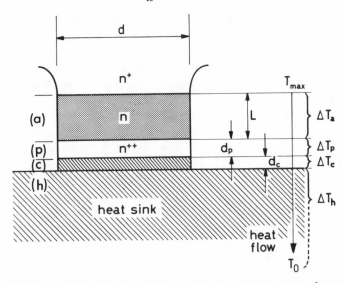

Fig. 6.18 Thermally significant part of mesa chip shown in Fig. 6.17, after [6.113], (not to scale). Heat flow into n⁺-substrate neglected; T_{max} maximum temperature in active region (a); T_0 ambient temperature at lower heat-sink surface; ΔT temperature rises in heat sink (h), contact region (c), passive n²⁺-layer (p), and active n-layer (a), respectively.

For the simple heat flow geometry considered, with a thermal conductivity κ (assumed to be independent of temperature) the ΔT_i's can be calculated from the heat-flow equation which is analogous to Ohm's law (e.g. [6.109a]) viz.

$$\Delta T_i = R_{th,i} P_0 \tag{6.9}$$

where P_0 is the d.c. power which for small d.c.-to-r.f. conversion efficiencies ($\eta \lesssim 10\%$) can be considered to represent the dissipated power, i.e. the total heat generated per unit time in region (a), and R_{th} is the appropriate thermal resistance ($i = h, c, p, a$). The one-dimensional model of heat flow through an area A (which here is the area of the mesa) yields for the heat generating region (a) (d.c. power dissipation assumed to be uniform)

$$R_{th, a} = \frac{L}{2\kappa_a A} \tag{6.10}$$

and for the regions that merely conduct heat

$$R_{th, p} = \frac{d_p}{\kappa_p A} \tag{6.11}$$

and

$$R_{th, c} = \frac{d_c}{\kappa_c A}. \tag{6.12}$$

Any thermal resistance at the contact interface between regions (c) and (h) can be included in eqn (6.12) by using an 'effective' thermal conductivity κ_c.

For the heat-sink region (h) in which lateral heat flow is also important, $R_{th,h}$ can be estimated from the maximum temperature rise* which is generated by a uniform heat flux P_0A^{-1} entering across an area A into the surface of a semi-infinite body [6.109b]. One obtains

$$R_{th,h} = \frac{g}{\kappa_h A^{1/2}} \tag{6.13a}$$

where g is a geometrical factor depending on the form of the area A:

$$\left.\begin{array}{l}
\text{circle } A = d^2\pi/4: \quad g = \pi^{-1/2} = 0\cdot564, \\[6pt]
\text{square } A = d^2: \quad\quad g = (2 \sinh^{-1} 1)\pi^{-1} = 0\cdot561, \\[6pt]
\text{rectangle } A = dw: \quad g = [(d/w)^{1/2} \sinh^{-1}(w/d) + (w/d)^{1/2} \sinh^{-1}(d/w)]\pi^{-1}, \\[6pt]
\quad\quad\quad\quad \text{e.g. for } d = 4w: g = 0\cdot492.
\end{array}\right\} \tag{6.13b}$$

Expressions (6.13b) indicate that for thermal reasons a rectangular active device area with as large a side ratio as possible is to be preferred to a circular one. By comparing eqn (6.13a) with (6.11) or (6.12) one may define an effective thermal thickness of the heat sink which is

$$d_h = gA^{1/2}. \tag{6.13c}$$

If the thermal conductivity depends on temperature, the thermal resistance R_{th} has to be calculated by using the average value

$$\bar{\kappa} = \int_{T_1}^{T_2} \frac{\kappa\, dT}{T_2 - T_1} \tag{6.14}$$

where T_1 and T_2 are the temperatures at the two endfaces of the body [6.109a]. For a metallic heat sink and contact layer, κ_h and κ_c can be regarded as temperature-independent. The thermal conductivity in GaAs follows, as in most semiconductors, a T^{-1} law (T is the absolute temperature) in the temperature range of interest, i.e.:

$$\kappa_a = \kappa_a^\star T^{-1} \text{ and } \kappa_p = \kappa_p^\star T^{-1}. \tag{6.15}$$

The average value can best be approximated by the value of eqn (6.15) at one half of the appropriate temperature rise ΔT. This yields simple rational equations for the ΔT's. Accordingly one has in eqns (6.10) and (6.11):

$$\kappa_a \approx \kappa_a^\star (T_{max} - \Delta T_a/2)^{-1} \tag{6.16a}$$

---

*The maximum temperature in the centre of the area A is chosen rather than the average across A since the first one determines also the highest temperatures actually present along the centre of the GaAs chip if $L \ll d$, a situation generally satisfied for mesa devices. In the model, these temperatures in the chip are approximated as being laterally uniform.

and

$$\kappa_p \approx \kappa_p^\star (T_{\max} - \Delta T_a - \Delta T_p/2)^{-1} \qquad (6.16b)$$

respectively, with $\kappa_a^\star = 150$ W cm^{-1} in the low-doped active region and $\kappa_p^\star = 120$ W cm^{-1} in the highly-doped passive region [6.110]; see also [6.111]. The approximation made by using eqns (6.16) is acceptable for most practical cases. This becomes obvious when comparing the results obtained to those derived by employing the rigorous expressions for $\bar{\kappa}$ according to eqns (6.14) and (6.15), as carried out in [6.111 to 6.113a]*. The system of eqns (6.8) to (6.13) with (6.16) can be used for either calculating the maximum sample temperature for given sample parameters or vice-versa.

A commonly used quantity for characterizing the quality of heat removal is the total thermal resistance

$$R_{\text{th,tot}} = (T_{\max} - T_0)P_0^{-1}. \qquad (6.17)$$

Because of the temperature dependence of the thermal conductivities involved $R_{\text{th,tot}}$ is a weak function of P_0. It can be measured simply if $T_{\max}$ is inferred from a temperature-dependent sample parameter such as low-field resistance $R_1(T)$ or threshold current $I_T(T)$ and these relationships have been pre-determined (cf. Fig. 7.18).

Using the rigorous $\bar{\kappa}$ value of GaAs, a calculation of the maximum possible sample dimensions of a square-area chip, assuming the total volume to be active, was carried out in [6.112] as a function of the tolerable dissipated power and/or power density for $T_{\max} = T_{\text{tol}} = 500$ K. The optimum case of negligible heat resistances $R_{\text{th,p}}$ and $R_{\text{th,c}}$ in the passive and contact layers, respectively, was assumed with copper of $\kappa_h = 3 \cdot 8$ W (cmK)$^{-1}$ as heat-sink material. The power density p_0 is only slightly temperature-dependent as long as n_0 stays constant and was considered to be temperature-independent in the calculations. It is preferably expressed in terms of the low-field electrical conductivity σ_1 of the active region (or carrier density n_0) and the applied electric field E_{h0}, viz.

$$p_0 = \frac{P_0}{AL} \approx 0 \cdot 75 E_p \sigma_1 E_{h0} = 0 \cdot 75 e \mu_1 E_p n_0 E_{h0}, \qquad (6.18a)$$

assuming as the bias current density $J_0 \approx 0 \cdot 75 \, \sigma_1 E_p$ (peak-to-valley ratio $J_p/J_v = v_p/v_v \approx 0 \cdot 5$). For a velocity peak field $E_p = 3 \cdot 4$ kV cm^{-1} and a low-field mobility $\mu_1 = 6\,250$ cm^2 (Vs)$^{-1}$ eqn (6.18a) becomes

$$p_0/\text{W cm}^{-3} \approx 2 \cdot 55 \times 10^3 \frac{\sigma_1 E_{h0}}{\text{A cm}^{-2}} \qquad (6.18b)$$

---

*In the references cited, $\bar{\kappa}$ has not been formulated explicitly, but the equations used there reduce to a system similar to eqns (6.8) to (6.13) with κ_a and κ_p replaced by the average value as defined by eqn (6.14) using expression (6.15). Note that eqns (6.16) are obtained if $\ln(T_2/T_1)$ is replaced by $2(T_2/T_1 - 1)(T_2/T_1 + 1)^{-1}$ which, in the range $0 \cdot 5 \leq T_2/T_1 \leq 2$, deviates from $\ln(T_2/T_1)$ by less than 5%.

with

$$\frac{\sigma_1 E_{h0}}{A\ cm^{-2}} = \frac{10^{-15} n_0 E_{h0}}{V\ cm^{-4}}.$$

(6.18c)

The results obtained for an ambient temperature of $T_0 = 300$ K are plotted in Figs 6.19 and 6.20. There, L is given as a function of the largest tolerable side length d_{tol} which causes T_{max} in the sample to reach $T_{tol} = 500$ K. In Fig. 6.19 the quantities P_0 (dash-dotted curves) and $\sigma_1 E_{h0}$ (solid curves) are fixed parameters. With a chosen E_{h0} the $\sigma_1 E_{h0}$ curves belong to fixed values of σ_1 and also of n_0 if μ_1 is known. The $\sigma_1 E_{h0}$ curves are specified for the particular bias field $E_{h0} = 10$ kV cm^{-1}, i.e.

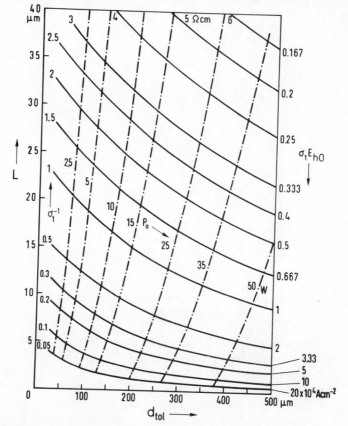

Fig. 6.19 Active-layer thickness L against thermally tolerable side length d_{tol} of a square GaAs chip (Fig. 6.6a with $d = w$) on a Cu heat sink with negligible thermal contact resistance $R_{th,c}$. The tolerable maximum temperature in the chip is set at $T_{tol} = 500$ K. Parameters are $\sigma_1 E_{h0}$ or σ_1^{-1} for fixed E_{h0}, here $E_{h0} = 10$ kV cm^{-1} (solid curves), and d.c. power P_0 (dash-dotted curves). After Becker and Bosch, [6.112].

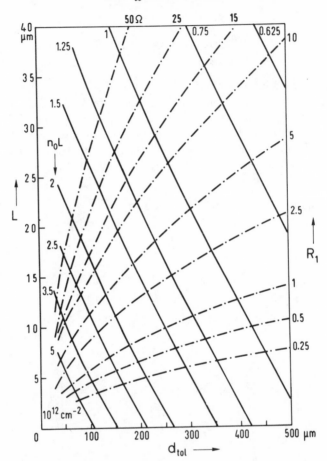

Fig. 6.20 Same diagram as in Fig. 6.19 but with the parameters n_0L product (solid curves) and low-field resistance R_1 (dash-dotted curves) for $E_{h0} = 10\,\text{kV cm}^{-1}$. After Becker and Bosch [6.112].

for about three times threshold. In this way curves are obtained, as shown in Fig. 6.20, with the low-field sample resistance $R_1 = L/\sigma_1 d_{\text{tol}}^2$ (dash-dotted) and the n_0L product (solid) as parameters.

As an example, we now discuss the design of a TE device constructed from $1\,\Omega$ cm GaAs that is to give the highest possible output power at 5 GHz when operated as an oscillator in the dipole-domain mode. With a mean domain drift velocity $\bar{v}_a \approx v_D = 10^7\,\text{cm s}^{-1}$ the relation $f_\tau = v_D/L$ requires $L = 20\,\mu\text{m}$. From the graph of Fig. 6.19 it follows that, at $E_{h0} = 10\,\text{kV cm}^{-1}$, we have $d_{\text{tol}} = 75\mu\text{m}$ with a power dissipation of $P_0 \approx 3$ W (total thermal resistance $R_{\text{th,tot}} = 67$ K W^{-1}). Assuming now that the efficiency is $\eta = 5\%$, a microwave output power of $P = \eta P_0 = 150$ mW is obtained. Further important sample data are found to have the following values: $R_1 = 38\Omega$, $V_T = E_T L = 7.8$ V, $V_0 = E_{h0} L = 20$ V, and eventually, with $\mu_1 = 6\,250$ cm^2 (Vs)$^{-1}$, a value $n_0 = 10^{15}$ cm^{-3} or $n_0 L = 2 \times 10^{12}$ cm^{-2}. Using GaAs with double the resistivity i.e. $\sigma_1^{-1} = 2\,\Omega$ cm, dramatically increases the achievable output power $P = \eta P_0$ to 1·3 W ($R_{\text{th,tot}} = 7.7$ K W^{-1}), i.e. by almost a factor of 10, since the dimension d_{tol} can be increased to 322 μm ($R_1 = 2.6\Omega$).

the passive n$^+$ layer ($R_{\text{th,c}}$ still neglected) is seen from Fig. 6.21 in which L is plotted

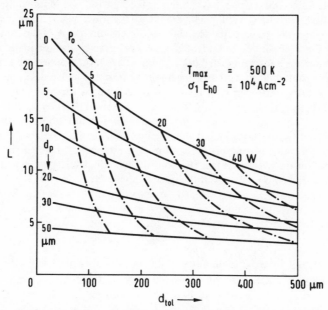

Fig. 6.21 Diagram similar to Fig. 6.20 but with thickness d_p of a finite passive n$^+$-layer between active layer and heat sink as parameter (solid curves). The d.c. power P_0 is again chosen as a second parameter (dash-dotted curves). After Becker and Bosch [6.112].

against d_{tol} for constant $\sigma_1 E_{h0} = 10^4$ A cm^{-1}, i.e. $\sigma_1^{-1} = 1\,\Omega$ cm at $E_{h0} = 10$ kV cm^{-1}. Taking our above example of $L = 20$ μm, an n$^+$ layer thickness of only $d_p = 2$ μm reduces d_{tol} to 36 μm and hence the available microwave power from $P = 150$ mW to $P = 33$ mW. This shows the importance of the 'up-side-down'-mounting technique of the mesa structure shown in Fig. 6.18.

If a second heat sink is attached to the top contact of the sample in such a way that a symmetrical heat flow into the two heat sinks results, the system of eqns (6.8) to (6.13) with (6.16) still remains valid if L and P_0 are replaced by $L/2$ and $P_0/2$. The same applies to Figs 6.19 to 6.21. In such an ideal situation the tolerable side length d_{tol} is increased and thus higher output powers P would be obtained. However, this case is difficult to achieve in practice. For asymmetrical heat flow, which is easier to realize [6.164], the system of equations (subscript 1) can be supplemented by a similar second set (subscript 2) in a straightforward way with the additional requirements of $L_1 + L_2 = L$ and $P_{01}/L_1 = P_{02}/L_2 = P_0/L = p_0 A$. In [6.112] also the asymmetrical case has been treated in the more rigorous manner of using for the GaAs sections the exact κ values according to eqn (6.14).

So far lateral heat flow has been neglected in the GaAs chip itself which is justified for sandwich structures like type (a), (d), or the up-side-down mounted mesa chip (c) of Fig. 6.6 where the different heat carrying layers are laterally exposed to air. This neglect is not always possible for the sandwich types (b), (e), or the right-side-up mounted mesa chip (c) of Fig. 6.6. Lateral heat flow is still negligible if the layer thickness L and d_p (d_p and d_s in case of Fig. 6.6e) are small compared to the linear dimensions d or w of the contact area A. (For structure Fig. 6.6b the condition $L < d, w$ always has to be ensured in order to obtain a sufficiently homogeneous field distribution in the active layer, see Section 6.4.) If this is not the case, particularly if $d_p > d$ or w, heat spread from the active device area A to the passive support area $A^\star$ (lower chip area) can be accounted for by using an electrical analogue [6.114] in which an equivalent capacitance is derived that can be transcribed into a thermal conductance. In this approximation the temperatures both at the interface between active and passive layer and at the interface between passive layer and heat sink are considered to be laterally uniform. Therefore, the estimate of the heat-sink resistance should be based on the average rather than on the maximum surface temperature in the lower area $A^\star$ of the chip [yielding, in the case of a circular area, $g = 8/(3\pi^{3/2}) = 0{\cdot}479$ in eqn (6.13a) instead of $g = \pi^{-1/2} = 0{\cdot}564$; in this situation, however, the heat resistance $R_{\mathrm{th,h}}$ approaches the value obtained for heat flow into a semi-infinite body with constant surface temperature across $A^\star$ which yields for a circular area: $g = (\pi^{1/2})/4 = 0{\cdot}442$ [6.109b])].

Now we consider briefly the planar sheet structure, shown in Fig. 6.6f. Here, the approach is basically the same as before, except that the heat flow is essentially perpendicular to the electrical current flow. If both the active layer (d) and the substrate (d_s) are thin enough, i.e. $d, d_s < L, w$, the system of eqns (6.8) to (6.13) can still be used with the following modifications: eqn (6.13a) is specified for a rectangular area; in all equations d and L are interchanged and the subscript p is replaced by the subscript s. The thermal conductivity of the active layer still follows from (6.16a) and that of the substrate from (6.16b) with $\kappa_s^\star \approx \kappa_a^\star = 150$ W cm^{-1} for semi-insulating GaAs. If the substrate is not GaAs but consists of BeO (hetero-epitaxy: [6.52a]) heat removal is highly improved since then $\kappa_s^\star \approx 670$ W cm^{-1} [$\kappa_{\mathrm{BeO}} = 2{\cdot}2$ W (cmK)$^{-1}$ at 300 K] according to [6.115]. An electrical-analogue approach [6.114] has to be used if the active layer or, particularly, the substrate is so thick that lateral heat flow becomes important.

An improvement in the power capability of TE devices can be achieved if the Cu heat sink is replaced by diamond of type IIa (see [6.116]) which has a room temperature thermal conductivity of $\kappa_{\mathrm{dia}} = 23$ W (cmK)$^{-1}$, i.e. about six times that of Cu. Even though κ_{dia} decreases strongly at elevated temperatures according to a $T^{-3/2}$ law, one still has about $\kappa_{\mathrm{dia}} = 11$ W (cmK)$^{-1}$ at 500 K [6.117]. Thus, with a diamond heat sink also κ_h in eqn (6.13a) has to be averaged according to eqn (6.14). To achieve the better heat-sink properties it would be sufficient to attach a diamond slab of the effective heat-sink thickness defined by eqn (6.13c) on top of the Cu heat sink. On the other hand, there are still technological difficulties in achieving good mechanical and thermal bonds with diamond.

Problems regarding the stability of the semiconductor-to-heat-sink bond may arise, particularly for large chip areas, due to the difference in the coefficient of expansion α; GaAs has a coefficient $\alpha = 6\cdot9 \times 10^{-6} \, \text{K}^{-1}$ [6.118] between -62 °C and 200 °C whereas for Cu one has $\alpha = 16\cdot8 \times 10^{-6} \, \text{K}^{-1}$ and for diamond $\alpha = 1\cdot2 \times 10^{-6} \, \text{K}^{-1}$. In this respect a diamond heat sink should again be superior to Cu since it exerts a compressional strain on the GaAs chip when heating up, whereas Cu tends to tear the chip apart.

To increase the power capability of TE devices the multi-contact approach [4.11, 6.85, 6.86] or the multi-chip approach ([6.119] and Fig. 6.10) are frequently used and/or the ambient temperature is reduced by forced cooling [6.120].

6.3.2 Thermal Device Behaviour

The thermal properties of the low-field mobility and the carrier density have already been discussed in Section 6.1.3 on GaAs material (see Fig. 6.4). For the power conversion efficiency η the temperature dependence of the velocity-field characteristic in the high-field range [6.121], (cf. Fig.2.10) is of special interest (Section 5.2.2). Experimentally a slight decrease in peak velocity v_p of about $0\cdot06\%$ per K was found with rising temperature [6.121], combined with a slight increase in the peak field, but no influence at fields beyond about $5 \, \text{kV cm}^{-1}$ (cf. also [2.36b]). This is, however, inconsistent with other results [6.122a] obtained for the temperature dependence of the device saturation current $I_s \approx 0\cdot75 \, Aen_0v_p$, cf. also [6.122b, 6.113b]. There, a more pronounced decrease of approximately $0\cdot25\%$ per K was experienced which better agrees with theoretical predictions [2.59]. At any rate, the theoretically expected influence on the efficiency is not excessive [3.52]. On the other hand, a strong influence of the temperature on the efficiency at constant bias condition has been found in practice [6.122a]. This can be explained by the existence of a parasitic resistance in series with the active GaAs region, this resistance having a negative temperature coefficient [3.52]. Such a behaviour of the contact resistance has, in fact, been confirmed (thermal activation energy $W_a \approx 0\cdot1$ eV; [6.105]. See p. 231).

The temperature dependence of the *total* low-field device resistance $R_{\text{tot}} = R_c + R_1$ is a combination of temperature influence on the low field mobility μ_1 (decreasing with temperature), the carrier concentration n_0 (constant of increasing with temperature), and the parasitic contact resistance R_c. Not surprisingly, the experimental findings differ very much; compare for example [6.105, 6.123]. Ideally, the contact resistance should be negligible and the low-field resistance vary like $R_1(T) \propto 1 \, \mu_1(T)$ (constant carrier concentration) yielding a *positive* temperature coefficient of the resistance and therefore a built-in protection against thermal breakdown.

The large negative temperature coefficient of $R_1(T)$ in bulk-grown material due to the carrier density increase with temperature (Fig. 6.4) can induce a current-controlled negative resistance in the d.c. current characteristic of the device even at relatively low device temperatures T_{max} which causes current filament formation with destructive breakdown as a consequence [6.38]. Such an increase of carrier density with temperature is avoided in most epitaxially-grown materials.

Now we compare the thermal properties of GaAs with those of two other semi-conducting materials which are of potential interest for TE devices, namely InP and the alloy $In_xGa_{1-x}Sb$ ($x \approx 0.5$), see Table 6.1. Of the three materials, InP has the highest thermal conductivity, namely about 1·5 times that of GaAs [6.110], whereas in $In_xGa_{1-x}Sb$ it is relatively low, only about 1/10 of that of GaAs [6.26]. InP possesses a high threshold field (≈ 1.7 times that of GaAs), a fact which at first look tends to compensate the advantage of the better thermal conductivity, because of an increased dissipated power density. Contrary to this, $In_xGa_{1-x}Sb$ has a low threshold field (≈ 0.18 times that of GaAs) and thus makes it attractive despite the low thermal conductivity. A meaningful comparison, however, has to be based on the bias-field values that lead to about the same efficiency for devices made from all three semiconductors, rather than on the threshold fields. We consider, then, the lowest d.c. power density that can be achieved in each material combined with the shortest possible length L for lowest heat resistance in the sandwich con-figuration, i.e. with the minimum n_0L product of domain-mode operation [6.124]. According to eqn (6.18a) the output power is

$$P = \eta P_0 = 0.75 \eta e n_0 L \mu_1 A E_p E_{h0}. \tag{6.19a}$$

Inserting $(n_0L)_{min}$ from eqn (4.51b) with $|\mu_{h0}'|_{max} = \mu_n$ at $v_{h0} \approx 0.6\, v_p$ one obtains

$$P \approx 1.5 \eta \varepsilon v_p (\mu_1/\mu_n) A E_p E_{h0}. \tag{6.19b}$$

For equal output power of two different semiconductors (subscripts I and II) which are biased so as to yield equal efficiencies, one thus has, using a permittivity ratio $\varepsilon_{II}/\varepsilon_I \approx 1$,

$$\frac{A_I}{A_{II}} = \left(\frac{E_{p,II}}{E_{p,I}}\right) \left(\frac{E_{h0,II}}{E_{h0,I}}\right) \frac{(\mu_1/\mu_n)_{II}}{(\mu_1/\mu_n)_I} \frac{v_{p,II}}{v_{p,I}}. \tag{6.20}$$

The mobility ratio μ_1/μ_n in InP and $In_xGa_{1-x}Sb$ is expected to be of the order of one half of that of GaAs [5.14, 5.36] (based on the three-level mechanism in InP and In_xGa_{1-x} Sb. More recently the mobility ratio is believed to be more like GaAs in the two-level model, cf. p. 152 ff.) Hence one has, using the bias-field values obtained for $\eta \approx 20\%$ from Figs 5.10 and 5.11 and the numerical parameters of Table 6.1,

$$\frac{A_{(InGa)Sb}}{A_{GaAs}} \approx 104 \quad \text{or} \quad \frac{(\kappa A)_{(InGa)Sb}}{(\kappa A)_{GaAs}} \approx 10$$

and

$$\frac{A_{InP}}{A_{GaAs}} \approx 0.71 \quad \text{or} \quad \frac{(\kappa A)_{InP}}{(\kappa A)_{GaAs}} \approx 1.1$$

where κA can be regarded as a figure-of-merit of the device for heat removal. Con-sequently, the room-temperature thermal resistance, eqns (6.10) to (6.13), can be made significantly smaller for an $In_xGa_{1-x}Sb$ device than for a GaAs one, whereas for InP devices it is roughly the same. However, this analysis does not take into account the differences in r.f. device impedance which should have such values that optimum matching to a practical microwave circuit is still possible (see Section 5.2.6). Allowing for this, $In_xGa_{1-x}Sb$ is still more suitable for high power application than GaAs, provided the smaller band gap does not reduce the tolerable maximum device temperature T_{tol} too much.

6.4 FAILURE MODES AND AGEING EFFECTS

In many non-TE device applications GaAs is known to be susceptible to degradation effects. Thus a decrease of efficiency during operation time has been observed for GaAs injection devices such as tunnel diodes, electroluminescent diodes and lasers: a non-radiative recombination path of the minority carriers leads to the creation of native Frenkel defects* in the crystal lattice and hence to unwanted additional energy levels within the band gap [6.138]. These effects are detrimental also in bipolar n-p-n transistors since the generated defects cause a mobility decrease of the injected electrons within the p-base and hence a lower cut-off frequency.

On the other hand, severe degradation of field-effect transistors which, like TE devices, are based on majority carrier flow only, has not been reported. For GaAs TE devices degradation appears to be absent, too, if the GaAs material and the applied ohmic contacts are 'good' and the bias fields not excessive. For example, GaAs devices made from n^{2+}-n-n^+ sandwiches have been operated on a C.W. basis for more than 20 000 h without any measurable change in their characteristics [1.35]. Life-test data at various temperatures even indicate an extrapolated life of 10^5 h at an operating device temperature of about 225 °C [6.139].

If failures occur they seem to be the result of *sudden* breakthroughs (short circuits), with no indication of prior drift, mostly during the first 24 to 170 h of operation [6.140]. *Slow* ageing effects, both during operation [6.98, 6.139, 6.141] and 'on the shelf' [6.98], have been observed to occur either at elevated *ambient* temperatures or in devices made of impure material or which have not been properly processed.

In order to develop a suitable technology to avoid failures, we have first to know their origin. For this reason, the underlying mechanisms causing failure modes and ageing effects in GaAs TE devices are discussed below. Generally, failures are induced either from the contacts, subsequently spreading to the active GaAs bulk, or they originate within the bulk itself. Device failures are also known to be caused by an imperfect GaAs surface.

Deterioration of contact properties occurs particularly in devices with metal contacts alloyed directly onto the active epitaxial GaAs layer. Such a contact can usually be operated only as the cathode since avalanching effects are observed when it is used as the anode (Fig. 6.22a) as explained further below. 'On-the-shelf' ageing

---

*Frenkel defects are pairs of a vacancy and a distant interstitional ion.

has been reported from Ag-based contacts that are not hermetically sealed in a protective gas atmosphere. If exposed to air, sulphur and moisture cause the formation and growth of Ag_2S whiskers which introduce additional strain at the contact.

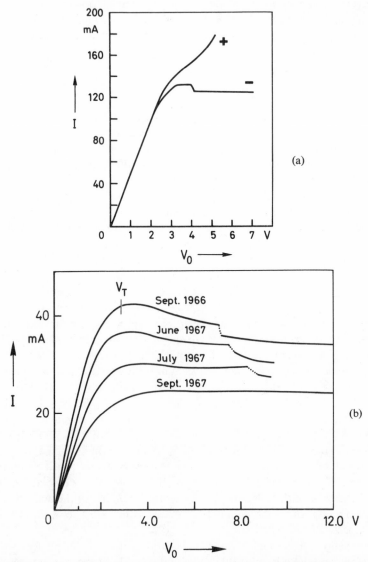

Fig. 6.22 Current-voltage characteristics of n-n$^+$ epitaxial TE devices with metal contact applied directly on top of epitaxial n-layer; (a) typical characteristic for both bias polarities with respect to metal contact, after Hasty *et al.* [3.3a]; (b) ageing effect of characteristic when negative bias polarity at metal contact, after Al-Moufti *et al.* [6.141].

This type of ageing can be accelerated artificially by exposing the device to a mixture of H_2S and H_2O [6.98]. A variety of metal-alloy contacts induce device degradation during C.W. operation as a consequence of the elevated device temperatures (200–300 °C) produced. In all cases a gradual decrease in threshold current combined with a slight increase in threshold voltage and a decrease in device efficiency is observed. As an example, Fig. 6.22b shows the result of one year's C.W. operation of an oscillator, decreasing the threshold current to almost one half of its original value [6.98, 6.141].

The degradation can be explained using a model which is based on the damaged layer of presumably higher resistivity formed during alloying underneath the metal contact (Section 6.2.2). This damaged layer produces a static high-field region at the contact which tends to decrease the threshold current density and the efficiency when biased as the cathode*. Such behaviour is predicted by a theoretical analysis [3.3, 3.5, 6.75], as seen by the $I(V)$ curves for negative polarity in Fig. 6.23. Ageing enforces the damage causing a further decrease in threshold current density and efficiency.

If opposite bias polarity is chosen so that the static high-field region is adjacent to the *anode*, another failure mechanism can take place, at least at higher bias voltages: impact ionization, cf. [6.167] causes an increase in current which usually leads to a destructive breakdown of the device due to formation of a channel of high current density (see below). This effect is again expected from theory [3.3a, 6.75], since, contrary to the case of negative bias, the static high-field region at the contact is tremendously increased when the differential mobility becomes negative beyond the threshold field. In this way breakdown fields ($\approx 10^5$ V cm^{-1}) are rapidly attained. The theoretical $I(V)$ curves in this case show a less pronounced saturation, as seen in Fig. 6.23 (curves with positive polarity). Here, avalanching effects are still neglected. Taking avalanching into account, a still stronger increase is expected, as found in the experiments (Fig. 6.22a).

A similar effect is observed if the static high-field region is a result of a particular contact geometry [6.143]. Thus the $I(V)$ characteristic becomes asymmetrical for a small contact (higher field) opposite to a large contact even though both contacts have been made by the same technology. For pure Sn sandwich contacts on bulk-grown chips (preform alloying) with a carrier drift length of $L = 50$ μm and a relatively large cathode contact area of more than 5×10^{-3} cm^2, substantial avalanching was observed when the (circular) small-area anode contact decreased below about 5×10^{-4} cm^2 (or $d/L < 5$ with $d = $ anode-contact diameter), whereas stable operation at reduced current densities was possible with the small contact biased as the cathode (see Fig. 6.24).

Another contact ageing effect during device operation is caused by the drift of

---

*Current saturation at relatively low values has been reported also for cases where the high-field region was trapped by some doping irregularity further away from the cathode within the GaAs bulk [6.142].

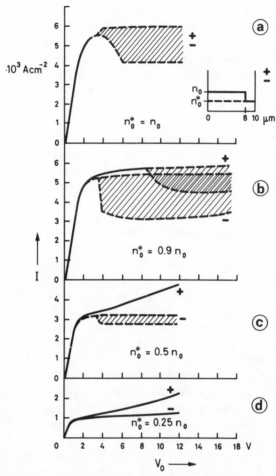

Fig. 6.23 Theoretical current-voltage characteristics with doping notch ($\delta n_0 = n_0 - n_0^\star$) at one contact, for both bias polarities; $n_0/n_0^\star$ being 1 in (a), 0·9 in (b), 0·5 in (c), and 0·25 in (d). Shaded area shows oscillation amplitude of instability for constant-voltage operation. After Suga and Sekido [6.75].

positive metal ions in the high domain fields with the anode metal acting as the source. In an investigation of migration of Sn, which on an As site acts as a donor (Table 6.3), a Sn movement of about one As-As lattice distance per transit of a high-field domain was found [6.32b]. Thus, operation of a device in the dipole modes at high bias fields with the anode contact containing Sn produces a lower-resistivity channel spreading from the anode towards the cathode.

Ageing effects originating from the GaAs bulk itself have been observed only in relatively impure compensated material fabricated usually by the bulk-crystal growth processes.

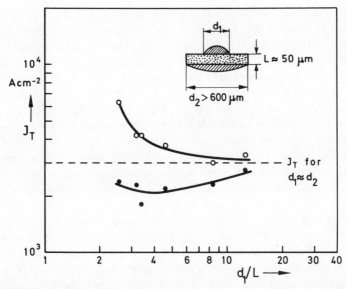

Fig. 6.24 Threshold-current density J_T for GaAs TE devices with Sn dot contacts showing influence of contact size. J_T is normalized to $\sigma_I^{-1} = 1\,\Omega\,\text{cm}$. After Engelmann [6.143].

Even at fields below threshold, with the device temperature increased due to self-heating, some type of drifting (presumably interstitial) impurity ion is electrolysed out at the contacts or possibly also at dislocations thus losing its donor or acceptor property. This leads to an increase [6.143] or decrease [6.4] of the GaAs bulk resistivity, the change usually saturating at longer times. In some cases the resistivity can be recovered by annealing at about 100°C without a bias field applied [6.143].

The most common device failure is a sudden destructive breakdown during operation. The exact mechanism of this breakdown appears to be complicated since the nature and geometry of the contacts, the GaAs bulk and surface properties, and also the thermal conditions play a role. Two main failure types seem to occur which both originate from the anode region: (1) excessive anode heating with contact melting induces growth of a conductive channel in the bulk or surface forming a short-circuit when the cathode is reached; (2) the existence of high average fields near the anode may lead to carrier multiplication which, at increased bias fields, causes avalanche breakdown and eventually destructive current-filament formation. Some observed failures may be due to a combination of both effects, cf. [6.140a].

Even in the absence of contact irregularities heat dissipation is increased near the anode by the field increase in the travelling domain during its extinction at the anode as first discussed in [6.145] (see also Section 3.3, Fig. 3.16). Contact melting usually starts at some hot spot dissolving the GaAs there, thus letting the molten metal penetrate further into the bulk until a short-circuit channel is formed, usually within a fraction of a second [6.108, 6.144a] but in some cases of pulsed operation within seconds, or, even minutes, depending on the applied bias relative to threshold voltage [4.51]. The breakdown is essentially thermal but undoubtedly is enhanced by ion-drift effects in the high field (of the kind already mentioned) since the conducting channel consists almost exclusively of one constituent of the contact alloy only, e.g. of Sn for Sn-Ag contacts and of In for In-Au contacts [6.108]. Because such a slow channel formation has been observed also for Au-Ge-Ni contacts [4.74], the presence of a low melting constituent is obviously not a necessity for its occurrence. TE devices with coplanar contacts are particularly susceptible to this type of failure since heat sinking at the anode contact is relatively poor. However, anode metal migration can be avoided and higher breakdown voltages (above 5 times threshold) are achieved if the anode contact is formed by selective etching and n^+ re-growth [6.144b].

Avalanching in the device may induce an S-type current voltage characteristic which can be explained by the strong carrier multiplication mechanism in the increased high field of the domain reaching the anode. After filling of any hole traps this establishes the situation of double injection [6.36, 6.145]; (for details on current injection see [6.156]). The device switches at a critical voltage to a high-current/low voltage state which is characterized by a small region of very high field ($\approx 10^5$ V cm^{-1}) near the anode, supporting the avalanche, and a low field in the rest of the sample [6.146]. As a result of the obtained S shape of the $I(V)$ characteristic this state is unstable and tends to form current filaments [1.11] which cause excessive heating and hence destructive breakdown*. Since this process takes a certain time to develop, the S-shaped characteristic can be used to excite relaxation oscillations without causing permanent deterioration of the device [6.147]. Avalanching is enhanced by any static high-field region at the anode as already mentioned.

Results of investigations on the breakdown (switching) voltage of sandwich devices as a function of the carrier density n_0 and the sample length L are shown in Fig. 6.25 [6.148]. The maximum field $E_{2,\text{bd}}$ of the travelling domain at the breakdown voltage V_{bd} was calculated from a simple triangular domain model yielding

$$\frac{V_{\text{bd}}}{V_{\text{T}}} = \frac{1}{2} + \frac{\varepsilon(E_{2,\text{bd}} - E_{\text{T}}/2)^2}{2en_0LE_{\text{T}}}. \tag{6.21}$$

The $E_{2,\text{bd}}$ values obtained seem to be somewhat low when compared with theoretical results on avalanche breakdown [6.163], but one has to keep in mind the enhancement of E_2 when the domain reaches the anode. The dependence of $E_{2,\text{bd}}$ on sample length L follows qualitatively from the dependence on the domain width b if breakdown fields are calculated in analogy to the case of an abrupt p-n junction [6.163]. When assuming a constant $E_{2,\text{bd}} \approx 150$ kV cm^{-2} eqn (6.21) can be approximated by $V_{\text{bd}} \approx 2 \cdot 5 \times 10^{13}$ cm^{-2} $(n_0L)^{-1}V_{\text{T}}$.

---

*A similar S-shaped $I(V)$ characteristic is obtained in C.W. operation of GaAs chips fabricated from material with a strongly negative temperature coefficient of resistivity solely because of self-heating. Here current-filament formation is also destructive [6.38]; p. 241.

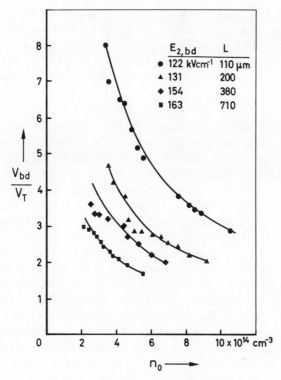

Fig. 6.25 Normalized breakdown voltage V_{bd}/V_T as a function of carrier concentration n_0 and sample length L. After Kennedy *et al.* [6.148].

The breakdown voltage V_{bd} is influenced also by the properties of the exposed GaAs surface. This seems to be one of the reasons for the low-voltage breakdown characteristics observed for coplanar-contact devices [6.108, 6.144a]. In the case of sandwich devices, a surface influence becomes apparent when comparing samples having a mechanically cleaved lateral active surface with mesa-etched structures from the same epitaxial wafer [6.79]. Mesa devices exhibit higher V_{bd} values (V_{bd} up to 30–35 V_T as measured with 20 ns pulses) and also have proved to be more reliable in C.W. operation (at 2·5 V_T).

Note that V_{bd} is not a constant device parameter but decreases slightly with increasing pulse width [6.79], probably due to hole-trapping effects [6.145, 6.148]. When V_{bd} is measured with very short pulses (<100 ns) no permanent device deterioration occurs [6.79]. Thus this method seems to be useful for a quantitative assessment of the reliability of a device.

In the LSA mode of oscillation the electric fields in the sample are considerably lower than in the domain modes. Consequently, the avalanche switching effect discussed previously is absent and higher reliability is expected. However, care has to be taken in the external circuit design so that the LSA device does not change to domain-mode operation, particularly since the applied bias voltages lie usually beyond V_{bd} in order to obtain efficient LSA operation [6.148].

For establishing the optimum in reliability of TE devices, further investigations of the ageing and failure mechanisms are necessary. A powerful tool for this seems to be the scanning electron microscope [6.149]. At any rate, special care has to be taken in the technology and geometry of the anode contact, in order to avoid any possible d.c.-field increase immediately adjacent to it. Reliability improvement in this respect has been demonstrated [6.150a] by increasing the cross-sectional area at the anode (lowering of field); see Fig. 8.19. A somewhat enhanced potential drop at the cathode, on the other hand, improves the reliability in the domain-mode operation without seriously influencing the efficiency [6.150b]. Judging from the theoretical findings of [3.3b] as compared to others [3.3a, 6.75], this could be due to a counteraction of the high-field region at the cathode against the field increase near the anode in the active bias range. Theoretically, a slightly enhanced cathode drop even improves the efficiency somewhat, as seen from the increased r.f. current amplitude in Fig. 6.23b as compared to Fig. 6.23a (negative polarity).

For minimizing the influence of the exposed GaAs surface on device failure or ageing, particularly in coplanar contact structures, surface passivation serves is an important technique. This can be achieved with SiO_2 [6.83] or Si_3N_4 [151], or possibly even with semi-insulating GaAs, either by growing it epitaxially [6.152] or by forming it by Cr diffusion on top of the active layer [6.153].

Degradation of TE devices as a result of nuclear γ radiation [6.154] and of electron/neutron irradiation [6.155] has also been reported. For example, a carrier density and mobility decrease (forming of acceptors) impairs device performance after exposures to about 10^{14} cm^{-2} neutrons.

In conclusion, it can be stated that careful fabrication of TE devices results in high reliability during r.f. operation. This careful fabrication should consist of strictly choosing the material on the basis of purity, device selection after an initial burn-in period, pulsed measurements of the break-down voltage V_{bd}, and measurements of the thermal resistance R_{th}. The origin of a number of failure modes and ageing effects observed in practice have been elucidated. However, a final optimization of the contact technology with respect to highly reliable r.f. performance is still lacking. In general, sandwich-type devices should preferably be of the mesa form, Fig. 6.6d, or possibly of the planar form, Fig. 6.6e, with the anode contact mounted to the heat sink as close as possible, i.e. without a substantial interface layer between the active n layer and the heat sink. For coplanar-contact devices, on the other hand, the form shown in Fig. 6.6f is most suited, with the n-n^{2+} contact fabricated to be as planar as possible and exhibiting a somewhat larger width w at the anode than at the cathode (cf. Fig. 8.19).

REFERENCES

6.1a Hilsum, C., Mullin, J. B., Prew, B. A., Rees, H. D. and Straughan, B. W., Instabilities of InP 3-level transferred-electron oscillators. *Electron. Lett.* **6** (1970), p. 307.

6.1b Colliver, D., Hilsum, C., Joyce, B. D., Morgan, J. R., Rees, H. D. and Knight, H. D., Microwave generation by InP 3-level transferred-electron oscillators. *Electron. Lett.* **6** (1970), p. 436.

6.2 Ludwig, G. W. and Aven, M., Gunn effect in ZnSe. *J. Appl. Phys.* **38** (1967), p. 5326.

6.3a Foyt, A. G., Halsted, R. E. and Paul, W., Evidence of impurity states associated with high-energy conduction band extrema in n-CdTe. *Phys. Rev. Lett.* **16** (1966), p. 55.

6.3b Ludwig, G. W., Gunn effect in CdTe. *IEEE Trans. Electron Dev.* **ED–14** (1967), p. 547.

6.4a Shyam, M., Allen, J. W. and Pearson, G. L., Effect of variation of energy minima separation on Gunn oscillations. *IEEE Trans. Electron Dev.* **ED–13** (1966), p. 63.

6.4b Califano, F. P., Gunn effect in $Ga_xAl_{1-x}As$ (in Italian). *Alta Frequenza* **38** (1969), p. 937.

6.5 McGroddy, J. C., Lorenz, M. R. and Plaskett, T. R., The Gunn effect and conduction band structure in $Ga_xIn_{1-x}Sb$ alloys. *Solid-State Commun.* **7** (1969), p. 901.

6.6 Fawcett, W., Hilsum, C. and Rees, H. D., Optimum semiconductor for microwave devices. *Electron. Lett.* **5** (1969), p. 313.

6.7a Allen, J. W., Shyam, M. and Pearson, G. L., Gunn oscillations in InAs. *Appl. Phys. Lett.* **9** (1966), p. 39.

6.7b Smith, Jr., J. E. and Camphausen, D. L., Gunn effect in InAs under hydrostatic pressure. *J. Appl. Phys.* **42** (1971), p. 2064.

6.8 Porowski, S., Paul, W., McGroddy, J. C., Nathan, M. I. and Smith, Jr., J. E., Electron-hole pair production and Gunn effect in InSb. *Solid-State Commun.* **7** (1969), p. 995.

6.9 McGroddy, J. C., Nathan, M. I., Paul, W., Porowski, S. and Smith Jr., J. E., The effects of hydrostatic pressure on hot electron phenomena in n-InSb. *IBM J. Res. Dev.* **13** (1969), p. 580.

6.10a McGroddy, J. C., Nathan, M. I. and Smith Jr., J. E., Negative conductivity effects and related phenomena in Ge. Part I. *IBM J. Res. Dev.* **13** (1969), p. 543.

6.10b Smith Jr., J. E., Nathan, M. L. and McGroddy, J. C., Negative conductivity effects and related phenomena in Ge. Part II. *IBM J. Res. Dev.* **13** (1969), p. 554.

6.11a Paige, E. G. S., Bulk negative differential conductivity in Ge: Theory. *IBM J. Res. Dev.* **13** (1969), p. 562.

6.11b Heinrich, H., Lischka, K. and Kriechbaum, M., Magnetoresistance and Hall effect of hot electrons in Ge and carrier transfer to higher minima. *Phys. Rev.* **B2** (1970), p. 2009.

6.12 Dumke, W. P., High-field conductivity of the $\langle 111 \rangle$ valleys of Ge. *Phys. Rev.* **B2** (1970), p. 987.

6.13a Kastal'skii, A. A. and Ryvkin, S. M., *Zh. ETF Pis. Red.* **7** (*1968*), p. 446. In Russian Translation: New type of current instability in n-Ge. *Sov. Phys.-JETP Lett.* **7** (1968), p 350.

6.13b De Biasi, R. S. and Yee, S. S., Bulk negative differential conductance in high-purity n-type Ge. *J. Appl. Phys.* **41** (1970), p. 3863.

6.13c De Biasi, R. S. and Yee, S. S., A bulk Ge LSA-mode oscillator. *Proc. IEEE* **58** (1970), p. 1404.

6.14 Smith Jr., J. E., Intervalley transfer and microwave current oscillations in strained n-type Ge. *Appl. Phys. Lett.* **12** (1968), p. 233.

6.15 Nuekermans, A. and Kino, G. S., The velocity-field characteristic of electrons in Ge under uniaxial pressure. *Solid-State Commun.* **8** (1970), p. 987.

6.16 Kastal'skii, A. A. and Ryvkin, S. M., *Fiz. Tekh. Poluprov.* **1** (1967), p. 622. In Russian. Translation: Gunn effect in uniaxially compressed Ge. *Sov. Phys. Semiconductors* **1** (1967), p. 523.

6.17a Smith Jr., J. E., Nathan, M. I., McGroddy, J. C., Porowski, S. A. and Paul, W., Gunn effect in n-type InSb. *Appl. Phys. Lett.* **15** (1969), p. 242.

6.17b Müller, E. and Perry, D. K., Effect of a magnetic field upon the Gunn effect in InSb. *Solid-State Commun.* **8** (1970), p. 855.

6.18 Madelung, O., *Physics of III–V Compounds.* (New York, John Wiley & Sons, 1964.)

6.19 Moll, J. L., III–V compounds and their applications. Paper at *Int. Electron Dev. Meeting, Washington, D.C., Oct. 1968.*

6.20 Enstrom, R. E. and Peterson, C. C., Vapor phase growth and properties of GaAs Gunn devices. *Trans. Metallurgical Soc. AIME* **239** (1967), p. 413.

6.21 Wasse, M. P., Lees, J. and King, G., The effect of pressure on Gunn phenomena in GaAs. *Solid-State Electron.* **9** (1966), p. 601.

6.22 Kosicki, B. B., Jayaraman, A. and Paul, W., Conduction band structure of GaSb from pressure experiments to 50 kbar. *Phys. Rev.* **172** (1968), p. 764.

6.23 Tietjen, J. J. and Amick, J. A., The preparation and properties of vapor-deposited epitaxial $GaAs_{1-x}P_x$ using arsine and phosphine. *J. Electrochem. Soc.* **113** (1966), p. 724.

6.24 Lorenz, M. R., McGroddy, J. C., Plaskett, T. R. and Porowski, S. A., Location of the $\langle 111 \rangle$ conduction band minima in the $Ga_xIn_{1-x}Sb$ alloy system. *IBM J. Res. Dev.* **13** (1969), p. 583.

6.25 Kudman, I. and Seidel, T. E., Conduction bands in GaSb-InSb alloys. *J. Appl. Phys.* **38** (1967), p. 4379.

6.26 Kudman, I., Ekstrom, L. and Seidel, T. E., High-temperature thermal and electrical properties of GaSb-InSb alloys. *J. Appl. Phys.* **38** (1967), p. 4641.

6.27 Herzog, A. H., Groves, W. O. and Craford, M. G., Electroluminescence of diffused $GaAs_{1-x}P_x$ diodes with low donor concentrations. *J. Appl. Phys.* **40** (1969), p. 1830.

6.28 Bube, R. H., *Photoconductivity of Solids.* (New York, John Wiley & Sons, 1960.)

6.29 Smith, R. A., *Semiconductors.* (Cambridge, University Press, 1961).

6.30 Oliver, M. R., McWhorter, A. L. and Foyt, A. G., Current runaway and avalanche effects in n-CdTe. *Appl. Phys. Lett.* **11** (1967), p. 111.

6.31 King, G., Wasse, M. P. and Sandbank, C. P., An assessment of epitaxial GaAs for use in Gunn effect devices. *Proc. Int. Symp. Gallium Arsenide, Reading, Sept. 1966,* p. 184 (Institute of Physics and Physical Society, London, 1967).

6.32a Heeks, J. S. and Woode, A. D., Localized temporary increase in material conductivity following impact ionization in a Gunn effect domain. *IEEE Trans. Electron Dev.* **ED–14** (1967), p. 512.

6.32b Guétin, P., Contribution to the experimental study of Gunn effect in long GaAs samples. *IEEE Trans. Electron Dev.* **ED–14** (1967), p. 522.

6.33a James, L. W., Van Dyke, J. P., Herman, F. and Chang, D. M., Band structure and high-field transport properties of InP. *Phys. Rev.* **B1** (1970), p. 3998.

6.33b Pitt, G. D., The conduction band structure of InP from a high pressure experiment. *Solid-State Commun.* **8** (1970), p. 1119.

6.34 Smith Jr., J. E., McGroddy, J. C. and Nathan, M. I., Bulk current instabilities in uniaxially strained Ge. *Ibid.* **186** (1969), p. 727.

6.35 Stratton, R , The influence of interelectronic collisions on conduction and breakdown in polar crystals. *Proc. Roy. Soc. (Lond.)* **A264** (1958), p. 406.

6.36a Southgate, P. D., Recombination processes following impact ionization by high-field domains in GaAs. *J. Appl. Phys.* **38** (1967), p. 4589.

6.36b Gelmont, B. L. and Shur, M. S., S-type current-coltage characteristic, current filament formation and stimulated emission in high doped Gunn diodes. *Proc. 1970 MOGA Conference,* p. 9/17 (Deventer, Kluwer, 1970).

6.37 Woodall, J. M., Crystal holds key to the future. *Electronics* **40** (1967), Nov. 13, p. 110.

6.38 Knight, S., Current runaway in n-GaAs bulk effect devices. *Proc. IEEE* **54** (1966), p. 1004.

6.39 Cronin, C. R. and Haisty, R. W., The preparation of semi-insulating GaAs by Cr doping. *J. Electrochem. Soc.* **111** (1964), p. 874.

6.40 Allen, J. W., Gallium arsenide as a semi-insulator. *Nature (London)* **187** (1960), p. 403.

6.41a Mehal, E. W. and Wacker, R. W., GaAs integrated microwave circuits. *IEEE Trans. Electron Dev.* **ED–15** (1968), p. 513.

6.41b Sandbank, C. P., An almost ideal substrate. *Electronics* **40** (1967), Nov. 13, p. 117.

6.42 Effer, D., Epitaxial growth of doped and pure GaAs in an open flow system. *J. Electrochem. Soc.* **112** (1965), p. 1020.

6.43 Knight, J. R., Effer, D. and Evans, P. R., The preparation of high purity GaAs by vapor phase epitaxial growth. *Solid-State Electronics* **8** (1965), p. 178.

6.44 Eddolls, D. V., Knight, J. R. and Wilson, B. L. H., The preparation and properties of epitaxial GaAs. *Proc. Int. Symp. Gallium Arsenide, Reading, Sept. 1966*, p. 3 (Institute of Physics and Physical Society, London, 1967).

6.45 Bolger, D. E., Franks, J., Gordon, J. and Whitaker, J., Preparation and characteristics of GaAs. *Proc. Int. Symp. Gallium Arsenide, Reading, Sept. 1966*, p. 16 (Institute of Physics and Physical Society, London, 1967).

6.46 Wolfe, C. M., Stillman, G. E. and Lindley, W. T., Tin doping of epitaxial GaAs. *Proc. 2nd Int. Symp. Gallium Arsenide, Dallas, Texas, Oct. 1968*, p. 43 (Institute of Physics and Physical Society, London, 1969).

6.47 Reid, F. J. and Robinson, L. B., Preparation of epitaxial GaAs for microwave applications. *Proc. 2nd Int. Symp. Gallium Arsenide, Dallas, Texas, Oct. 1968*, p. 59 (Institute of Physics and Physical Society, London, 1969).

6.48 Wolfe, C. M., Foyt, A. G. and Lindley, W. T., Epitaxial GaAs for high-efficiency Gunn oscillators. *Electrochem. Technol.* **6** (1968), p. 208.

6.49 Gramann, W., Ge-doped epitaxial GaAs layers for Gunn elements (abstract, in German). *Verhandl. Deutsche Physikal. Gesellsch.* **4/VI** (1969), p. 206. (European Meeting "Semiconductor Device Research", Munich, March 24–27, 1969).

6.50a Wolfe, C. M. and Stillmann, G. E., High purity GaAs. *Proc. 3rd Int. Symp. Gallium Arsenide and Related Compounds, Aachen, Oct. 1970*, p. 3 (Institute of Physics and Physical Society, London, 1971).

6.50b Wolfe, C. M., Stillman, G. E. and Lindley, W. T., Electron mobility in high-purity GaAs. *J. Appl. Phys.* **41** (1970), p. 3088.

6.50c Wolfe, C. M., Stillman, G. E. and Owens, E. B., Residual impurities in high-purity epitaxial GaAs. *J. Electrochem. Soc.* **117** (1970), p. 129.

6.51a Berson, B. E., Enstrom, R. E. and Reynolds, J. E., High-power L- and S-band transferred electron oscillators. *RCA Rev.* **31** (1970), p. 20.

6.51b Lawley, K. L., Film-making: a delicate job performed under pressure. *Electronics* **40** (1967), Nov. 13, p. 114.

6.52a Manasevit, H. M., Single-crystal GaAs on insulating substrates. *Appl. Phys. Lett.* **12** (1968), p. 156.

6.52b GaAs on sapphire. *Electronics* **41** (1968), July 8, p. 51.

6.52c Owens, J. M., GaAs on sapphire Gunn effect devices. *Proc. IEEE* **58** (1970), p. 930.

6.52d Gutierrez, W. A., Pommerring, H. D., Jasper, M. A. and Mantzouranis, A. P., Epitaxial growth of GaAs on insulating substrates using HCl-H_2 vapor transport. *Solid-State Electron.* **13** (1970), p. 1199.

6.52e Thorsen, A. C. and Manasevit, H. M., Heteroepitaxial GaAs on aluminium oxide: electrical properties of undoped film. *J. Appl. Phys.* **42** (1971), p. 2519.

6.53 Joyce, B. A., Growth and perfection of chemically-deposited epitaxial layers of Si and GaAs. *J. Crystal Growth* **3** (1968), p. 43.

6.54 Wolff, G., Keck, P. H. and Broder, J. D., Preparation and properties of III–V compounds. *Phys. Rev.* **94** (1954), p. 753.

6.55 Nelson, H., Epitaxial growth from the liquid state and its application to tunnel and laser diodes. *RCA Rev.* **24** (1963), p. 603.

6.56 Solomon, R., Factors influencing the electrical and physical properties of high quality solution grown GaAs. *Proc. 2nd Int. Symp. Gallium Arsenide, Dallas, Texas, Oct. 1968*, p. 11 (Institute of Physics and Physical Society, London, 1969).

6.57 Beneking, H. and Vits, W., Properties of GaAs luminescence and laser diodes prepared by liquid-phase epitaxy (abstract, in German). *Verham dl. Deutsche Physikal. Gesellschaft* **2/VI** (1967), p. 57. (European Meeting "Semiconductor Device Research", Bad Nauheim, April 17–22, 1967).

6.58 Grobe, E. and Salow, H., Preparation of epitaxial GaAs layers from the liquid phase (in German). *Zeitschr. Angew. Physik* **32** (1972), p. 381.

6.59 Hicks, H. G. B. and Manley, D. F., High purity GaAs by liquid phase epitaxy. *Solid-State Commun.* **7** (1969), p. 1463.

6.60 Kinoshita, J., Stein, W. W., Day, G. F. and Mooney, J. B., Solution epitaxy of GaAs with controlled doping. *Proc. 2nd Int. Symp. Gallium Arsenide, Dallas, Texas, Oct. 1968*, p. 22 (Institute of Physics and Physical Society, London, 1969).

6.61 Goodwin, A. R., Dobson, C. D. and Franks, J., Liquid phase epitaxial growth of GaAs. *Proc. 2nd Int. Symp. Gallium Arsenide, Dallas, Texas, Oct. 1968*, p. 36 (Institute of Physics and Physical Society, London, 1969).

6.62 Kang, C. S., Greene, P. E., Tin and tellurium doping characteristics in GaAs epitaxial layers grown from Ga solution. *Proc. 2nd Int. Symp. Gallium Arsenide, Dallas, Texas, Oct. 1968*, p. 18 (Institute of Physics and Physical Society, London, 1969).

6.63 Harris, J. S. and Snyder, W. L., Homogeneous solution grown epitaxial GaAs by tin doping. *Solid-State Electron.* **12** (1969), p. 337.

6.64a Van der Pauw, L. J., Method for measuring specific resistivity and Hall effect of disks of arbitrary shape. *Philips Res. Repts* **13** (1958), p. 1.

6.64b Kane, P. F. and Larrabee, G. B., *Characterization of Semiconductor Materials.* (Texas Instruments Electronics Series; New York, McGraw-Hill, 1970).

6.65 Jervis, T. R. and Johnson, E. F., Geometrical magnetoresistance and Hall mobility in Gunn effect devices. *Solid-State Electron.* **13** (1970), p. 181.

6.66a Fitzpatrick, J. R. and Conn, G. K. T., Non-destructive measurements of carrier concentration in epitaxial GaAs. *Brit. J. Appl. Phys. (J. Phys. D)* **2** (1969), p. 1407.

6.66b Decker, D. R., Measurement of epitaxial resistivity versus depth. *J. Electrochem. Soc.* **115** (1968), p. 1085.

6.67a Copeland, J. A., A technique for directly plotting the inverse doping profile of semiconductor wafers. *IEEE Trans. Electron Dev.* **ED–16** (1969), p. 445.

6.67b Copeland, J. A., Diode edge effect on doping-profile measurements. *IEEE Trans. Electron Dev.* **ED–17** (1970), p. 404.

6.68 Meyer, N. I. and Guldbrandsen, T., Method for measuring impurity distributions in semiconductor crystals. *Proc. IEEE* **51** (1963), p. 1631.

6.69 Sah, C. T., Rosier, L. L. and Forbes, L., Low-temperature high frequency capacitance measurements of deep- and shallow-level impurity center concentrations. *Appl. Phys. Lett.* **15** (1969), p. 161.

6.70a Spears, D. L. and Bray, R., Optical probing of inhomogeneities in n-GaAs with application to the acoustoelectric instabilities. *J. Appl. Phys.* **39** (1968), p. 5093.

6.70b Christensson, S., Woodward, D. W. and Eastman, L. F., High peak-power LSA operation from epitaxial GaAs. *IEEE Trans. Electron Dev.* **ED–17** (1970), p. 732.

6.71a Herzog, A. H. and Groves, W. O., Preparation and evaluation of the properties of GaAs, Part 2. *Semiconductor Products* **5** (1962), Dec., p. 25.

6.71b Abe, T., Nishi, Y., Goto, K. and Konaka, M., Thickness measurement of n on n⁺ one-sided lapped wafers by the far infrared interference method. Abstracts, *133rd Meeting of the Electrochem. Soc., Boston, Ma., May 5–9, 1968*, p. 14 (New York, Electrochem. Soc., 1968).

6.72   Larrabee, R. D., Hicinbothem Jr., W. A. and Steele, M. C., A rapid evaluation technique of functional Gunn diodes. *IEEE Trans. Electron Dev.* **ED–17** (1970), p. 271.

6.73a   Hilsum, C. and Morgan, J. R., The selection of GaAs epitaxial layers for CW X band Gunn diodes. *IEEE Trans. Electron Dev.* **ED–14** (1967), p. 532.

6.73b   Colliver, D. J., Gibbs, S. E. and Taylor, B. C., Material selection for efficient transferred-electron devices at Q band. *Electron. Lett.* **6** (1970), p. 353.

6.74   Cohen, L. D., Drago, F., Shortt, B., Socci, R. and Urban, M., Epitaxial GaAs Gunn effect oscillators: influence of material properties on device performance. *Proc. 2nd Int. Symp. Gallium Arsenide, Dallas, Texas, Oct. 1968*, p. 153 (Institutes of Physics and Physical Society, London, 1969).

6.75   Suga, M. and Sekido, K., Effect of doping profile upon electrical characteristics of Gunn diodes. *IEEE Trans. Electron Dev.* **ED–17** (1970), p. 275.

6.76   Copeland, J. A., Doping uniformity and geometry of LSA oscillator diodes. *IEEE Trans. Electron Dev.* **ED–14** (1967), p. 497.

6.77a   Hobson, G. S., Effect of Johnson noise on L.S.A. oscillators. *Electron. Lett.* **4** (1968), p. 230.

6.77b   Hobson, G. S., Three dimensional limitations on space charge growth in transferred electron devices. *Brit. J. Appl. Phys. (J. Phys. D)* **2** (1969), p. 1203.

6.78   Bott, I. B., Colliver, D. J., Hilsum, C. and Morgan, J. R., The design and performance of transferred electron oscillators (Gunn diodes). *Proc. Int. Symp. Gallium Arsenide, Reading, Sept. 1966*, p. 172 (Institute of Physics and Physical Society, London, 1967).

6.79   Takeuchi, M., Sekido, K., Mitsuhata, T. and Aono, Y., Surface effects on breakdown characteristics of GaAs bulk diodes. *IEEE Trans. Electron Dev.* **ED–15** (1968), page 748.

6.80a   Zettler, R. A. and Cowley, A. M., Batch fabrication of integral heatsink IMPATT diodes. *Electron. Lett.* **5** (1969), p. 693.

6.80b   Hoefflinger, B., Recent developments on avalanche diode oscillators. *Microwave J.* **12** (1969), p. 101.

6.80c   Di Lorenzo, J. V., C.W. performance of vapor grown GaAs transferred electron oscillators. *17th Int. Electron Devices Meeting, Washington, D.C., Oct. 1971*; Abstracts p. S5 (Inst. Electrical Electron. Engrs, New York, 1971).

6.81   Cox, R. H. and Mehal, E. W., Planar Gunn oscillator for microwave integrated circuits. *Trans. Metallurgical Soc. AIME* **242** (1968), p. 461.

6.82   Dienst, J. F., Dean, R. H., Enstrom, R. and Kokkas, A., Coplanar-contact Gunn-effect devices. *RCA Rev.* **27** (1967), p. 585.

6.83   Sekido, K., Takeuchi, T., Hasegawa, F. and Kikuchi, S., CW oscillations in GaAs planar-type bulk diodes. *Proc. IEEE* **57** (1969), p. 815, and *IEEE Trans. Electron Dev.* **ED–16** (1969), p. 256.

6.84   Koyama, J., Ohara, S., Kawazura, S. and Kumabe, K., Bulk GaAs travelling-wave amplifier. *Proc. 2nd Int. Symp. Gallium Arsenide, Dallas, Texas, Oct. 1968*, p. 167 (Institute of Physics and Physical Society, London, 1969).

6.85   Berson, B. E., Collard, J. R. and Narayan, S. Y., Epitaxial GaAs Gunn oscillators. *Proc. 1967 Cornell University Conf. on High Frequency Generation and Amplification*, p. 78 (Ithaca, N.Y., School of Electrical Engng, Cornell University, 1967).

6.86   Harris, K. J., 100 mW C.W. Gunn diode. *Electron. Lett.* **5** (1969), p. 123.

6.87   Enstrom, R. E., Reynolds, J. F. and Berson, B. E., Vapour growth of multilayered GaAs structures for series operation of transferred electron oscillators. *Electron. Lett.* **5** (1969), p. 714.

6.88a   Heeks, J. S., King, G. and Sandbank, C. P., Transferred-electron bulk effects in GaAs. *Electron. Commun. (USA)* **43** (1968), p. 334.

6.88b   King, G. and Heeks, J. S., Bulk effect modules pave way for sophisticated uses. *Electronics* **42** (1969), Feb. 3, p. 94.

6.89 Bott, I. B., Hilsum, C. and Smith, K. C. H., Construction and performance of epitaxial transferred electron oscillators. *Solid-State Electron.* **10** (1967), p. 137.

6.90 Salow, H. and Grobe, E., Ohmic metal contacts for Gunn elements (in German). *Zeitschr. Angew. Phys.* **25** (1968), p. 137.

6.91 Hakki, B. W. and Knight, S., Microwave phenomena in bulk GaAs. *IEEE Trans. Electron Dev.* **ED–13** (1966), p. 94.

6.92 Petzel, B., The coherence of Gunn oscillations. *Phys. Stat. Sol.* **33** (1969), p. K59.

6.93 Braslau, N., Gunn, J. B. and Staples, J. L., Metal-semiconductor contacts for GaAs bulk effect devices. *Solid-State Electron.* **10** (1967), p. 381.

6.94 Cox, R. H. and Strack, H., Ohmic contacts for GaAs devices. *Solid-State Electron.* **10** (1967), p. 12.

6.95 Ramachandran, T. B. and Santosuosso, R. P., Contacting n-type high-resistivity GaAs for Gunn oscillators. *Solid-State Electron.* **9** (1966), p. 733.

6.96a Engelmann, R. W. H. and Zizelmann, W., Contacting of GaAs, Patent pending.

6.96b Becker, R., AEG-Telefunken, Research Institute, private communication (1968).

6.97a Knight, S. and Paola, C. R., Ohmic contacts for GaAs bulk effect devices. *Proc. Conference Ohmic Contacts to Semiconductors, Montreal, Canada, Oct. 1968*, p. 102 (New York, Electrochem. Soc., 1969).

6.97b Paola, C. R., Metallic contacts for GaAs. *Solid-State Electron.* **13** (1970), p. 1189.

6.98 Cox, R. H. and Hasty, T. E., Metallurgy of alloyed ohmic contacts for the Gunn oscillator. *Proc. Conference Ohmic Contacts to Semiconductors, Montreal, Canada, Oct. 1968*, p. 88 (New York, Electrochem. Soc., 1969).

6.99a Harris, J. S., Nannich, Y., Pearson, G. L. and Day, G. F., Ohmic contacts to solution-grown GaAs. *J. Appl. Phys.* **40** (1969), p. 4575.

6.99b Saito, T. and Hasegawa, F., Cause of the high resistance region of vapour epitaxial GaAs layer-substrate interface. *Jap. J. Appl. Phys.* **10** (1971), p. 197.

6.99c Muñoz, E., Snyder, W. L. and Moll, J. L., Effect of arsenic pressure on the heat treatment of liquid epitaxial GaAs. *Appl. Phys. Lett.* **16** (1970), p. 262.

6.100 Brady, D. P., Knight, S., Lawley, K. L. and Uenohara, M., Recent results with epitaxial GaAs Gunn effect oscillators. *Proc. IEEE* **54** (1966), p. 1497.

6.101 Lawley, K. L., Heilig, J. A. and Klein, D. L., Preparation of ohmic contacts for n-type GaAs. *Electrochem. Techn.* **5** (1967), p. 374.

6.102 Bass, J. C., Eldridge, A. L. and Knight, J. R., Gunn effect oscillators with vapour-grown contact layers. *Electron. Lett.* **3** (1967), p. 24.

6.103 Duraev, V. P., Kubetskii, G. A., Pugach, M. K. and Shveikin, V. I., *Pribory Tekh. Eksperim.* **11** (1968), p. 214. In Russian. Translation: Ohmic contacts to GaAs. *Sov. Instrument. Exp. Techn.* **11** (1968), p. 469.

6.104 Klohn, K. L. and Wandinger, L., Variation of contact resistance of metal-GaAs contacts with impurity concentration and its device implication. *J. Electrochem. Soc.: Solid-State Sci.* **116** (1969), p. 507.

6.105 Bolton, R. M. G. and Jones, B. F., Effects of temperature on contact resistance of Gunn diodes. *Electron. Lett.* **5** (1969), p. 662.

6.106 Engelbrecht, R. S., Solid-state bulk phenomena and their application to integrated circuits. *IEEE J. Solid-State Circuits* **SC–3** (1968), p. 210.

6.107 Macrae, A. U., Low energy electron diffraction study of the polar III-surfaces of GaAs and GaSb. *Surface Sci.* **4** (1966), p. 247.

6.108 Jeppsson, B. and Marklund, I., Failure mechanism in Gunn diodes. *Electron. Lett.* **3** (1967), p. 213.

6.109a Carslow, H. S. and Jaeger, J. C., *Conduction of Heat in Solids*, p. 92 ff. (London, Oxford University Press, 1959).

6.109b Ibid., pp. 216 ff. and 265.

6.110 Maycock, P. D., Thermal conductivity of silicon, germanium, III–V compounds and III–V alloys. *Solid-State Electron.* **10** (1967), p. 161.

6.111 Knight, S., Heat flow in n^{++}-n-n^{+} epitaxial GaAs bulk effect devices. *Proc. IEEE* **55** (1967), p. 112.

6.112 Becker, R. and Bosch, B. G., Design of Gunn devices for CW operation. *IEEE Trans. Electron Dev.* **ED–14** (1967), p. 615.

6.113a Clorfeine, A. S., Power dissipation limits in solid-state oscillator diodes. *Microwave J.* **11** (1968) March, p. 93.

6.113b Bravman, J. S. and Eastman, L. F., Thermal effects of the operation of high average power Gunn devices. *IEEE Trans. Electron Dev.* **ED–17** (1970), p. 744.

6.114 Ramachandran, T. B., Sheet Gunn oscillator and thermal considerations. *Proc. IEEE* **56** (1968), p. 336.

6.115 *Handbook of Chemistry and Physics*, 47th ed. (Cleveland, Ohio, The Chemical Rubber Co., 1966–1967).

6.116a Migitaka, M., Miyazaki, M. and Saito, K., High-power Gunn oscillator diodes on type IIA diamond heat sinks. *IEEE Trans. Microwave Theory Techniques* **MTT–18** (1970), p. 1004.

6.116b Swan, C. B., Improved performance of silicon avalanche oscillators mounted on diamond heat sinks. *Proc. IEEE* **55** (1967), p. 1617.

6.117 Berman, R., *Physical Properties of Diamond*, p. 387 (New York/Oxford, Clarendon Press, 1965).

6.118a Pierron, E. D., Parker, D. L. and McNeeley, J. B., Coefficient of expansion of GaAs, GaP, and Ga(AsP) compounds from $-62°$ to 200°C. *J. Appl. Phys.* **38** (1967), p. 4669.

6.118b Feder, R. and Light, T., Precision thermal expansion measurements of semi-insulating GaAs. *J. Appl. Phys.* **39** (1968), p. 4870.

6.119 Tang, D. D., High power varactor package for multichip application. *Proc. IEEE* **57** (1969), p. 799.

6.120 Plevyak, T. J., Vapor phase cooling of semiconductor circuit components. IEEE Industry and General Applications Group, Conference Record of 1968 Annual Meeting, Chicago, Ill., Sept./Oct., p. 541.

6.121 Bostock, P. A. and Walsh, D., Variation of velocity/field curve of GaAs in the temperature range 40–180°C. *Electron. Lett.* **5** (1969), p. 623.

6.122a Bott, I. B. and Holliday, H. R., The performance of X-band Gunn oscillators over the temperature range 30°C to 120°C. *IEEE Trans. Electron Dev.* **ED–14** (1967), p. 522.

6.122b Higashisaka, A., Temperature dependence of Gunn effect in GaAs over the range 77°K to 545°K. *Jap. Jl Appl. Phys.* **9** (1970), p. 583.

6.123 Berson, B. E. and Narayan, S. Y., L-band epitaxial Gunn oscillators. *Proc. IEEE* **55** (1967), p. 1078.

6.124 McGroddy, J. C., IBM Thomas J. Watson Research Center, private communication (1969).

6.125 Sato, H. and Jida, S., A thin GaAs n on n^{+} epitaxial film with abrupt interface in carrier concentration profile. *Jap. J. Appl. Phys.* **9** (1970), p. 156.

6.126 Clarke, G. M., Edridge, A. L. and Bass, J. C., Planar Gunn effect oscillators with concentric electrodes. *Electron. Lett.* **5** (1969), p. 471.

6.127 Jeppsson, B., Marklund, I. and Olsson, K., Voltage tuning of concentric planar Gunn diodes. *Electron. Lett.* **3** (1967), p. 498.

6.128a Vorob'ev, Y. V., Karhanin, Y. I. and Tretyak, O. V., Electrical instability in half-insulating GaAs. *Phys. Stat. Sol.* **36** (1969), p. 499.

6.128b Heath, D. R., Selway, P. R. and Tooke, C. C., Photoconductivity and infra-red quenching in Cr-doped semi-insulating GaAs. *Brit. J. Appl. Phys.* (*J. Phys. D*) **1** (1968), p. 29.

6.129 Weisberg, L. R., Rosi, F. D. and Hekart, P. G., Materials research on GaAs and InP. In: *Properties of Elemental and Compound Semiconductors, Proc. Techn. Conf., Boston, Mass., 1959* (Ed.: Gatos, H. C.), p. 25 (New York, Interscience, 1960).

6.130 Von Münch, W., *Technology of Gallium Arsenide Devices* (in German; Berlin, Springer, 1969).

6.131 Constantinescu, C. R. and Petrescu-Prahova, I., Acceptor behaviour of Ge in GaAs. *J. Phys. Chem. Solids* **28** (1967), p. 2397.

6.132 Copeland, J. A., Semiconductor impurity analysis from low frequency noise spectra. *IEEE Trans. Electron Dev.* **ED–18** (1971), p. 50.

6.133 Hickey Jr., J. E., Gallium arsenide: what is its status? *Electron. Ind.* **22** (1963), Feb., p. 47.

6.134 Honig, R. E., Vapor pressure data for the solid and liquid elements. *RCA Rev.* **23** (1962), p. 567.

6.135 Copeland, J. A. and Niehaus, W. C., Traps and low-field breakdown in n-GaAs devices, 1970, unpublished.

6.136a Stillman, G. E., Wolfe, C. M. and Dimmock, J. O., Magnetospectroscopy of shallow donors of GaAs. *Solid-State Commun.* **7** (1969), p. 921.

6.136b Stillman, G. E., Wolfe, C. M. and Dimmock, J. O., Effect of donor density on extrinsic GaAs photodetectors. *Solid-State Res. Rpt* No. 3, 1969, Lincoln Laboratory, MIT, Lexington, Mass,, p. 6 (Oct. 23, 1969).

6.137 Wolfe, C. M., Stillman, G. E. and Dimmock, J. O., Ionized impurity density in n-type GaAs. *J. Appl. Phys.* **41** (1970), p. 504.

6.138 Steiner, S. A. and Anderson, R. L., Degradation of GaAs injection devices. *Solid-State Electron.* **11** (1968), p. 65.

6.139 Kuru, I., Performance degradation of Gunn diodes at elevated temperature. *Proc. 2nd Conference on Solid State Devices, Tokyo, Japan, 1970: J. Jap. Soc. Appl. Phys.* **40** (1971), Suppl., p. 137.

6.140 Chilton, R. H., Failure mechanisms and life test of Gunn and avalanche devices. *Proc. 1970 MOGA Conference*, p. 9/4 (Deventer, Kluwer, 1970).

6.141 Al-Moufti, M. N., Jaskolski, S. V. and Ishii, T. K., Ageing effects of bulk GaAs devices. *Proc. IEEE* **56** (1968), p. 236.

6.142 Yamashita, A. and Nii, R., A mechanism of current saturation in GaAs. *IEEE Trans. Electron Dev.* **ED–13** (1966), p. 196.

6.143 Engelmann, R. W. H., Hewlett-Packard Ass., unpublished investigation (1965).

6.144a Colliver, D. J. and Fray, A. F., Limitation to the performance of planar Gunn effect devices. *Solid-State Electron.* **12** (1969), p. 671.

6.144b Clarke, G. M. and Griffith, I., An investigation of X and J band transverse Gunn oscillators. *Proc. 1970 MOGA Conference*, p. 2/21 (Deventer, Kluwer, 1970).

6.145 Christensson, S., Lundström, I., Marklund, I. and Jeppsson, B., Simple model for electron-hole generation and switching mechanisms in n-type GaAs diodes. *Electron. Lett.* **3** (1967), p. 507.

6.146 Thim, H. W. and Knight, S., Carrier generation and switching phenomena in n-GaAs devices. *Appl. Phys. Lett.* **11** (1967), p. 83.

6.147 Acket, G. A. and Scheer, J. J., Relaxation oscillations due to impact ionization in epitaxial sheet-type Gunn oscillators. *Electron. Lett.* **5** (1969), p. 160.

6.148 Kennedy Jr., W. K., Eastman, L. F. and Gilbert, R. J., LSA operation of large volume bulk GaAs samples. *IEEE Trans. Electron Dev.* **ED–14** (1967), p. 500.

6.149a Thornton, P. R., Sulway, D. V. and Shaw, D. A., Scanning electron microscopy in device diagnostics and reliability physics. *IEEE Trans. Electron Dev.* **ED–16** (1969), p. 360.

6.149b Gopinath, A., On scanning electron microscope conduction-mode signals in bulk semiconductor devices: linear geometry. *J. Phys. D.: Appl. Phys.* **3** (1970), p. 467.

6.149c Gopinath, A. and de Monts de Savasse, T., On scanning electron microscope conduction-mode signals in bulk semiconductor devices: annular geometry. *J. Phys. D.: Appl. Phys.* **4** (1971), p. 2031.

6.150a Shoji, M. and D'Alessio, F. J., Improvement of reliability of Gunn diodes. *Proc. IEEE* **57** (1969), p. 250.

6.150b Adams, R. F., CW operation of GaAs planar Gunn diodes with evaporated contacts. *Proc. IEEE* **57** (1969), p. 2164.

6.151 Seki, H. and Moriyama, K., Vapor deposition of silicon nitride on GaAs by $SiCl_4$-NH_4N_2 system. *Jap. J. Appl. Phys.* **6** (1967), p. 1345.

6.152a Hoyt, P. L. and Haisty, R. W., Preparation of epitaxial semi-insulating GaAs by iron doping. *J. Electrochem. Soc.* **113** (1966), p. 296.

6.152b Mizuno, O., Kikuchi, S. and Seki, Y., Epitaxial preparation of semi-insulating GaAs. *Jap. J. Appl. Phys.* **10** (1971), p. 208.

6.153 Selway, P. R. and Nicolle, W. M., Negative resistance in chromium-doped GaAs p-i-n diodes. *J. Appl. Phys.* **40** (1969), p. 4087.

6.154a Anderson, W. A., The performance of a Gunn diode under the influence of γ radiation. *Proc. IEEE* **57** (1969), p. 1198.

6.154b Brehm, G. E. and Pearson, G. L., Effects of gamma radiation on Gunn diodes. *IEEE Trans. Electron Dev.* **ED–17** (1970), p. 475.

6.155 Stein, H. J., Electrical studies of low-temperature neutron- and electron-irradiated epitaxial n-type GaAs. *J. Appl. Phys.* **40** (1969), p. 5300.

6.156 Lampert, M. A. and Mark, P., *Current Injection in Solids.* (New York, Academic Press, 1970.)

6.157 Omelianovski, E. M., Deep levels in semi-insulating GaAs doped with Cr. *Bull. Amer. Phys. Soc.* **15** (1970), p. 1615.

6.158a Champlin, K. S., Erlandson, R. J., Glover, G. H., Hauge, P. S. and Lu, T., Search for resonance behaviour in the microwave dielectric constant of GaAs. *Appl. Phys. Lett.* **11** (1967), p. 348.

6.158b Braslau, N., On the dielectric constant of GaAs at microwave frequencies. *Appl. Phys. Lett.* **11** (1967), p. 350.

6.158c Rogers, C. B., Thompson, G. H. B. and Antell, G. R., Dielectric constant and loss tangent of semi-insulating GaAs at microwave frequencies. *Appl. Phys. Lett.* **11** (1967), p. 353.

6.159 Minden, H. T., Recent advances in gallium arsenide materials technology. *Solid-State Technol.* **12** (1969), April, p. 25.

6.160 Hilsum, C., Smith, K. C. H., Taylor, B. C. and Knight, J. R., C.W. X and K band radiation form GaAs epitaxial layers. *Electron. Lett.* **1** (1965), p. 178.

6.161 Foxell, C. A. P., Summers, J. G. and Wilson, K., Surface-orientated Gunn-effect oscillators. *Electron. Lett.* **1** (1965), p. 217.

6.162 Hubbard, H. D. and Meggers, W. F., *Periodic Chart of Atoms.* (Skokie, Ill., The Welch Scientific Company, 1965).

6.163a Owens, J. and Kino, G. S., Theoretical study of Gunn domains and domain avalanching. *J. Appl. Phys.* **42** (1971), p. 5006.

6.163b Owens, J. and Kino, G. S., Experimental study of Gunn domains and avalanching. *J. Appl. Phys.* **42** (1971), p. 5019.

6.164 German Fed. Patent Nr. 1,614,826 (inventor: R. Becker, priority: June 19, 1967).

6.165 Orton, J. W., *Material for the Gunn Effect* (London, Mills and Boon Ltd., 1971).

6.166 Pilkuhn, M., Stuttgart University, private communication, 1972.

6.167 Williams, R., Avalanche and tunneling currents in gallium arsenide. *RCA Rev.* **27** (1966), p. 336.

6.168 Hilsum, C., Band structure engineering. Invited paper presented at European Semiconductor Device Research Conference, Munich, March 16–19, 1971.

7
Microwave Devices and Applications

This chapter is intended to provide a survey of the performance of TE devices for microwave applications. Whereas Chapter 5 was devoted to the theoretical analysis of the inherent TE-device capabilities as oscillators and amplifiers, we now consider practical circuit design, measured performance data and suitability for microwave systems.

7.1 OSCILLATORS

7.1.1 Circuit Considerations

In Chapters 3 to 5 mainly simple lumped-element RLC parallel circuits of a fairly high quality factor Q were considered for establishing a sinusoidal voltage waveform across the TE device. In practice transmission-line or waveguide resonators of various designs are employed. In addition, the device package introduces parasitic circuit elements. The exact analysis of such circuits requires a high degree of sophistication and, in general, can be performed only by making use of a digital computer. The circuits have to be characterized not only with regard to their behaviour at the fundamental frequency but also at higher harmonics, since these higher frequencies play an important role in device performance as outlined in Section 5.2. For this reason circuit analysis in the time domain [5.101] may yield faster results than frequency-domain techniques which have briefly been described in Chapter 5. In the following we intend to familiarize the reader with several microwave circuits frequently employed for TE devices without going into particular details of their full mathematical description. The interested reader is referred to the original literature.

From all resonator circuits the TEM transmission-line resonator offers the most straightforward analysis. Both coaxial and strip-line types can be employed up to about 10 GHz. Beyond this frequency TEM waves are subjected to heavy losses in these circuits and better performance is obtained in waveguide resonators.

A versatile coaxial resonator is shown in Fig. 7.1a. The TE device is mounted in series with the centre conductor at an appropriate location which can be adjusted by the two sliding r.f. short circuits at the two ends of the resonator. One of the r.f.

short circuits is capacitive for supplying the d.c. or pulsed bias to the TE device between the centre and outer conductors of the resonator. The other (galvanic) short circuit is provided with a coupling loop for extracting the r.f. power into a coaxial line with matched termination. An r.f. equivalent circuit of the resonator is shown in Fig. 7.1b. The coaxial TEM transmission line of real characteristic impedance [7.1]

$$Z_0 = \frac{1}{2\pi} \sqrt{\frac{\mu_0}{\varepsilon_0}} \sqrt{\frac{\varepsilon_0}{\varepsilon}} \ln \frac{r_2}{r_1} = 60 \sqrt{\frac{\varepsilon_0}{\varepsilon}} \ln \frac{r_2}{r_1} \, \Omega \tag{7.1}$$

(assuming a permeability ratio $\mu/\mu_0 = 1$) and length l is interrupted at the position $x_2 = 1 - x_1$ for connecting the terminals of the TE device chip between 1 and 2. For determining the load admittance $Y_L = G_L + jB_L$ between the points 1 and 2 as

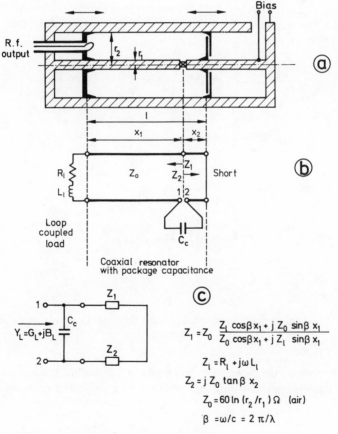

Fig. 7.1 Coaxial-line TEM-mode resonator with loop output coupling: (a) cross-sectional view, (b) TEM transmission-line equivalent circuit, (c) lumped-element equivalent circuit. After Pollmann *et al.* [7.2].

seen by the device, the incorporation of several lumped elements into the trans-mission-line circuit is important. The r.f. coupling loop in the galvanic sliding short is considered to be represented by an impedance Z_1 consisting of a series combination of a frequency dependent resistance R_1 (ω) and a frequency dependent inductance L_1 (ω). The resistance R_1 describes the *external* r.f. load resistance as transformed to the position of the coupling loop and/or, in an approximate way, any additional ohmic losses in the resonator cavity. A frequency dependent capacitance $C_c(\omega)$ between terminals 1 and 2 describes the parasitic reactance of the device case and its mount in the centre-conductor gap.

A simple equivalent circuit for the device case or package (Fig. 6.13) with frequency inde-pendent lumped elements valid at X-band frequencies is shown in Fig. 7.2a. L_1 is the lead inductance, L_2 the inductance of the mounting stud, C_1 the stud capacitance, and C_2 the capacitance of the cylindric insulating spacer [3.41]. For representing $C_c(\omega)$ over wider frequency bands, more elements [7.22] or even transmission-line sections [7.23] have to be incorporated into the circuit. For the packaged device the symbol sketched in Fig. 7.2b will be used from now on.

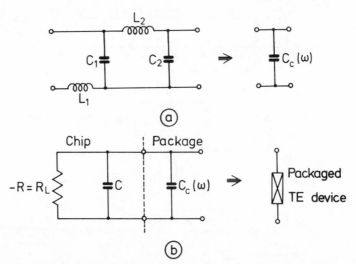

Fig. 7.2 (a) Equivalent circuit of device package, (b) symbol for packaged device used in later figures.

Figure 7.1c shows the details for calculating the load admittance $Y_L = G_L + jB_L$ of the coaxial resonator as presented to the device chip. The resonance condition reads

$$B_L + \omega C = 0 \qquad (7.2)$$

where C is the capacitance of the TE chip in the oscillating state. In general this chip capacitance is a function of the frequency ω, the generated r.f. power P, and the harmonic content of the oscillation. Equation (7.2) determines the oscillation frequency at which the device load conductance $G_L = 1/R_L$ is to be calculated for obtaining the expected r.f. output according to the discussion in Section 5.2. Relation (7.2) is plotted in Fig. 7.3a [7.2] for a given TEM-mode resonant frequency at a resonator wavelength λ: the resonator length l is shown as a function of the

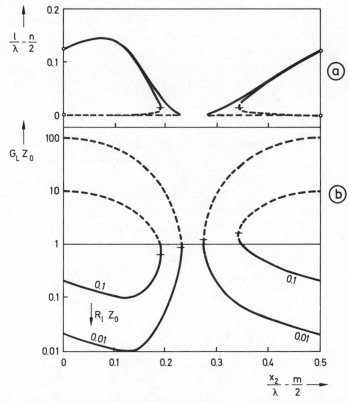

Fig. 7.3 Mode chart of coaxial resonator of Fig. 7.1 for a given resonator wavelength $\lambda = c/f$: (a) normalized resonator length l/λ and (b) normalized load conductance $G_L Z_0$ as functions of TE-device position x_2/λ. $n \geq 0$ and $m \leq n$ are positive integers. After Pollmann *et al.* [7.2].

TE-chip position x_2, both normalized to the resonator wavelength $\lambda = 2\pi c/\omega$ of the r.f. signal (c is the TEM wave velocity along the transmission line). Owing to the periodicity of the resonances with respect to length l and position x_2, integer order numbers $n \geq 0$ and $m \geq 0$ with $m \leq n$ can be defined for each resonance. Figure 7.3b shows the behaviour of the load conductance along the curves of Fig. 7.3a. As can be seen, strong variations of G_L occur when changing the TE-device position x_2 within the resonator if the resonator length l is adjusted to keep the oscillation frequency constant. This technique of varying the load conductance G_L allows a simple experimental determination of the maximum output power of the TE oscillator. The curves shown in Fig. 7.3 were calculated assuming $L_1 = 0$ and $\omega(C_c + C) = Z_0^{-1}$ for two degrees of loop coupling determined by the parameter R_1/Z_0.

Coaxial resonator structures with loop coupling into coaxial lines are widely used for TE devices [7.3–7.9]. In addition to mechanical tuning by sliding shorts [7.2–7.4], mechanical tuning by sliding dielectrics [7.5] or by capacitive screws [7.6, 7.7] is employed, and electrical tuning by varactor diodes [7.7, 7.8] or by yttrium-iron garnet (YIG) spheres [7.9] (see below) inserted into the resonator at an appropriate location. An example for a possible arrangement of coupling a varactor diode to the resonator is shown in Fig. 7.4. The varactor is mounted in parallel to the TE

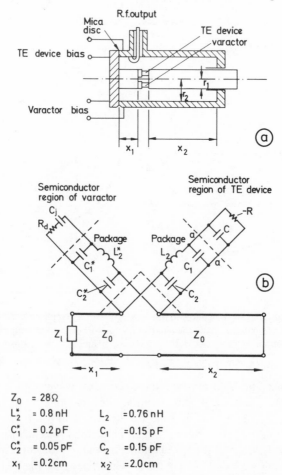

$$
\begin{array}{ll}
Z_0 & = 28\,\Omega \\
L_2^* & = 0.8\,\text{nH} \qquad L_2 = 0.76\,\text{nH} \\
C_1^* & = 0.2\,\text{pF} \qquad C_1 = 0.15\,\text{pF} \\
C_2^* & = 0.05\,\text{pF} \qquad C_2 = 0.15\,\text{pF} \\
x_1 & = 0.2\,\text{cm} \qquad x_2 = 2.0\,\text{cm}
\end{array}
$$

Fig. 7.4 Coaxial-line TEM-mode resonator with varactor tuning (cf. Section 7.1.2.2). Adapted from Smith and Crane, [7.8a]. (a) Cross-sectional view, (b) equivalent circuit.

device on a split-centre conductor allowing separate bias connections to be made. In the equivalent circuit the varactor package is represented by a similar network as used for the TE device package (Fig. 7.2). The resonant frequency is affected by changing the varactor junction capacitance C_j. The varactor series resistance R_d introduces losses which reduce the r.f. output power and limit the tuning range. Hence, varactors with high quality factor $(R_d \omega C_j)^{-1}$ are desirable.

Instead of loop-coupling the external load resistance to the resonator it can be connected also via movable low-impedance slugs [5.49a, 5.97, 7.10] as sketched in Fig. 7.5, or via a transmission-line tuner [6.87].

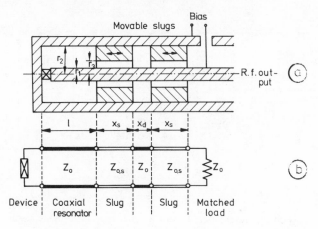

Fig. 7.5 Coaxial-line TEM-mode resonator with output coupling across low-impedance slugs: (a) cross-sectional view, (b) TEM transmission-line equivalent circuit. The load impedance Z_L can easily be calculated from the dimensions of the slugs and the resonator by using coaxial transmission-line formulas. After Camp Jr. [7.10a].

Short coaxial resonators can be excited both in TEM-mode resonances and radial TM-mode resonances [7.11, 7.12]. The different nature of the electric field pattern and of the equivalent circuit is shown in Fig. 7.6. The possibility of the two types of resonances has to be taken into account when interpreting actual oscillator performance [7.12]. If the centre conductor is removed from the structure a pure radial resonator results [7.13, 7.14, 5.59]. Various forms of such resonators are shown in Fig. 7.7. In examples (a) and (b) r.f. power is extracted from the resonator into a coaxial line by galvanic coupling which results in low loaded-circuit Q for obtaining highest efficiency [7.14, 5.59]. Capacitive coupling, example (c), is used for increasing the circuit Q if low noise performance is required [7.13]. Tuning of the resonator can be effected with a sliding dielectric rod (Fig. 7.7c), with a turning screw (Fig. 7.8) or, electronically, with a varactor diode [7.13]. An engineering drawing of a practical radial resonator is presented in Fig. 7.8.

For microwave integrated circuits TEM stripline resonators have proved to be

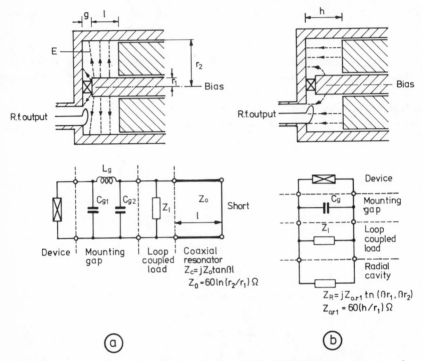

Fig. 7.6 Short coaxial resonator: (a) with 'coaxial-line' TEM-mode resonance and (b) with 'radial-line' TM-mode resonance. The lumped-element equivalent circuits with the coupling loop neglected are shown in each case. tn $(\beta r_1, \beta r_2)$ is the 'small radial tangent function' and $Z_{0,r1}$ the radial characteristic impedance at r_1 as defined in [7.15]. After Foulds [7.12].

suitable in triplate [7.36] and microstrip versions [6.41a, 6.88b, 7.7, 7.37–7.41]. Satisfactory performance has been achieved up to X-band frequencies. However, losses are higher than in coaxial circuits. On the other hand, adaptability to easy mass production is a great advantage. A typical X-band microstrip module and its equivalent circuit is shown in Fig. 7.9. The circuit pattern is etched into an Au-Cr film evaporated on an Al_2O_3 ceramic substrate. For higher performance the ceramic is replaced by single-crystal sapphire.

For frequencies at or beyond X band, most frequently rectangular waveguide circuits are employed. A widely used resonator is shown in Fig. 7.10a [7.16a]. The TE device is mounted on a cylindrical post across the waveguide. The resonator structure is formed by a sliding short on one side and an iris output coupler at the other. (Load coupling can, in principle, be achieved also by using an E-H tuner [4.5, 5.55] or a series of tuning screws [7.7].) In the r.f. equivalent circuit for the propagating TE_{10} (H_{10}) waveguide mode (electric field parallel to post axis) the post is represented by a T-network of an inductance L_p and two capacitances C_p,

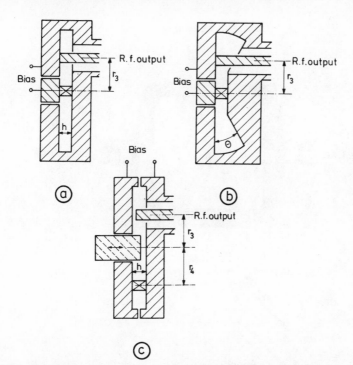

Fig. 7.7 Radial TM-mode resonators of various designs: (a) of constant height h with position (r) dependent radial characteristic impedance in air $Z_{0,r} = 60\ (h/r)\ \Omega$; (b) of constant angle θ with constant radial characteristic impedance in air $Z_{0,\theta} = 60\ \theta\Omega$; (c) with excentric position of the TE device, ceramic rod tuning and capacitive output coupling; (a) and (b) after Eastman [5.95], (c) after Olfs [7.13a].

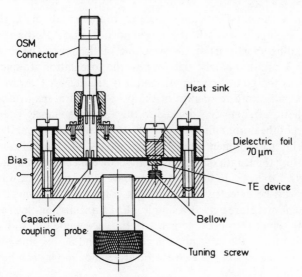

Fig. 7.8 Engineering drawing of a radial TM_{010} resonator cavity (cavity height 0·5 cm; inner diameter 3 cm for J band, 1·8 cm for X band).

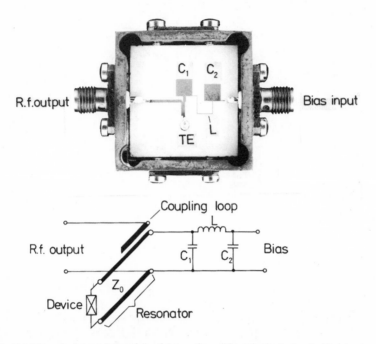

Fig. 7.9 X-band microstrip module: (a) top view, (b) equivalent circuit. (By courtesy of H. Shah, AEG-Telefunken.)

see Fig. 7.10b [7.16]. A fringing-field capacitance C_g represents the mounting gap. Z_1 is the equivalent load impedance of the iris output portion, and Z_w is the waveguide impedance of the sliding-short portion at the plane of the post. The impedance $Z_w = jZ_0 \tan(2\pi x_1/\lambda_g)$ approximately determines the oscillation frequency by $x_1 \approx n\lambda_g/2$ where λ_g is the guide wavelength and n is an integer. However, if the sliding short is tuned close to a resonance of the post structure, major deviations occur since the oscillation frequency is locked to this 'post resonance' [7.16–7.19]. Both the reactive part of the output impedance Z_1 [7.16b] and the device and package reactances [7.16a, 7.17, 7.18] influence the post resonance. Mechanical tuning can be achieved by adjusting the length of the post with a coaxial extension as indicated in Fig. 7.11a [7.17, 7.20, 7.21]. If the post resonance falls below the waveguide cut-off frequency, it can be detected by inserting a coupling loop connected to a coaxial output line. The structure acts as a simple TEM transmission-line resonator in this case (Fig. 7.10c) since the post together with the waveguide side walls form a slab line and the influence of the waveguide portions Z_1 and Z_w become negligible [5.101]. In addition, resonances of the evanescent waveguide modes can be excited in such a configuration [7.33–7.35].

A vane or stub extension in the post mounting structure (Fig. 7.11b) proved to be

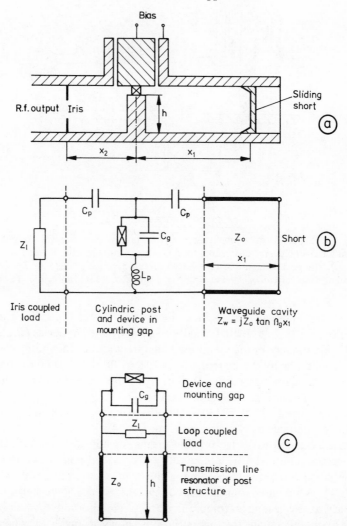

Fig. 7.10 Rectangular waveguide resonator with post structure for mounting TE device and with iris output coupling: (a) cross-sectional view, (b) equivalent circuit for TE_{01} waveguide mode (resonances above the waveguide cut-off frequency), (c) equivalent circuit for TEM-mode resonance for the post structure below the waveguide cut-off frequency. (a) and (b) after Tsai *et al.* [7.16a]; (c) after Jeppesen and Jeppsson [5.101].

particularly useful for LSA or hybrid-mode operation by producing a primary resonance with a loading delay for fast build-up of the oscillations [4.5, 5.55, 5.97, 7.24–7.27]. Fast build-up of the LSA oscillation prevents the formation of dipole domains with their high fields which may destroy the device. Mounting structures

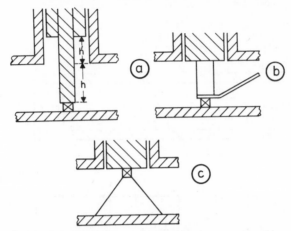

Fig. 7.11 Different device mounting structures: (a) post structure with coaxial extension for second degree-of-freedom to obtain proper tuning, after Hansen and Rowe [7.20]; (b) post structure with vane to establish primary TEM resonance with loading delay, after Copeland and Spiwak [4.5]; (c) ridge structure, after Taylor and Howes [7.19].

other than a cylindrical post include ridges (Fig. 7.11c) [7.19, 7.25] or specially formed irises [5.102, 7.28, 7.29]. If small size is important, the device may be operated solely at the resonance of the mounting structure and thus the sliding-short portion of the waveguide cavity can be removed [5.102, 6.70b, 7.30]. On the other hand, the influence of the mounting structure can largely be eliminated by using reduced-height [5.55, 7.31, 7.32] or ridge waveguide cavities [7.32]. For achieving high resonant frequencies (beyond about 20–30 GHz) the device package needs special consideration to minimize its parasitic inductances [3.41c].

The circuits described so far provide means for adjusting the fundamental-frequency termination. Only the circuit with the extended post structure (Fig. 7.11a) possesses a further degree of freedom for improving the output power of the oscillations [7.21]. As pointed out in Section 5.2.4.1 independent harmonic tuning, particularly at the second harmonic frequency, is important to achieve high efficiencies. A suitable circuit for this purpose is shown in Fig. 7.12, where a coaxial resonator is coupled to a section of a reduced-height waveguide [7.2, 7.42]. The former establishes the fundamental resonance whereas the latter, being designed for a cut-off frequency between the fundamental and second harmonic frequency, takes charge of the tuning at the second harmonic without changing the load conditions at the fundamental. Of course, the coaxial resonator can be replaced by a waveguide with the proper dimensions to support the fundamental frequency [7.44]. In purely coaxial circuits the two frequencies can simply be separated by filters [7.2, 7.42, 7.43]. A circuit consisting of two coupled bars that support TEM-wave propagation proved to be also very suitable for harmonic tuning [7.6, 7.45]. It has an especially simple structure compatible with microstrip technology.

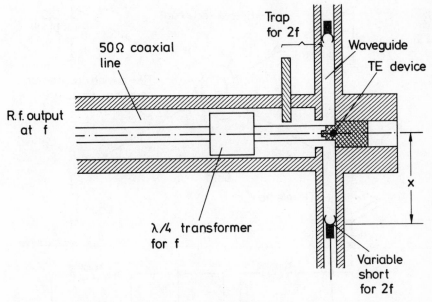

Fig. 7.12 Circuit for independent tuning of second-harmonic frequency 2*f*. After Frey *et al.* [7.42].

An attractive microwave resonator of a nature completely different from the structures discussed so far is formed by a small sphere of yttrium iron garnet (YIG), a high-quality ferrimagnetic crystal [7.46–7.50]. The sphere, having typically a diameter of 0·5 to 1 mm, resonates in a d.c. magnetic field approximately at the natural precession frequency of the electron spins in the material [7.46c], viz.

$$\omega_p \approx (e/m_0)B_0 \qquad (7.3)$$

where $B_0 = \mu_0 H_0$ denotes the external d.c. magnetic induction field. Equation (7.3) predicts for an applied magnetic induction of, for example, $B_0 = 0\cdot4$ T a precession frequency of $f_p = 11\cdot2$ GHz. The resonance is excited by r.f. magnetic fields perpendicular to the d.c. magnetic field. Thus only properly adjusted inductive circuit elements are able to couple to the YIG resonator. Figure 7.13a shows a possible arrangement for coupling the TE device and the r.f. output circuit to the sphere [7.46a]. A fairly accurate and a simplified equivalent circuit of this arrangement are given in Fig. 7.13b. The tank circuit L_y, C_y, R_y represents the YIG resonance, including internal losses. The additional reactive elements in the circuit lead to an actual oscillation frequency that is pulled away somewhat from the electron spin precession frequency of eqn (7.3). An important advantage of the YIG resonator is its simple electronic frequency tunability with an electro-magnet. According to eqn (7.3) the tuning follows linearly the magnetic field and hence the magnet current. However, the saturation magnetization sets a lower limit on the frequency, being

271

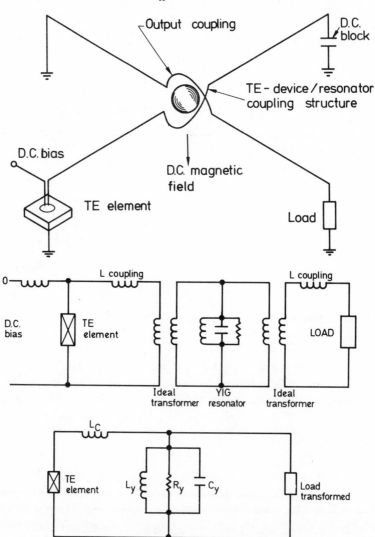

Fig. 7.13 YIG sphere in a d.c. magnetic field as a microwave resonator: (a) schematic representation, (b) equivalent circuits. After Clark and Swartz [7.46a].

about 3·5 GHz for pure YIG. With doped YIG crystals lower frequencies can be reached. YIG resonators allow multiple octave-bandwidth tuning with satisfactory operation up to about 18 GHz. Unloaded Q values of the YIG resonator exceed those of mechanical microwave cavities, reaching up to 10^4 in X band [7.46].

YIG spheres can also be employed in conventional microwave resonators, particularly for wide-range electronic tuning [7.9, 7.51]. Electronic tuning with varactor diodes, however, is more common if an extremely wide tuning range is not required [7.7, 7.8, 7.13, 7.31, 7.36–7.38]. Details on the tuning performance of TE oscillators are presented in Section 7.1.2.2.

7.1.2 Performance

In this section we will present performance data of *GaAs* TE oscillators only. Recently, TE oscillators made of *InP* have also received attention, but have not yet reached the mature state of GaAs oscillators. Qualitatively, however, the behaviour of the InP devices appear to be rather similar [5.94, 7.113, 7.114].

7.1.2.1 *Output Power and Efficiency*

The theoretical analysis of output power and efficiency of TE oscillators as presented in Section 5.2 did not take into account any self-heating effects. In practical oscillators the dissipated d.c. power will always raise the temperature of the active device, as we have seen in Section 6.3. Hence, efficiency is reduced (Fig. 5.12), limiting the r.f. output power particularly in C.W. operation. Figure 7.14 shows the typical decrease

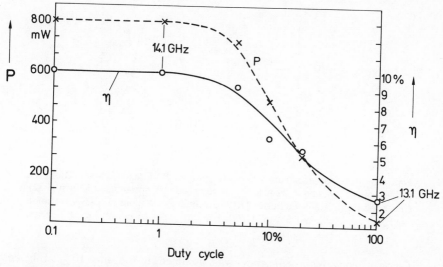

Fig. 7.14 R.f. output power P and efficiency η of a GaAs Gunn oscillator as a function of duty cycle. Sample data: mesa structure (Fig. 6.6c) with $d = 110\ \mu m$, $L = 8\ \mu m$, $n_0 \approx 2\text{–}3 \times 10^{15}\ cm^{-3}$, $V_0/V_T = 10$. After Narayan and Gobat [7.55].

of efficiency and output as a function of duty cycle up to C.W. operation (duty cycle 100%). Figure 7.15 compares experimental r.f. output powers of dipole-domain (Gunn), and LSA, single-device oscillators in C.W. operation and in low duty-cycle pulsed operation. Some efficiency values are also indicated. In Fig. 7.16 theoretically estimated C.W. power limits are shown for a single square-chip sandwich-type GaAs device mounted with the active n-layer directly on a Cu heat sink. The dissipated power was calculated according to eqn (6.18) with $E_{h0} = 10\ kV\ cm^{-1}$ and

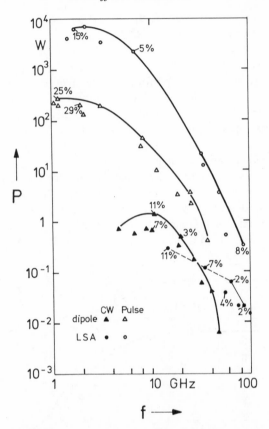

Fig. 7.15 Measured maximum r.f. output powers P of sandwich-type GaAs TE oscillators as a function of frequency f, as compiled by Copeland and Knight [1.32], Unger and Harth [1.36] and Berson [7.196], including C.W. points for LSA operation from Barrera [7.26] and Shyam [7.73]. Numbers refer to conversion efficiencies.

the r.f.-power conversion efficiency was assumed to be $\eta = 10\%$. Efficiency reduction at higher frequencies was taken into account using eqn (5.51) with $\tau_{\text{eff}} = 0.8$ ps and $\bar{\mu}/\bar{\mu}' = -1.43$ (cf. Fig. 5.28). The curves were based on the following constraints for the device [7.52]:

(a) Maximum tolerable temperature 500 K (at 300 K heat-sink temperature, neglecting any thermal resistance between active layer and heat sink, cf. Section 6.3.1).
(b) Doping-density × length product $n_0 L \geq 10^{12}$ cm^{-2}.
(c) Minimum low-field device resistance $R_1 \geq 1\Omega$. ·
(d) Skin effect limitation $d = w \leq 2.45\delta$, where δ is the skin depth assuming an r.f. device resistance of $|R| \approx 3R_1$ (cf. Fig. 7.2b).

274

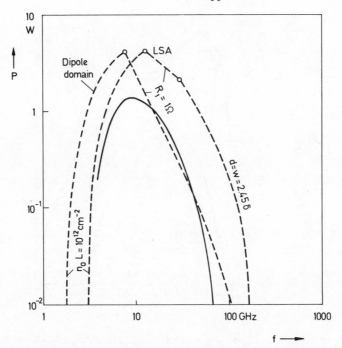

Fig. 7.16 Theoretically estimated r.f. power limits for C.W. operation in Gunn-domain and LSA modes (dashed curves) of a single GaAs sandwich device tolerating a device temperature of 500 K at a bias field of $E_{h0} = 10$ kV cm^{-1}; after R. Becker and B. G. Bosch [7.52]. Solid curve corresponds to best experimental data for Gunn mode; after R. J. Becker [7.53a], and M. Omori [7.53b].

(e) Frequency $f = f_\tau \approx v_D/L \approx (10^7/L)$ cm s^{-1} for the dipole-domain mode and $f = (n_0/6) \, 10^{-4}$ cm^3 s^{-1} for the LSA mode.

Figure 7.16 also shows measured performance of dipole-domain oscillators [7.53a, b]. As can be seen, the theoretically estimated limit is about 4 W at X-band frequencies, whereas experimentally somewhat above 1 W has so far been achieved for a single device. Power reduction at the low-frequency side is due to the limitation set by the lower limit of the $n_0 L$ product, on the high-frequency side it is caused by the impedance limitation $R_1 \geq 1 \, \Omega$. In case of the LSA mode the skin-effect limitation, however, becomes more severe at the highest frequencies [7.54]. The power reduction is enhanced beyond about 40 GHz by the frequency dependence of the efficiency (Section 5.2.5). Using coplanar devices (Fig. 6.6f) yields similar power envelopes, though shifted to somewhat lower frequencies [7.52].

In Fig. 7.17 C.W. values of the output power obtained experimentally are compared for different single microwave semiconductor devices [7.53]. Silicon IMPATT diodes are obviously superior to TE oscillators operated in the Gunn mode but have the disadvantage of exhibiting a higher noise level (see Fig. 7.34). Thus

275

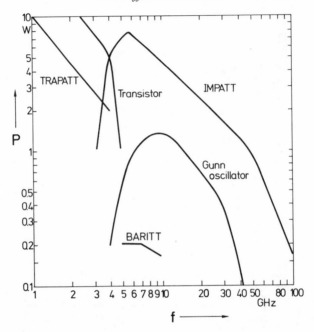

Fig. 7.17 Experimental C.W. values of r.f. output power P against frequency f for various semiconductor microwave devices. After R. J. Becker [7.53a].

both types of devices are found useful at X-band frequencies and beyond, depending on the requirements of the application desired.

Some curves representative for the bias dependence of Gunn-oscillator performance are presented in Figs 7.18–7.20. Figure 7.18 shows a d.c. current-voltage characteristic of a GaAs device superimposed on a series of curves obtained in low-duty cycle pulsed operation at different ambient temperatures. This procedure allows one to determine the operating temperature of the active crystal and, hence, the thermal resistance of its mount. In the example shown one finds $R_{th} = 32$ W/°C leading to a crystal temperature of about 235 °C at 10 V d.c. bias. The bias dependence of frequency f, r.f. output power P, and efficiency η of a C.W. GaAs oscillator is sketched in Fig. 7.19. The slight decrease in frequency is not typical, rising or bell-shaped curves are also observed [7.7, 7.29]. As to be expected theoretically from Fig. 5.15, output power and efficiency increase first with bias but they both show a sharp drop at higher bias values, commencing near two times threshold [7.7, 7.56]. This early drop is a consequence of self-heating [7.56b] since in pulsed operation the output power decreases generally only at relatively high bias values [3.52, 7.57, 7.58]. Figure 7.20 shows a particularly good result for a pulsed Gunn oscillator: a continuous rise in power and efficiency is observed up to 14 times threshold voltage. Curves of the type of Figs 7.19 and 7.20 exhibit sometimes discontinuities due to a

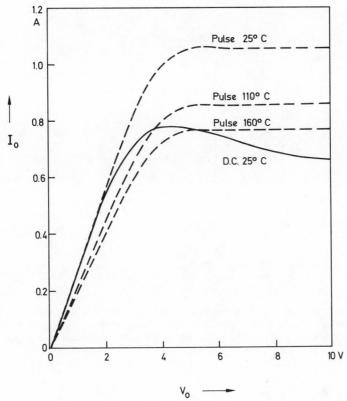

Fig. 7.18 Current-voltage characteristics of GaAs Gunn oscillator with pulsed and d.c. bias showing effect of temperature. After Murakami *et al.* [7.56a].

sudden jump from one oscillation mode to another, e.g. [7.21a, 7.72, 7.112, 7.129]. A similar bias performance as shown for the sandwich devices was also reported for coplanar device structures [6.126, 7.59, 7.60].

A statistical investigation of measured efficiency data of Gunn oscillators shows that values obtained on the average are still far from optimum. Figure 7.21 gives the dipole-domain mode results for a large number of similar devices fabricated from different GaAs wafers and studied in the same coaxial resonator at L- and S- band frequencies under pulsed-bias conditions. About 50% of the devices yielded less than 9% efficiency whereas the maximum efficiency observed was beyond 30%. A pronounced increase of efficiencies was obtained for the best wafer, demonstrating the strong influence exerted by the material properties on efficiency performance. In Fig. 7.22 the efficiency data are plotted against the $n_0 L$ product. Different symbols belong to different GaAs wafers. The scatter is considerable, but high efficiencies seem to crowd around $n_0 L \approx 6 \times 10^{12} \, \text{cm}^{-2}$. Finally, Fig. 7.23 shows the efficiency performance of LSA oscillators which were C.W.-operated at V-band frequencies

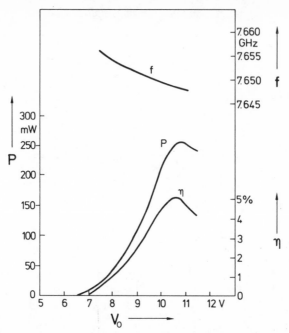

Fig. 7.19 Frequency f, r.f. output power P, and efficiency η measured as functions of bias voltage V_0 for C.W. operated GaAs Gunn oscillator (waveguide circuit). Sample data: $L = 15\ \mu m$, $n_0 = 2 \times 10^{15}\ cm^{-3}$. After Murakami *et al.* [7.56a].

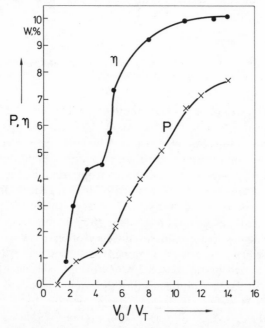

Fig. 7.20 Measured output power P and efficiency η as functions of bias voltage V_0 for pulsed operation of a GaAs Gunn oscillator. (Waveguide circuit, $f = 14$ GHz, duty cycle 5×10^{-5}; sample data: $L = 5 \cdot 5\ \mu m$, $n_0 = 3 \times 10^{15}\ cm^{-3}$.) After Edridge *et al.* [7.57].

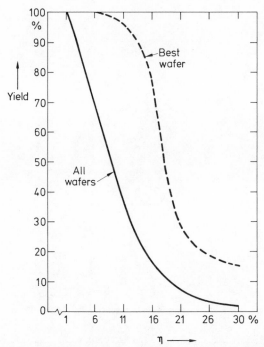

Fig. 7.21 Plot of device yield as a function of efficiency η in the Gunn-domain mode. Pulsed bias (duty cycle $0\cdot01\%$); sample data: $L = 75-100$ μm, $d = w = 250-625$ μm, $n_0 = 10^{14}-10^{15}$ cm^{-3}; frequency $f = 1-3$ GHz (coaxial circuit). After Reynolds *et al.* [5.56].

in a waveguide circuit of the type sketched in Fig. 7.11b. The dependence on $n_0 L/f$ is qualitatively to be expected from theory [4.9, 5.97]; cf. also Figs 5.16, 5.17. A distinct maximum of $\eta \approx 2\%$ appears near $n_0/f = 1\cdot1 \times 10^5$ s cm^{-3}.

Combining the r.f. power of individual TE devices operated in a phase-locked mode (see Section 7.1.2.4) is a way to realize, in principle, high-power solid-state microwave sources [7.62, 7.63]. Successful operation of direct parallel [7.64, 7.65] or series connection [7.66–7.68] of several GaAs chips has been reported. As an example, by operating four Gunn devices in parallel a C.W. output power of $2\cdot1$ W was achieved at $12\cdot8$ GHz with a paralleling efficiency of 70% [7.65], see also [7.53b]. For large-scale power combination heat sinking considerations call for substantial spatial separation of individual chips. This is achieved by inserting the chips in a microwave circuit at points of equal impedance [7.69–7.71]. Since both series and parallel connecting methods can be employed simultaneously, a satisfactory impedance match should not pose severe problems even if the number of combined devices is high.

7.1.2.2 *Frequency Tuning*

In Chapter 5 we have seen that the intrinsic tuning ranges for the quenched dipole-

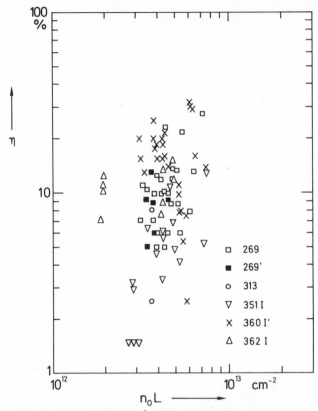

Fig. 7.22 Measured Gunn mode efficiencies η as a function of n_0L product. Operation and sample data same as in Fig. 7.21. After Berson *et al.* [6.51a].

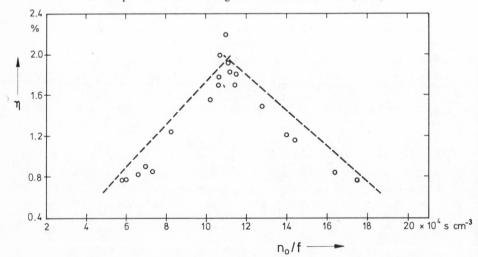

Fig. 7.23 Measured LSA efficiencies η of C.W. GaAs oscillators operating between 58 and 64 GHz (waveguide circuit of Fig. 7.11b) as a function of the n_0/f ratio (sample data: $n_0 = 3\cdot7 \times 10^{15}$–$1\cdot1 \times 10^{16}$ cm^{-3}, $L = 5$–15 μm, $d = 50$–125 μm). After Barrera [7.26].

domain, the hybird, and LSA, modes are relatively large. According to Fig. 5.20, for instance, we expect tuning ranges of up to 4:1. In practice, however, the tuning range is limited by the circuit employed. As we have already mentioned in Section 7.1.1 frequency tuning is effected either mechanically, or electronically using varactor diodes or ferrimagnetic material, particularly YIG spheres. Limited electronic tuning is also possible by changing the effective TE chip capacitance with the applied bias. The method of frequency tuning to be selected will, naturally, depend on the desired application.

Typical curves obtained for mechanical frequency tuning are shown in Fig. 7.24 referring to a low-power X-band Gunn oscillator and a high-power R-band LSA oscillator. In the first case the power difference between optimized loading at each frequency and fixed loading is indicated. Up to octave-bandwidth tuning can be achieved by mechanical means if the circuits are properly designed [6.100, 7.7]. Figure 7.25 demonstrates the power increase obtainable by optimizing the device termination at the second harmonic (harmonic tuning) by mechanical resonator adjustment.

Electronic frequency tuning by altering the bias voltage is usually rather small for sinusoidal voltage oscillations. In the example presented in Fig. 7.19 for a C.W. Gunn oscillator at 7·65 GHz the variation is only $-2·3$ MHz/V. Moreover, the amount varies from one device to another and might even change sign [7.7]. The influence of the bias voltage on the effective device capacitance and, hence, the frequency tuning is certainly influenced by the voltage and temperature dependence of the transit-time frequency [7.84]. However, changes in space-charge growth and decay, and in harmonic content of the voltage waveform also play a role. A relatively strong and consistent frequency increase with bias is observed in Gunn-mode oscillations if the domain capacitance is tightly coupled to the circuit [7.72a] and in LSA relaxation oscillators due to the high harmonic content [5.97, 5.101, 5.102, 6.70b, 7.72b]. An example of a relaxation-oscillator tuning curve is shown in Fig. 7.26.

Varactor tuning is relatively simple and straight-forward. The varactor diode acts as a variable capacitance in the load circuit thereby changing its resonant frequency (Fig. 7.4). Because it needs no tuning power and is relatively fast (sweep rates in the MHz range [7.74] this type of tuning is preferred practically for many applications. The tuning range is limited by the quality factor of the varactor diode [7.7, 7.8b, c] which is connected either in parallel (as in Fig. 7.4) or in series with the TE device [7.74]. A typical frequency tuning characteristic for parallel connection is shown in Fig. 7.27. Series coupling increases the tuning range as follows from a detailed circuit analysis [7.8c]: in a carefully designed circuit an octave bandwidth was achieved (Fig. 7.28). Presently available values of the varactor quality factor, which varies about inversely with frequency, set an upper frequency limit at about 13 GHz [7.74].

Magnetic frequency tuning is possible with ceramic ferrite material inserted into the microwave resonator [7.75, 7.76]. However, tuning ranges are relatively small

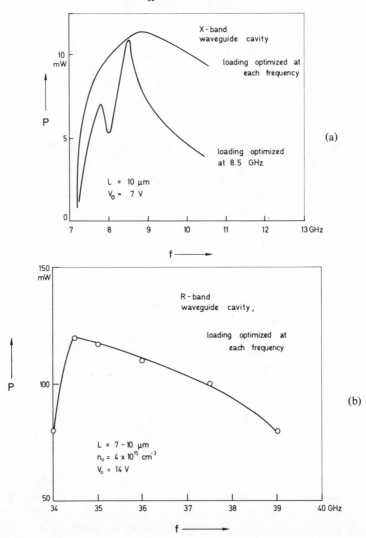

Fig. 7.24 Experimental r.f. output power P of C.W. GaAs oscillators as function of frequency f, with frequency variation effected by mechanical resonator tuning (waveguide circuits): (a) Gunn mode, after Wilson [7.7]; (b) LSA mode, after Shyam [7.73].

(800 MHz at X band, [7.76]) and speeds are slow (sweep rate < 400 Hz; [7.76]). Much better results are obtained by employing small YIG-sphere resonators (Section 7.1.1). Achieving a tuning range of $3:1$ presents no difficulty as demonstrated by Fig. 7.29. Thus YIG tuning is particularly suited for wide-band applications as in swept-frequency oscillators. Commercially available YIG-tuned TE

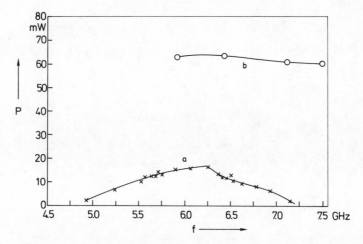

Fig. 7.25 Improvement in r.f. output power P of a C.W. Gunn oscillator obtained by optimizing device load at second harmonic (coaxial circuit). Curve (a) for sinusoidal r.f. voltage; curve (b) for fundamental plus second-harmonic r.f. voltage. After Troughton [7.43].

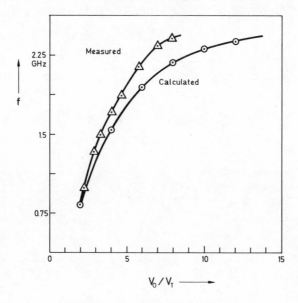

Fig. 7.26 Bias dependence of frequency f for GaAs LSA relaxation oscillator operated in waveguide circuit (pulsed bias). Sample data: $L = 560\ \mu$m, $n_0 = 6 \cdot 3 \times 10^{14}\ \text{cm}^{-3}$. Calculated curve from simple analytical investigation. After Jeppesen and Jeppsson [5.101].

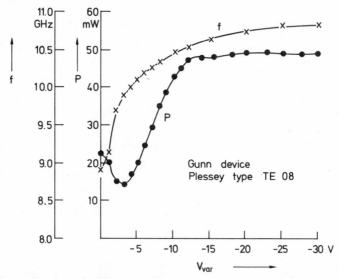

Fig. 7.27 R.f. output power P and frequency f of a varactor-tuned C.W. Gunn oscillator against varactor bias voltage V_{var}. Varactor diode connected in parallel to GaAs Gunn device (X-band waveguide circuit). After Downing and Myers [7.31].

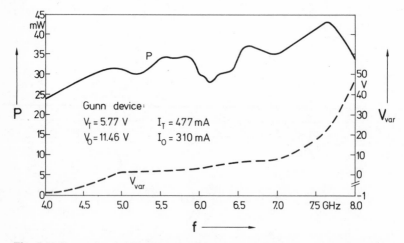

Fig. 7.28 Octave-bandwidth frequency tuning of a C.W. Gunn oscillator by two varactor diodes in series connection (C-band waveguide circuit). Shown is r.f. output power P as function of frequency f with indication of associated varactor bias V_{var}. After Large [7.74a].

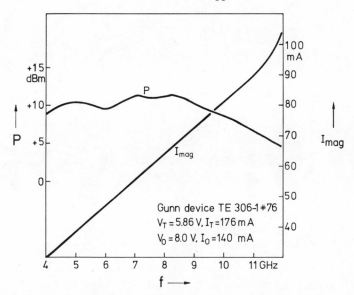

Fig. 7.29 Relation between r.f. output power P, frequency f and magnet d.c. current I_{mag} of a YIG-tuned C.W. GaAs Gunn oscillator in microstrip circuit. After Hanson [7.77].

oscillators cover X to Ku bands (6–18 GHz) [7.46a]. Sweep rates range from 100 Hz to 1 kHz [7.78].

7.1.2.3 *Modulation Properties*

The simplest way of modulating TE oscillators is by superimposing the modulation signal on the bias voltage [7.79–7.86]. According to the bias dependence of power and frequency (Fig. 7.19) both amplitude (AM) and frequency modulation (FM) is achieved, the latter sometimes expressed as phase modulation (PM). The sensitivity depends strongly on the bias point and/or circuit loading and frequency-tuning conditions selected [7.79, 7.80]. It can be optimized either for AM or for FM. Figure 7.30 shows typical performance data of a C.W. Gunn oscillator in X band as a function of oscillation frequency. The AM sensitivity is given in percent power deviation per volt, the FM sensitivity in frequency deviation per volt. Note that the FM sensitivity increases with decreasing loaded quality factor of the oscillator, Q_{L}, while the AM sensitivity decreases. The linearity of FM modulation is very good up to modulation-voltage amplitudes of about 0·7 V if the oscillator quality factor is low enough, typically $Q_{\text{L}} \approx 10$, but degrades rapidly for higher Q_{L} [7.79]. At low modulation frequencies f_{m} the modulation sensitivity generally decreases with increasing f_{m} [7.82, 7.83]. Sensitivity enhancement because of thermal effects is observed at $f_{\text{m}} \lesssim 100$ kHz [7.80, 7.83, 7.84]. At modulation frequencies $f_{\text{m}} \gtrsim 1$ MHz, the finite time constant of energy storage in the resonator leads to phase delay of the

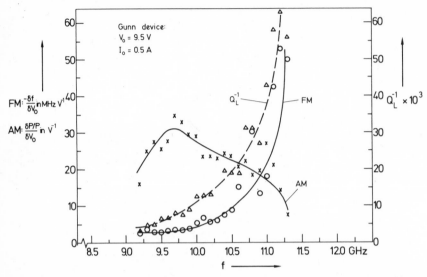

Fig. 7.30 AM and FM sensitivities of an X-band C.W. Gunn oscillator (waveguide circuit) expressed as per cent r.f. power deviation per volt ($\delta P/P\delta V_0$) and frequency deviation per volt ($-\delta f/\delta V_0$), respectively, as functions of oscillation frequency f. The change in reciprocal quality factor Q_L^{-1} is also given. Modulation frequency $f_m = 500$ kHz. After Tsai and Rosenbaum [7.79].

amplitude modulation, which in turn causes r.f. phase modulation and enhances the FM sensitivity [7.85]. The various modulation sensitivities of an X-band oscillator at optimum output-power coupling are plotted as a function of the modulation frequency f_m in Fig. 7.31. A modulation cut-off frequency appears close to 1 GHz corresponding to the relaxation frequency of the resonator energy storage. However, modulation is still feasible at a frequency as high as $f_m = 3$ GHz.

Frequency modulation is achieved also by varactor tuning of the microwave resonator [3.47, 7.87, 7.88]. For digital frequency modulation, p-i-n diodes which are able to rapidly switch the circuit resonant frequency can be employed [7.89], or one may simply utilize hysteresis properties [7.112] of a TE oscillator in certain microwave circuits [7.90].

The external negative conductance of a TE oscillator (Section 5.1.4.1) can be used for exciting self-modulating bias-circuit oscillations [5.20, 5.21]. Modulation frequencies in the range of 300 to 550 MHz have been reported for X-band oscillators [5.21].

7.1.2.4 *Injection Phase Locking and Large-Signal Amplification*
TE oscillators belong to the type of the self-excited oscillator which shows a pronounced dependence of the oscillation frequency on the external load impedance. Such oscillators can readily be phase-locked to an injected external r.f. signal if the frequency of the locking signal is close enough to that of the free-running oscillator. The injected signal can be thought to simulate a load impedance change in such a

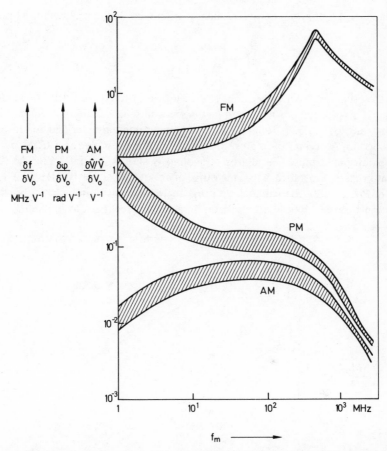

Fig. 7.31 Modulation sensitivities for AM, FM and PM of an X-band C.W. Gunn oscillator as functions of modulating frequency f_m. Curves FM and PM represent the same results expressed either as frequency or as phase modulation, respectively ($\delta f = f_m \delta \varphi$). Contrary to Fig. 7.30, AM sensitivity is given as fractional *voltage-amplitude* deviation per volt. After Martin and Hobson [7.85a].

way as to shift the oscillator frequency f towards the frequency f_{ex} of the injected signal until the phase difference φ between the two signals reaches a constant value, i.e. the two frequencies are equal [7.91, 7.92]. For a frequency shift of $\Delta f = f_{ex} - f$ one has

$$\sin \varphi = 2Q_e \left(\frac{P}{P_{ex}} \right)^{1/2} \frac{\Delta f}{f} \tag{7.4}$$

where P is the oscillator power, P_{ex} the power of the injected external signal, and Q_e an effective loaded quality factor of the locked oscillator [7.13, 7.94]. According to eqn (7.4) the phase difference φ increases with increasing frequency difference Δf

until at $\varphi = \pi/2$ the maximum frequency difference $\Delta f_{\max}$ for locking is reached, viz.

$$\frac{\Delta f_{\max}}{f} = \frac{1}{2Q_e} \left(\frac{P}{P_{ex}}\right)^{-1/2}. \qquad (7.5)$$

Since the oscillator frequency can be shifted upwards and downwards, depending on whether f_{ex} is larger or smaller than the free-running oscillator frequency f, the locking range is $2\Delta f_{\max}$. Relation (7.5) was experimentally verified for resonator-controlled GaAs Gunn oscillators [7.95–7.101]. An example for a C.W. oscillator at X-band and a pulsed oscillator at S-band is shown in Fig. 7.32. The X-band oscillator has, for instance, a locking range of about $2\Delta f_{\max} = 10$ MHz at a 'locking gain' of $P/P_{ex} = 30$ dB. The locking range increases substantially by stabilizing the bias supply of the TE oscillator with a phase-sensing feedback loop [7.99].

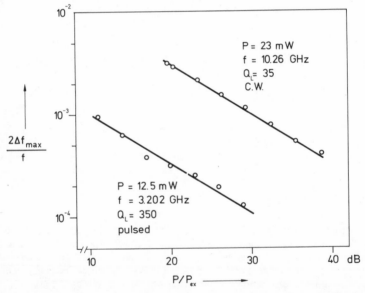

Fig. 7.32 Normalized locking range $2\,\Delta f_{\max}/f$ as a function of the locking 'gain' P/P_{ex} for two GaAs Gunn oscillators. Q_L is loaded quality factor of circuit. After Pollmann [7.101b].

As a consequence of eqn (7.4), locked oscillators can readily be phase-modulated by changing the frequency f of the originally free-running oscillator [7.85c, 7.102]. A convenient way of achieving this is by varying the bias voltage V_0. A typical phase modulation characteristic is shown in Fig. 7.33. Qualitatively it follows directly from eqn (7.4) assuming a linear bias/frequency relationship.

Another type of phase-locking of TE oscillators may occur at high enough levels of the injected signal both for Gunn-domain [3.45, 3.47, 7.98, 7.103], and for LSA oscillations [7.104]. It is related to the self-synchronizing mechanism of the TE oscillator in a resonant circuit as described in Sections 3.5 and 4.2. The voltage swing of the injected signal across the TE device controls dipole-domain, or accumulation layer, nucleation and extinction and thereby synchronizes the oscillations. Frequency division [3.47] with power gain is, for instance, possible by this technique [7.98, 7.103]. On the other hand, phase-locking TE oscillators with signals at subharmonic frequencies (ratios as low as 1:6) has been achieved [7.105]. The locking range, however, is then strongly reduced.

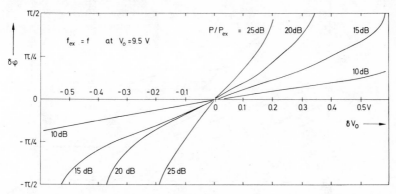

Fig. 7.33 Phase change $\delta\varphi$ plotted against deviation of bias voltage δV_0 for locked X-band GaAs Gunn oscillator. After Holliday [7.102a].

A phase-locked oscillator can be regarded as a non-linear amplifier (7.104, 7.106–7.108]. High-power microwave amplification is possible particularly with the LSA-mode oscillator [7.104, 7.108]. Amplification of FM signals with locked oscillators in X band has proved to be feasible and was used in a miniature television link [7.97]. There, at a frequency deviation of 1 MHz, up to 46 dB power gain was achieved, with typical values varying between 20 and 38 dB.

High-power microwave generation is possible by phase-locking several individual TE oscillators [7.14, 7.95, 7.109] as we have already mentioned in Section 7.1.2.1. The locking signal may or may not be generated by a TE oscillator itself.

Related to the phase-locking mechanisms described here is the reduction of delay times at the onset of r.f. oscillations from pulsed devices and the improvement of their pulse-to-pulse coherence by injection priming [7.101a, 8.11b]. In both cases phase-locking is effective only during the build-up period of the oscillations.

7.1.2.5 *Noise Properties and Frequency Stability*
As we have seen in Section 5.3.2, noise in TE oscillators is dominated by up-converted low-frequency base-band noise, at least at the smaller modulation frequencies. Thus noise and frequency stability are closely related since both are influenced by the bias and temperature dependence of frequency and/or output power. Con-

sequently techniques developed for reducing noise also improve frequency stability and vice versa.

The noise performance of TE oscillators was studied extensively [4.12, 5.79, 5.84a, 7.4, 7.7, 7.56, 7.116–7.122]. Figure 7.34 summarizes FM noise spectra of

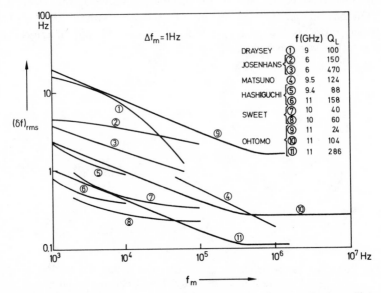

Fig. 7.34 FM noise level (r.m.s. frequency deviation) of GaAs Gunn oscillators measured by various workers at X-band frequencies $f_{(-)}^{+} f_{m}$; normalized to noise-measuring bandwidth $\Delta f_{m} = 1$ Hz. Note the different loaded quality factors Q_{L}. After Tanimoto *et al.* [8.84].

GaAs TE oscillators of the Gunn-domain type (cf. Fig. 5.33) as measured by several workers [8.84]. In Fig. 7.35 typical AM and FM noise spectra of a GaAs Gunn oscillator are compared with those of a klystron and an IMPATT diode oscillator (for noise measuring techniques see, e.g. [5.81, 7.115]). AM noise is given as the ratio of noise power $P_{N,AM}$ to oscillator ('carrier') power P, and the FM noise as root-mean-square frequency deviation $(\delta f)_{rms}$ (cf. Section 5.3.2). The noise from a single side band is normalized to a noise-measuring bandwidth of $\Delta f_{m} = 1$ Hz in both cases. The Gunn oscillator reaches the low noise level of the klystron, particularly at the higher modulation frequencies f_{m}, whereas IMPATT oscillators do not perform as good. Since a frequency deviation of $(\delta f)_{rms} = 0.1$ Hz, at a modulation frequency of $f_{m} = 10$ kHz, corresponds to a single-side-band noise power ratio of $P_{N,FM}/P = -103$ dB [see eqn (5.85)], AM noise of TE oscillators is practically negligible as compared to the FM component. Ways for reducing FM noise will be considered below.

Figure 7.36 demonstrates the influence of the FM modulation sensitivity ('voltage pushing factor') $\delta f / \delta V_{0}$ on the FM noise spectrum of a Gunn oscillator ($\delta f / \delta V_{0}$

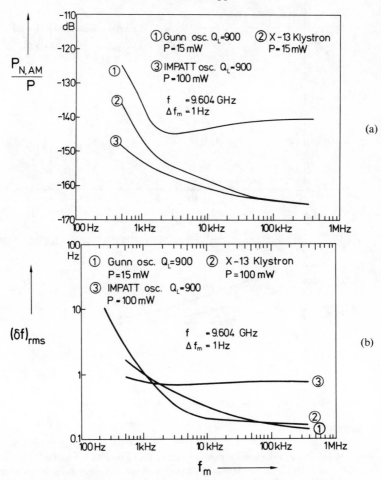

Fig. 7.35 (a) AM and (b) FM noise levels of a GaAs Gunn oscillator, a klystron, and an IMPATT-diode oscillator measured at X-band frequencies $f_{(-)}^{+} f_m$; noise-measuring bandwidth $\Delta f_m = 1$ Hz. After Gilden *et al.* [7.119].

can be changed e.g. by varying the bias voltage V_0). Smaller modulation sensitivity results in reduced noise in the sloping range ('flicker noise') of the spectrum close to the carrier frequency. However, it has no effect on the white-noise portion at higher modulation frequencies f_m. This supports the contention that the white noise results from intrinsic r.f. fluctuations (Section 5.3.2).

Another important noise-determining parameter is the loaded quality factor, Q_L, of the circuit [5.84a, 7.4] [cf. eqns (5.86) and (5.87)]. Figure 7.37a shows the reduction in FM noise obtained by increasing Q_L. As sketched in Fig. 7.37b, a simultaneous decrease in the voltage pushing factor is observed. Interestingly, this

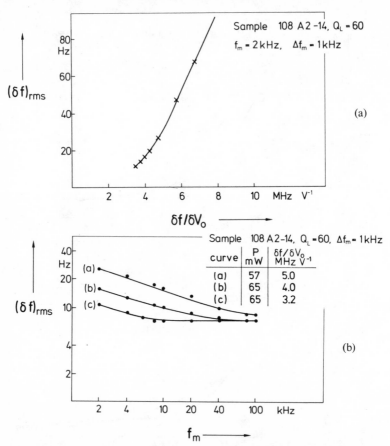

Fig. 7.36 Influence of the voltage pushing factor $\delta f/\delta V_0$ on FM noise of a GaAs Gunn oscillator at X-band: (a) Variation of r.m.s. frequency deviation close to the carrier frequency ($f_m = 2$ kHz), (b) change in FM noise spectrum with pushing factor. Note different noise-measuring bandwidth $\Delta f_m = 1$ kHz as compared to Figs 7.34 and 7.35. After Sweet and MacKenzie [7.120].

particular Gunn device does not show FM noise saturation at high modulation frequencies f_m, suggesting very low intrinsic r.f. noise in this case.

Unfortunately, increasing the circuit Q of the oscillator itself results generally in a strong decrease in r.f. output power. However, FM noise close to the carrier frequency (up to modulation frequencies $f_m \approx 1$ MHz) can be reduced without substantial power loss, as known from other oscillators [7.123–7.127], by coupling an external high-Q stabilizing cavity to the oscillator [7.128–7.131] or by injection phase locking [7.117, 7.132 and 7.133]. Figure 7.38 shows the FM noise reduction achieved by employing an external cavity with a quality factor of $Q_s \approx 10\,000$. Noise suppression is strongest when the resonant frequency f_s of the stabilizing

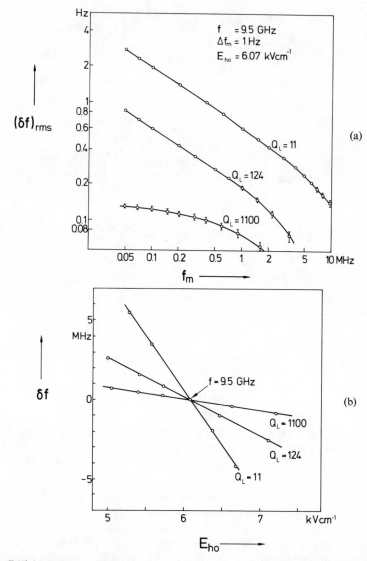

Fig. 7.37 (a) Influence of the loaded quality factor of the circuit, Q_L, on the FM noise spectrum of an X-band GaAs Gunn oscillator. (b) Simultaneous change in voltage pushing factor $\delta f/\delta V_0$ ($V_0 = E_{ho}L$, $L = 13\ \mu m$). After Matsuno [5.84a].

cavity equals the oscillation frequency f. Too large a frequency difference, $\Delta f = f_s - f$, on the other hand, may even increase the noise. The r.f. power reduction in the case of $\Delta f = 0$ was only 1·5 dB [7.128]. An example of FM noise reduction by injection phase locking is shown in Fig. 7.39. The Gunn oscillator was driven first with a regulated (a) and then with a noisy (b) d.c. power supply. The locking signal

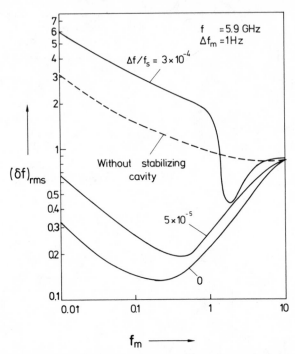

Fig. 7.38 FM-noise reduction in GaAs Gunn oscillator by employing an external stabilizing cavity ($Q_s \simeq 10{,}000$). $\Delta f = f_s - f$ is difference between resonant frequency f_s of stabilizing cavity and oscillation frequency f with modulation frequency f_m (abscissa) given in MHz. Adapted from Müller [7.128].

was generated by a varactor-chain oscillator and its frequency was chosen to be identical to the frequency of the free-running TE oscillator ($\Delta f = 0$). For high enough injection levels P_{ex}/P (reciprocal 'locking gain') the noise characteristic of the varactor-chain oscillator is approached.

Which of the above-sketched techniques for noise reduction is to be chosen depends largely on the type of application envisaged for the oscillator. For example, space systems and doppler radar applications require a modulation range of a few Hz up to 1 MHz, whereas telecommunications FM applications employ modulation frequencies between 10 kHz and 10 MHz [7.134]. Consequently, in the first case external stabilizing cavities or phase-locking to low-noise oscillators is adequate for FM noise reduction. In the second case intrinsic noise reduction techniques (low voltage pushing, high circuit Q) must be adopted.

The long-term frequency stability ('drift') of TE oscillators is mainly a question of the bias, load, and temperature, stability. Obviously, the frequency dependencies on bias voltage ('pushing'), load impedance ('pulling'), and temperature [6.51a, 7.7, 7.56, 7.72b, 7.135–7.137] have to be kept small. As already pointed out, techniques similar to those for noise reduction can be used here [3.47, 7.4, 7.129, 7.137–7.139].

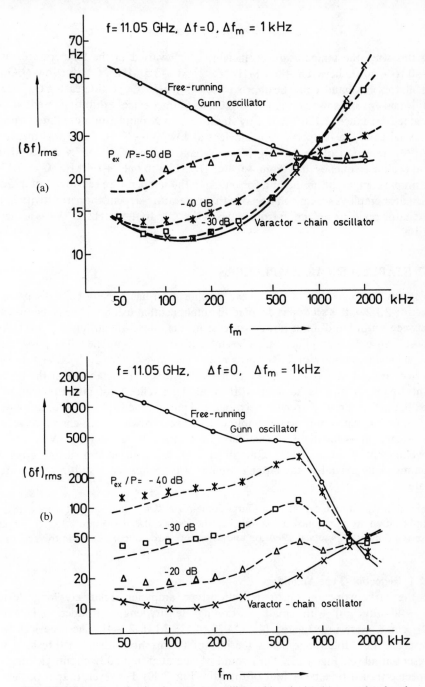

Fig. 7.39 FM-noise reduction in a GaAs oscillator, biased with (a) a regulated and (b) a noisy d.c. power supply, by injection phase locking from varactor-chain source. Parameter is the relative injection power level P_{ex}/P. Symbols are from experiment, dashed curves are calculated from the noise characteristics of both oscillators. After Sugiura and Sugimoto [7.132].

In this way, the temperature coefficient, which varied in the temperature range -40 to $+70\,°C$ between $+0.5\,\text{MHz}\,°C^{-1}$ and $-3.5\,\text{MHz}\,°C^{-1}$ for low-Q Gunn oscillators in X band, could be improved to $-170\,\text{kHz}\,°C^{-1}$ by increasing Q_L [7.135]. Still smaller values were obtained with GaAs Gunn devices exhibiting essentially no contact resistance [7.136, 7.137]. By stabilizing an X-band Gunn oscillator with an external cavity, a temperature coefficient as low as $-7\,\text{kHz}\,°C^{-1}$ was measured [7.129]. Injection phase locking of an X-band Gunn oscillator by a second one reduced the frequency drift from $-0.46\,\text{MHz}\,°C^{-1}$ to only $+1.4\,\text{kHz}\,°C^{-1}$ [7.139]. A simple compensation method for reducing the temperature sensitivity of a Gunn oscillator employs ceramic material with temperature-dependent permittivity in the oscillator cavity; a reduction from $-0.6\,\text{MHz}\,°C^{-1}$ to $+54\,\text{kHz}\,°C^{-1}$ was achieved [7.140].

7.2 STABLE LINEAR AMPLIFIERS

Whereas a special type of non-linear amplification has already been discussed in Section 7.1.2.4, this section is devoted to amplifiers that exhibit a linear relationship between output and input power, at least in their small-signal range [7.141]. As a consequence these amplifiers must be stable if no signal is applied. The possibilities for achieving stability of TE devices have been investigated in Section 4.4. Such devices are characterized by a two-terminal negative conductance near the transit-time frequency and its harmonics, thus allowing reflection-type amplification to be obtained as familiar from tunnel diode and parametric varactor-diode amplifiers. Additionally, with appropriate coupling structures, growing space-charge waves can be excited and unilateral transmission amplification be obtained. Whereas the development of the first class of amplifier is fairly advanced and various types are commercially available, the latter travelling-wave amplifier is still in a laboratory state.

Reflection-type amplification with Gunn-domain *oscillators* and *reciprocal* transmission amplification are mentioned in Section 7.3 (the former also in Section 5.1.4). They have received only little practical attention and, therefore, are not considered in this section.

7.2.1 Reflection-Type Amplifiers
In the early work on GaAs reflection-type amplifiers relatively long devices ($L = 40$–$120\,\mu\text{m}$) with small net-doping density/length product, $n_0L \lesssim 5 \times 10^{11}\,\text{cm}^{-2}$, were investigated [4.26, 4.39, 4.51, 7.141–7.144]. These devices were biased in the stable range below the instability threshold if such a threshold was present at all (cf. Fig. 4.13). They could be operated in a $50\,\Omega$ circuit yielding the expected transit-time behaviour of the gain (Fig. 7.40). However, output powers at gain saturation were relatively low [7.144].

GaAs TE amplifiers became practical when high-n_0L product devices ($n_0L \approx 1$–$4 \times 10^{12}\,\text{cm}^{-2}$), designed for C.W. operation with a drift length L small enough

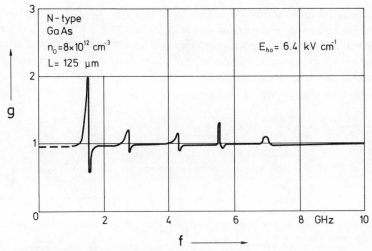

Fig. 7.40 Experimentally obtained power gain g as a function of frequency f for a GaAs TE device with $n_0 L = 10^{11}$ cm^{-2} showing transit-time behaviour of the gain. After Thim [4.51].

to yield a fundamental transit-time gain peak at C-band frequencies and above, could be stabilized at high bias voltages *beyond* the instability range of Fig. 4.13 [4.44, 4.46, 5.75, 6.139, 7.145–7.154]. A typical current/voltage characteristic of such a 'supercritically'-doped device is plotted in Fig. 7.41. Stabilization was possible

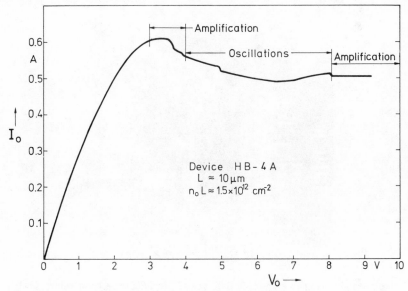

Fig. 7.41 D.c.-current/voltage characteristic $I_0(V_0)$ of a 'supercritically'-doped GaAs TE amplifier of the reflection type. After Perlman *et al.* [5.75].

297

beyond about 2·5 times threshold voltage in coaxial resonant-circuit structures.

A schematic diagram of the amplifier circuit is sketched in Fig. 7.42 and Fig. 7.43 reproduces a photograph of such an amplifier in integrated-circuit form. A three-port circulator with a characteristic impedance of $Z_0 = 50\ \Omega$ is used for isolating signal output and input. The network N couples the active device to the circulator. It is designed both for stabilizing the TE device and for matching it to the load impedance

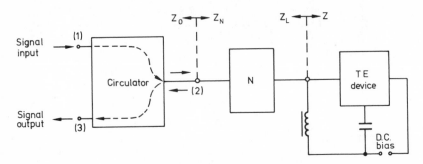

Fig. 7.42 Basic circuit of reflection-type TE amplifier. After Perlman [7.148].

Fig. 7.43 Photograph of GaAs reflection-type TE amplifier for X band (centre: GaAs chip, disk: ferrite circulator). (By courtesy of Messrs H. Huang and R. Paglione, RCA Corp.)

$Z_0 = 50 \ \Omega$ as to obtain high power gain g [5.75, 7.148] given by

$$g = \left| \frac{Z_N - Z_0}{Z_N + Z_0} \right|^2 = \left| \frac{Z_L - Z^\star}{Z_L + Z} \right|^2. \tag{7.6}$$

The impedance quantities of eqn (7.6) are defined in Fig. 7.42 and the star denotes the complex conjugate quantity. The available maximum bandwidth $\Delta f = f_2 - f_1$ of such a tuned amplifier is determined by the quality factor of the TE device, $Q = \mathrm{Im}\, Z / \mathrm{Re}\, Z$, viz.

$$\Delta f = \frac{2\pi \sqrt{f_1 f_2}}{|Q| \ln g_f} \tag{7.7}$$

if one assumed a flat gain g_f across Δf, and $\mathrm{Re}\, Z$ is considered as frequency independent in this range [7.148]. Equation (7.7) is a consequence of the fact that the reactive part of the sample impedance, $\mathrm{Im}\, Z$, has to be tuned out for maximum gain according to eqn (7.6), i.e. $\mathrm{Im}(Z_L + Z) \approx 0$. In general, a practical 'equalizing' network N never reaches the ideal properties to achieve relation (7.7), particularly if it has only one single resonance. Carefully designed multiple-resonant networks may approach the ideal bandwidth more closely. A typical measured gain response of an X-band GaAs TE amplifier in a triple-resonant coaxial circuit (three-section Chebychev-filter type) is shown in Fig. 7.44. Similarly broad amplifier bandwidths were achieved in C and Ku bands [5.75, 7.147–7.150]. Besides coaxial equalizing networks [5.75, 7.147, 7.148, 7.152] micro-stripline [7.149] and wave-guide circuits [7.151, 7.154] have been successfully employed.

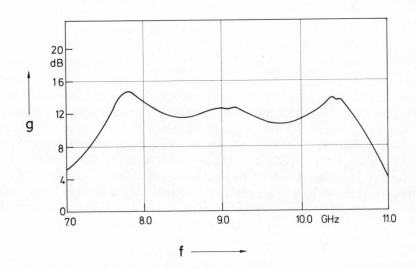

Fig. 7.44 Gain-frequency response $g(f)$ of a GaAs reflection-type TE amplifier in a three-section Chebychev filter-type circuit (X band). After Perlman [7.148].

299

The dynamic behaviour of a C- and an X-band GaAs TE amplifier is shown in Fig. 7.45. The linear range reaches beyond 100 mW output power, and the saturation power at 3 dB gain approaches 1 W, being the output power of the device when

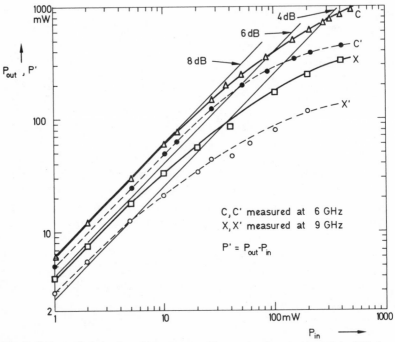

Fig. 7.45 Dynamic behaviour (output power P_{out} against input power P_{in}) of GaAs TE reflection-type amplifiers C.W.-operated in C and X band, respectively. After Perlman *et al.* [5.75].

operated as an oscillator. The linear dynamic range extends over 60 to 90 dB for a measured noise figure F of 15 to 20 dB [5.75]. The noise performance is thus superior to that of IMPATT-diode amplifiers which, however, possess higher power capabilities (cf. Fig. 7.17). When cascading amplifiers it is, therefore, advantageous to use TE amplifiers for the input stages and an IMPATT amplifier for the final stage [7.153]. A summary of performance data of GaAs TE amplifiers in C, X, and Ku bands is presented in Table 7.1. Considering these data it is anticipated that TE amplifiers might well replace travelling-wave tubes in some of their present-day applications [7.149]. The highest frequency achieved with stable TE amplifiers was in the 30 GHz range for fundamental operation [5.76b] and in the 50 GHz range for second-harmonic operation [7.155].

Performance data on InP reflection-type TE amplifiers are still scarce. However, they seem to offer substantially lower noise figures F than GaAs devices [5.72] with a preliminary experimental value of $F = 7.5$ dB [5.76a].

Table 7.1—Performance of GaAs TE reflection-type amplifiers
(two stages in brackets)*

Frequency band	3 dB-bandwidth over range GHz	Small-signal gain g dB	P_{out} at -1 dB gain compression W	P_{out} at 3 dB gain ($\approx P_{sat}$) W	Efficiency $\eta = P_{sat}/P_0$ %	Gain bandwidth $g^{1/2}\Delta f$ GHz	Noise figure F dB
C	4·5 – 8·0	8	0·25	1	3	8·75	15
	(5·0 – 7·5)	(22)	(0·20)	(1)	(1·5)	(28)	—
X	7·5 –10·75	12	0·20	0·55	2·3	13·0	—
	(7·65–10·25)	(24)	(0·20)	(0·5)	(1·2)	(41·6)	—
	8·0 –12·0	6	0·25	0·6	2·5	8·0	15
	7·0 –10·0	7	0·35	0·5	2·1	6·75	15
Ku	12·0 –16·0	6	0·25	0·5	2·5	8·0	—
	13·0 –15·0	8	0·05	0·12	2·0	5·0	—

*After Perlman and Upadhyayula [7.149].

7.2.2 Unilateral Travelling-Wave Amplifiers

The unilateral travelling-wave amplifier [7.156] relies on the growth of space-charge waves in stable TE devices as described in Section 4.3. The basic principle of this 'transmission-type' of amplification is sketched in Fig. 7.46. R.f. fields are coupled into the TE semiconductor at the cathode side $(C - K_1)$ where a space-charge wave is excited which grows exponentially in amplitude towards the anode. There, the electrodes $A - K_2$ pick up the amplified r.f. fields and energize the output gap. Assuming ideal lossless coupling between the r.f. fields and the fundamental space-charge-wave mode $m = 0$ (Fig. 4.10) at the cathode (input) and anode (output), the power gain is given by the growth coefficient $\alpha = \text{Re}\gamma$, as defined by eqns (4.42) and (4.44), according to

$$g = \exp 2\alpha L \qquad (7.8)$$

where L is the distance between input and output coupler. Referring to Fig. 4.12 and eqn (4.23) one may expect amplification from almost zero frequency up to a cut-off frequency determined by diffusion. However, the frequency characteristics of real couplers cause losses which reduce the bandwidth of the external gain. In addition, the boundary condition at the output coupler gives rise to fast electromagnetic waves which introduce internal feedback, and hence a further reduction of the external gain bandwidth may occur.

Whereas for reflection-type amplifiers n_0L-stabilization is generally used, for transmission amplifiers n_0d-stabilization is more appropriate since long structures are necessary to accommodate several space-charge wavelengths for sufficient amplification and also to simplify the way of coupling r.f. power in and out of the active

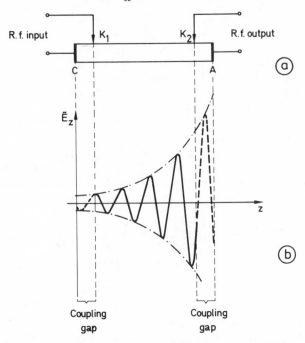

Fig. 7.46 Illustration of travelling-wave amplification in a stable TE semiconductor. (a) TE-semiconductor arrangement, C = cathode, A = anode (bias connections not shown); C-K_1 input coupler, A-K_2 output coupler. (b) Exponentially growing electric field E of space-charge wave.

semiconductor region. As pointed out in Section 4.4.2, several difficulties have to be overcome in stabilizing long structures and, at the same time, achieving a satisfactorily homogeneous d.c. field distribution such that wave growth in the total drift region is ensured. A number of early investigations on travelling-wave amplifiers were conducted with pulsed-bias operation [5.77, 6.84, 7.156–7.160]. Wideband gain was observed in the frequency region between 700 MHz and 4 to 5 GHz with saturation powers near 100 mW at the low-frequency boundary [5.77]. The gain/frequency response exhibited typical undulations, periodic in the transit-time frequency, indicating electro-magnetic feedback from output to input. This gain undulation is particularly pronounced in thin epitaxial GaAs devices on a semi-insulating GaAs substrate (Section 6.2.1) designed for C.W. operations at X-band frequencies or beyond [5.78, 7.161, 7.162]. An example, as measured on the structure of Fig. 7.47* in a 50 Ω circuit, is shown in Fig. 7.50. Gain peaks appear at the 3rd,

---

*The GaAs chip in Fig. 7.47 was provided with Schottky-barrier contacts except for the (shielded) ohmic cathode contact. By choosing Schottky contacts in this way, boundary conditions can be established at the cathode and the anode which approximately lead to the homogeneous electric-field distribution along the specimen as required for obtaining optimum gain [4.95, 7.162]; cf. Section 4.4.2, and [4.92, 4.93, 5.78].

4th and 5th harmonic of the transit-time frequency. The reverse attenuation a_r, which is also indicated, generally exceeds the cold loss a_0 leading to a high degree of unilaterality (about 40 dB at the gain peaks). The gain can be improved by tuning the source and load impedances, for example with a triple-stub tuner, to an optimum value as shown by the dash-dotted curve g'. As much as 30 dB gain was obtained in X band; the saturation output power [7.163] is low, however, (up to 0·2 mW measured). Another deficiency of the device is the relatively high noise figure obtained so far ($F \geq 25$ dB for GaAs device [5.78]) as compared to the reflection-type amplifier. This seems mainly to be a consequence of the still high coupling losses (>10 dB) at the cathode (cf. Section 5.3.1). A realization of the GaAs travelling-wave amplifier structure of Fig. 7.47 is shown in Figs 7.48 and 7.49.

An interesting property of the TE travelling-wave amplifier is the reasonably linear dependence of its electronic length l_e on the bias voltage. This is a direct consequence of the velocity/field characteristic of TE semiconductors: the decreasing drift velocity (roughly equal to the phase velocity of the space-charge waves, Section 4.3) at increasing bias field introduces a phase delay $\Delta\varphi$ which changes the electrical length by $\Delta l_e = (\Delta\varphi/2\pi)\,(c/f)$, where c is the velocity of light in vacuum.

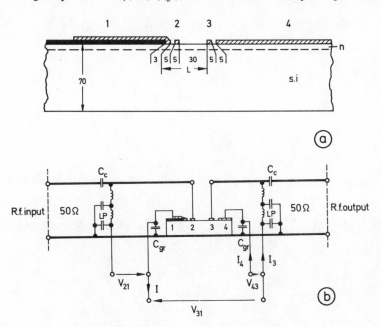

Fig. 7.47 GaAs travelling-wave amplifier for X-band frequencies (dimensions in μm). (a) GaAs chip: n = epitaxial n-layer (several μm thick); s.i. = semi-insulating substrate; black: ohmic contact; hatched: Schottky barriers. 1 cathode, 4 anode, 2 and 3 r.f. electrode. (b) GaAs chip connected to circuit: C_c = r.f. coupling capacitances; C_{gr} = r.f. grounding capacitances; LP = low-pass filter for bias application. After Frey *et al.* [7.162].

303

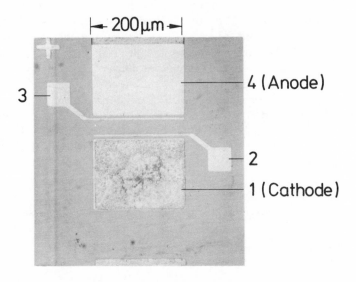

Fig. 7.48 Top-view photograph showing GaAs chip of an X-band C.W. travelling-wave amplifier as depicted in Fig. 7.47a. After Frey *et al.* [7.162].

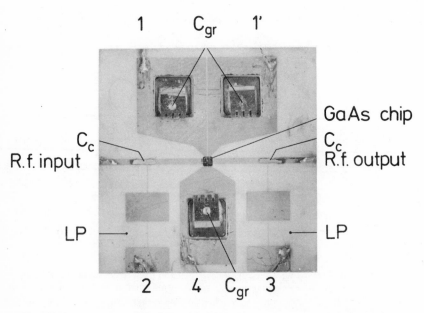

Fig. 7.49 Top-view photograph of microstrip travelling-wave amplifier circuit as depicted in Fig. 7.47b (but with one additional control electrode on GaAs chip, terminal 1'). After Frey *et al.* [7.162].

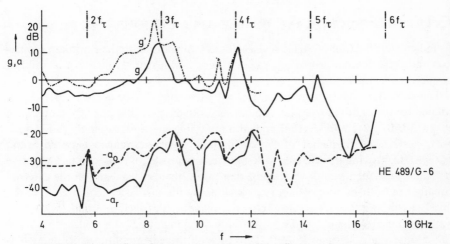

Fig. 7.50 Measured gain-frequency response $g(f)$ of C.W. GaAs travelling-wave amplifier (Fig. 7.49) with 50 Ω source and load impedances, and response $g'(f)$ for condition of tuned source and load impedances. Also indicated are reverse attenuation $a_r(f)$ and cold loss $a_0(f)$. Bias settings (Fig. 7.47b): $V_{31} = 27$ V, $V_{21} = -7.8$ V, $V_{43} = 8.0$ V ($I_3 = 2.15$ mA, $I_4 = 8.2$ mA). After Frey *et al.* [7.162].

Figure 7.51 shows a typical phase variation at X-band frequencies and the simultaneous change in external gain. A phase shift of $\pi/2$ is possible with gain changes of less than 3 dB. Application of this property for phase-code modulation or phased-array systems might be anticipated.

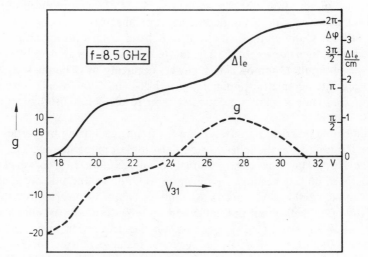

Fig. 7.51 Bias dependence of electrical length l_e and simultaneous change in gain for C.W. GaAs travelling-wave amplifier (Fig. 7.49). Bias settings (Fig. 7.47b): $V_{21} = -8.6$ V, $V_{43} = 8.0$ V. After Frey *et al.* [7.162].

7.3 FURTHER MICROWAVE DEVICE APPLICATIONS

As indicated in Section 5.1.4 TE devices possess, in addition to their primary oscillation and amplification properties, various secondary features which can be utilized in microwave applications.

The secondary negative resistance of an oscillating TE device was extensively studied for its usefulness in generating lower-frequency oscillations of various wave shapes [5.20, 5.21, 7.164–7.166] and in amplifying microwave signals [4.12, 5.22, 5.23, 7.167]. The frequency of the oscillations generated ranged between 5 and 1 300 MHz, proving the extreme wide-band characteristic of the secondary negative resistance. Frequency and wave shape depend on the d.c. bias and on the external circuit seen both by the secondary as well as by the primary oscillations (i.e. the 'bias' circuit and the 'microwave' circuit). Wave shapes include sinusoids, square waves, damped sinusoids, and exponentially decaying pulses. The pulses may have sub-nanosecond rise times [5.21a]. Typical peak-to-peak amplitudes of the oscillations are of the order of the bias voltage. Linear amplification of r.f. signals with an oscillating Gunn device was achieved at signal frequencies as high as 9 GHz [7.167]. Power gain reached 17 dB and saturated output power 17·5 dBm; the latter corresponds roughly to the optimum output power obtainable at the oscillation frequency which lay a few GHz above the frequency of the amplified r.f. signal. Noise figures were reported to be somewhat below 20 dB [5.23a]. The practical interest in the oscillating TE amplifier diminished because of the possibility of stabilizing high-n_0L product TE devices as described in Section 7.2.1 which showed superior amplifier performance.

Parametric amplification with TE oscillators [5.25–5.32] has received only little practical attention so far. A non-degenerate amplifier for signals at X-band frequencies with the Gunn pump oscillations in K band yielded power gains of up to 30 dB and a noise figure down to 18 dB; saturated output power approached 10 mW, whereas the optimum Gunn-oscillator power, occurring in Q band, was 20 mW [5.29]. Degenerate parametric amplification with X-band Gunn oscillators was also reported [5.28a]; power gains were greater than 20 dB, and noise figures ranged from 15 to 20 dB.

Of considerable practical importance are the frequency mixing properties of TE oscillators resulting from the inherent non-linearities of an oscillator [4.12, 7.168–7.173]. The TE oscillator has the additional advantage of yielding conversion *gain* because of its secondary negative conductance. Both down and up conversion are possible with such 'self-excited' mixers. A Gunn-effect down-converter from 33 GHz to 60 MHz worked with input powers as low as −80 dBm (10 pW) and a usable bandwidth of 32·5 MHz [7.169]. The conversion gain dropped from 4 dB at 10 pW input power to 0·2 dB at 1 μW. By proper input matching, mixer noise figures of 10 to 15 dB should be realizable with high-Q oscillators [7.170]. Highest conversion gain achieved so far was somewhat beyond 20 dB [7.171]. On the other hand, with *up*-converters no conversion gain is generally observed, at least in the linear mixing

306

range [4.12, 7.172, 7.173]. A gain of 6·5 dB was reported, however, for high input powers (>3 dBm) when converting 600 MHz to 9·38 GHz with a 9·98 GHz Gunn oscillator [7.172]. Up-conversion *with* gain by using the subharmonic phase-locking properties of a Gunn oscillator (Section 7.1.2.4) combined with its secondary negative conductance has also been proposed [7.174].

The non-linear impedance characteristic of TE devices can in principle be used by itself for harmonic generation [7.175] or for achieving variable attenuation [7.176–7.178] at microwave frequencies. With X-band input frequencies harmonics up to the fourth were generated, yet the second harmonic showed a conversion efficiency of only 1 % [7.175]. Reflection-type attenuation with a Gunn device could be varied by the bias voltage between +20 and −5 dB, negative values occurring in the active bias region [7.176]. By using two Gunn devices in a hybrid ring, r.f. amplitude modulation with an on-off range from 30 to 0·6 dB was possible [7.178].

The utilization of negative-conductivity semiconductor material in distributed form within a waveguide or transmission line for realizing active microwave devices has received major theoretical attention [4.75, 7.179–7.188]; practical realization, however, has made only little progress so far [4.75, 7.183–7.185]. Both TEM-wave transmission amplification [4.75, 7.179–7.181] and self-resonant oscillations [7.182–7.188] are possible. In the first case stabilization of the TE material against dipole-domain formation is necessary (Section 4.4), in the second case the TE material can be considered as oscillating in a kind of LSA mode (Section 4.2). The advantage of such distributed TE devices as compared to the convential ones described in Sections 7.1 and 7.2 are their easy adaptability to microwave integrated circuit techniques and, possibly, their high microwave-power capabilities. However, impedance matching problems seem to be severe [7.188].

Other microwave applications of TE devices may result from magnetic-field effects [7.189, 7.190], from the possibility of generating high-frequency acoustic waves by the piezo-electric effect in the high-field domains [7.191], from the modulation properties of an additional field-effect type terminal attached to the device [7.192–7.194, 8.40], and from the potential possibility of AM-to-FM conversion by employing a Gunn-domain oscillator [3.47, 7.195].

A further application concerns microwave modulation of infra-red light transmission through a TE sample in which dipole domains propagate [7.224, 7.225]. This modulation relies on the shift of the band edge for fundamental absorption in high electric fields towards longer wavelengths (Franz–Keldysh effect). If the optical wavelength corresponds to the fundamental edge, the high electric field of a passing-by domain causes this edge to shift and the absorption to increase suddenly. Using n-type GaAs samples and a GaAs laser as the light source, modulations of up to 66 % at a wavelength of 9 050 Å have been achieved experimentally in this way [7.225].

7.4 SYSTEMS APPLICATIONS

TE oscillators and amplifiers have found their way into a number of electronic systems, and it is anticipated that entirely new products may become feasible with their help [7.195]. However, in many applications they have to compete with other microwave devices, such as IMPATT, TRAPATT or BARITT diodes, bipolar and field-effect transistors, or electron tubes. To provide an idea of the prospects of TE devices, their strengths and weaknesses against the competitors are summarized in Table 7.2. Such a survey should be helpful for choosing the most suitable device for a particular application considered. As can be seen the decided strengths of TE devices are their low noise properties and low bias-voltage requirements in moderate power C.W. applications, and their high peak powers available in pulsed operation, particularly in the LSA mode. A number of systems applications suitable for TE oscillators and amplifiers are listed in Table 7.3.

TE Gunn oscillators have proved to be particularly advantageous in small portable C.W. doppler radar equipment [7.198–7.202] in which the TE oscillator itself may directly be utilized for the mixing process [7.201, 7.202]. Important commercial applications are foreseen in using such systems for intrusion alarms [7.203], automobile safety equipment [7.203–7.205], police radar [7.203], tanker docking radar [7.206], and other types of traffic control including radar for pleasure boats [7.207].

Table 7.2—Comparison of TE oscillators (TEO) and reflection-type amplifiers (TEA) with other microwave devices*

(a) State of the Art of TEOs

		Strengths	Weaknesses
vs	IMPATT diodes	Lower noise Higher pulsed power Higher pulsed efficiency Lower bias voltage	Lower C.W. efficiency and power
vs	TRAPATT diodes	Lower noise Higher duty-cycle operation Higher frequency operation Lower bias voltage	Lower pulsed power and efficiency Less temperature stable GaAs *vs* Si
vs	Transistors	Higher frequency operation Higher pulsed power Less complicated structure	No C.W. L & S band devices GaAs *vs* Si
vs	Tubes	Longer life, more reliable Equivalent noise performance Lower bias voltage Simpler power supplies	Lower power and efficiency Less temperature stable
vs	BARRIT diodes	Higher power and efficiency	GaAs *vs* Si

*By courtesy of B. Berson [7.196].

308

Table 7.2 (contd)

(b) State of the Art of TEAs (reflection-type)

		Strengths	Weaknesses
vs			
	IMPATT amplifiers	Greater bandwidth (Higher gain-bandwidth product) More linear Lower noise Wider dynamic range	Lower C.W. efficiency and power
vs			
	TRAPATT amplifiers	Much greater bandwidth (Much higher gain-bandwidth product) Much more linear Much lower noise Much wider dynamic range Higher frequency operation	Much lower pulsed power and efficiency No UHF or L-band operation GaAs *vs* Si
vs			
	Transistor amplifiers	Greater bandwidth (Higher gain-bandwidth product) Higher power in C, X, and Ku bands	Two-terminal device GaAs *vs* Si
vs			
	Tubes	Longer life. More reliable Low bias voltages Simpler power supplies	Less power and efficiency Less temperature stable
vs			
	BARITT-diode amplifiers	Greater bandwidth (Higher gain-bandwith product) Higher power and efficiency	GaAs *vs* Si

In pulsed radar systems TE oscillators have satisfactorily operated both as local oscillators [7.200] and as transmitters [7.208, 7.209]. TE oscillators have been investigated also as elements for phased-array aerials [7.210–7.212] and as transmitters and local oscillators in radio-link equipment [7.213]. For C.W. transmitter applications requiring higher power levels (in the watt range) and simultaneously low-noise performance, a series combination of a Gunn oscillator with a phase-locked IMPATT oscillator might be advantageous [7.213, 7.214].

Other TE oscillator applications in microwave systems described in the literature are parametric-amplifier pumps [7.200, 7.215, 7.216], signal generators [7.217], self-oscillating mixers [7.218, 7.219], microwave-biased photo detectors [7.220, 7.221], and equipment for microwave spectroscopy [7.222, 7.223] or absorption measurement techniques [7.222].

Table 7.3—Systems applications*

(a) Commercial applications

	Frequency	Power
Intrusion alarms	10·525 GHz	5–15 mW
Automobile systems	Ku and up	10–100 mW
Radar braking		
Anti-skid		
Lane change ind.		
Pre-impact sensor		
Rear warning		
Police radar	10·525 GHz	10 mW
Small boat radar	C, X, Ku	0·1–1 kW (P)
Telecommunications		
L.O.	C, X	5–10 mW ⎫ FM-CW
Trans.†		1, 10 mW ⎭
Aircraft systems		⎧ 10–100 W(P)
Collision avoidance	C	⎩ 2 W
Transponders†	C, X, Ku	100 W(P)
Altimeters	9·3/13·325 GHz	10 W(P)
Instruments		
Swept sources	C, X, Ku	100 mW
Signal sources	C, X, Ku	100 mW
Noise sources		

(b) Military applications

Fuses		
L.O.	S-Ku	5–15 mW
Trans.	S-Ku	10 W(P)
Electronic Warfare		
Driver Osc. for power amp.	S-Ku (T)	100–500 mW
Noise gen.	X-Ku	100–200 mW
L.O.	S-Ku (T)	5–15 mW
Satellite/Comm.		
Osc.	S-Ka (T)	100–500 mW
Collision avoidance	L-Ku	0·5–1 W
Airborne radar		
L.O.	X-Ka (T)	5–50 mW
Paramp pumps	X-Ka	100–300 mW

*After Rosenbaum [7.197].

†Potential amplifier application

  (T) Electronically tunable

  (P) Peak power.

REFERENCES

7.1 Altman, J. L., *Microwave Circuits*, p. 37 (Princeton, N.J., D. van Nostrand, 1964).

7.2 Pollmann, H., Engelmann, R., Frey, W. and Bosch, B. G., Load dependence of Gunn oscillator performance. *IEEE Trans. Microwave Theory and Techniques* **MTT–18** (1970), p. 817.

7.3   Hakki, B. W. and Irvin, J. C., CW microwave oscillations in GaAs. *Proc. IEEE* **53** (1965), p. 80.

7.4   Court, W. P. N., Herman, P., Hilsum, C., Holliday, H. R. and Warner, F. L., Reduction of frequency modulation noise from Gunn oscillators. *Electron. Lett.* **3** (1967), p. 567.

7.5a   Warner, F. L. and Herman, P., Miniature X band Gunn oscillator with a dielectric tuning system. *Electron. Lett.* **2** (1966), p. 467.

7.5b   Warner, F. L. and Herman, P., Miniature X band Gunn oscillators. *Proc. 1st Biennial Cornell Conference, 1967, "High Frequency Generation and Amplification"*, p. 206 (Ithaca, N.Y., School of Electrical Engineering of Cornell University, 1967).

7.6   Reynolds, J. F., Berson, B. E. and Enstrom, R. E., Microwave circuits for high-efficiency operation of transferred electron oscillators. *IEEE Trans. Microwave Theory and Techniques* **MTT–18** (1970), p. 827.

7.7   Wilson, K., Gunn effect devices and their applications. Mullard Technical Communications No. 100 (July 1969), p. 286.

7.8a   Smith, R. B. and Crane, P. W., Varactor-tuned Gunn-effect oscillator. *Electron. Lett.* **6** (1970), p. 139.

7.8b   Cawsey, D., Varactor-tuned Gunn-effect oscillators. *Electron. Lett.* **6** (1970), p. 246.

7.8c   Cawsey, D., Wide-range tuning of solid-state microwave oscillators. *IEEE Journal of Solid-State Circuits* **SC–5** (1970), p. 82.

7.9   James, D. A., Wide-range electronic tuning of a Gunn diode by an yttrium-iron garnet (Y.I.G.) ferrimagnetic resonator. *Electron. Lett.* **4** (1968), p. 451.

7.10a   Camp, O. W., Jr, Experimental observations of relaxation oscillator waveforms in GaAs from less than transit-time frequency to several times transit-time frequency. *Proc. IEEE* **59** (1971), p. 1248.

7.10b   Camp, O. W., Jr, Computer simulation of multifrequency LSA oscillations in GaAs. *Proc. IEEE* **57** (1969), p. 220.

7.11   Dienst, J. F. and Thomas, J. J., An oversize cavity for exciting the LSA mode in n-GaAs. *IEEE Trans. Electron Dev.* **ED–15** (1968), p. 615.

7.12   Foulds, K. W. H., L.S.A. oscillation in coaxial-line circuit. *Electron. Lett.* **6** (1970), p. 189.

7.13a   Olfs, P., Gunn oscillators with EO_{10}-resonators—Possible applications today. *Microwave J.* **14** (6), June 1971, p. 18B.

7.13b   Olfs, P., A mechanically and electronically tunable resonator for the operation of IMPATT and Gunn diodes (in German). *Proc. 7th Int. Conference MOGA, Hamburg, 1968, Nachrichtentechnische Fachberichte* **35** (1968), p. 479.

7.14   Eastman, L. and Wilson, Jr, W. L., Optimization of performance and control of LSA oscillators. *Proc. 3rd Biennial Cornell Electrical Engineering Conference "High Frequency Generation and Amplification: Devices and Application", 1971*, p. 361 (Ithaca, N.Y., School of Electrical Engineering of Cornell University, 1971).

7.15   Marcuvitz, N., *Waveguide Handbook*. MIT Radiation Laboratory Series Vol. 10, pp. 31 ff. (New York, McGraw-Hill, 1951).

7.16a   Tsai, W. C., Rosenbaum, F. J. and Mackenzie, L. A., Circuit analysis of waveguide-cavity Gunn oscillator. *IEEE Trans. Microwave Theory and Techniques* **MTT–18** (1970), p. 808.

7.16b   Jethwa, C. P. and Gunshor, R. L., Circuit characterization of waveguide-mounted Gunn-effect oscillators. *Electron. Lett.* **7** (1971), p. 433.

7.16c   White, J. F., Simplified theory for post coupling diodes to waveguide. *Proc. European Microwave Conf., Stockholm, 1971*, p. A4/2: 1 (Stockholm, The Royal Swedish Academy of Engineering Sciences, 1971).

7.17a   Taylor, B. C., Fray, S. J. and Gibbs, S. E., Frequency saturation effects in transferred electron oscillators. *IEEE Trans. Microwave Theory and Techniques* **MTT–18** (1970), p. 799.

7.17b   Fray, S. J. and Taylor, B. C., Frequency-tuning characteristics of waveguide-mounted transferred-electron oscillators. *Electron. Lett.* **6** (1970), p. 708.

7.18   Howes, M. J., Circuit considerations in the design of wide-band tunable transferred-electron oscillators. *IEEE Trans. Electron Dev.* **ED–17** (1970), p. 1060.

7.19   Taylor, B. C. and Howes, M. J., LSA operation of GaAs layers in large-scale tunable microwave circuits. *IEEE Trans. Electron Dev.* **ED–16** (1969), p. 928.

7.20   Hanson, D. C. and Rowe, J. E., Microwave circuit characteristic of bulk GaAs oscillators. *IEEE Trans. Electron Dev.* **ED–14** (1967), p. 469.

7.21a   Lazarus, M. J., Novak, S. and Bullimore, E. D., Use of the voltage controlled cap resonance to obtain higher power and higher frequencies for millimeter-wave Gunn oscillators. *Proc. IEEE* **59** (1971), p. 716.

7.21b   Novak, S., Double cavity tuning of Gunn oscillators at millimeter wavelengths. *Proc. IEEE* **59** (1971), p. 1026.

7.22a   Pence, I. W. and Khan, P. J., Broad-band equivalent-circuit determination on Gunn diodes. *IEEE Trans. Microwave Theory and Techniques* **MTT–18** (1970), p. 784.

7.22b   Ito, Y., Komizo, H., Meguro, T., Daido, Y. and Umebo, I., Experimental and computer simulation analysis of a Gunn diode. *IEEE Trans. Microwave Theory and Techniques* **MTT–19** (1971), p. 900.

7.23a   Owens, R. P. and Cawsey, D., Microwave equivalent-circuit parameters of Gunn-effect-device packages. *IEEE Trans. Microwave Theory and Techniques* **MTT–18** (1970), p. 790.

7.23b   Owens, R. P. and Cawsey, D., The influence of Gunn effect package reactances on circuit. *Proc. 1971 European Microwave Conference, Stockholm*, p. A4/1:1 (Stockholm, The Royal Swedish Academy of Engineering Sciences, 1971).

7.24   Copeland, J. A., The LSA oscillator: Theory and applications. *Proc. 1st Biennial Cornell Conference, 1967, "High Frequency Generation and Amplification"*, p. 4 (Ithaca, N.Y., Cornell University, 1967).

7.25   Taylor, B. C. and Gibbs, S. E., Fundamental microwave oscillations in epitaxial GaAs by control of space-charge growth. *Electron. Lett.* **4** (1968), p. 471.

7.26   Barrera, J. S., GaAs LSA V-band oscillators. *IEEE Trans. Electron Dev.* **ED–18** (1971), p. 866.

7.27   Huang, H. C., Enstrom, R. E. and Narayan, S. Y., High efficiency operation of transferred electron devices in the hybrid mode. *Electron. Lett.* **8** (1972), p. 271.

7.28   Spiwak, R. R., A step-iris resonator for LSA operation. *IEEE Trans. Microwave Theory and Techniques* **MTT–18** (1970), p. 973.

7.29   Bybokas, J. and Farrell, B., The Gunn flange—a building block for low-cost microwave oscillators. *Electronics* **44**, March 1, 1971, p. 47.

7.30   Schneider, H. M. and Kennedy, W. K., Cavity design for millimeter-wavelength gallium arsenide devices. *Proc. IEEE* **57** (1969), p. 1213.

7.31   Downing, B. J. and Myers, F. A., Broadband (1·95 GHz) varactor-tuned X-band Gunn oscillator. *Electron. Lett.* **7** (1971), p. 407.

7.32   Narayan, S. Y., Huang, H. C. and Gobat, A. R., Operation of transferred-electron oscillators in the ridge wave guide circuit. *Electron. Lett.* **7** (1971), p. 31.

7.33   Ivanek, F., Shyam, M. and Reddi, V. G. K., Investigation of waveguide-below-cutoff resonators for solid state active devices. *Electron. Lett.* **5** (1969), p. 214.

7.34   Ivanek, F., Reddi, V. G. K. and Shyam, M., Mode indentification for an oscillator using solid state active devices in a waveguide-below-cutoff resonator. *Electron. Lett.* **6** (1970), p. 151.

7.35   Craven, G., Waveguide bandpass filters using evanescent modes. *Electron. Lett.* **2** (1966), p. 251.

7.36   Ellis, D. J. and Gunn, M. W., Stripline Gunn oscillators are compatible with microwave i.c.'s. *Electronic Engng* **43**, May (1971), p. 50.

7.37   Brehm, G. E. and Mao, S., Varactor tuned integrated Gunn oscillators. *IEEE J. Solid-State Circuits* **SC–3** (1968), p. 217.

7.38   Wasse, M. P., Pearson, A. and King, G., A microstrip circuit module for the Gunn oscillator. *Proc. 7th Int. Conference MOGA, Hamburg, 1968, Nachrichtentechn. Fachberichte* **35** (1968), p. 470.

7.39   Jones, S., Frequency-stable microstripline X-band Gunn oscillator. *Proc. IEEE* **57** (1969), p. 364.

7.40   Quine, J. P., An LSA-mode TED oscillator for microstrip. *Proc. IEEE* **58** (1970), p. 1291.

7.41   Monroe, J. W. and Camp, Jr, W. O., LSA operation of a gallium arsenide device in microstrip. *IEEE Trans. Electron Dev.* **ED–18** (1971), p. 69.

7.42   Frey, W., Pollmann, H., Engelmann, R. W. H. & Bosch, B. G., Influence of 2nd harmonic frequency termination on Gunn oscillator performance. *Electron. Lett.* **5** (1969), p. 691.

7.43   Troughton, P., An evaluation circuit for fundamental and harmonically tuned GaAs devices. *Proc. IEEE* **58** (1970), p. 1165.

7.44   Eddolls, D. V., Ward, F. S. and Whitehead, A. J., High power high-efficiency c.w. Gunn oscillators in X band. *Electron. Lett.* **7** (1971), p. 472.

7.45   Reynolds, J. F., Rosen, A., Berson, B. E. and Huang, H. C., A coupled TEM bar circuit for solid state microwave oscillators. *IEEE Int. Solid-State Circuits Conference, Philadelphia, Pa., 1970;* Digest of Technical Papers, p. 12.

7.46a   Clark, Jr, R. J. and Swartz, D. B., Combine YIG's with bulk-effect diodes. *Microwaves* **11** (3), March (1972), p. 46.

7.46b   Clark, Jr, R. J. and Swartz, D. B., Take a fresh look at YIG-tuned sources. *Microwaves* **11** (2), Feb. (1972), p. 40.

7.46c   Marriott, S. P. A., Microwave devices using spheres of mono-crystalline garnet materials. *Marconi Rev.* 1st quarter 1970, p. 79.

7.47   Chang, N. S., Hayamizu, T. and Matsuo, Y., YIG tuned Gunn effect oscillator. *Proc. IEEE* **55** (1967), p. 1621.

7.48   Dydyk, M., Ferrimagnetically tunable Gunn effect oscillator. *Proc. IEEE* **56** (1968), p. 1363.

7.49a   Omori, M., Octave electronic tuning of a CW Gunn diode using a YIG sphere. *Proc. IEEE* **57** (1969), p. 97.

7.49b   Omori, M., The YIG-tuned Gunn oscillator, its potentials and problems. *GMTT International Microwave Symposium, Dallas, Texas, 1969*, Digest of Technical Papers, p. 176.

7.50   Easson, R. M., Design and performance of YIG-tuned Gunn oscillators. *Microwave J.* **14** (2), Feb. (1971), p. 53.

7.51   Magarshack, J. and Spitalnik, R., Magnetically tunable Gunn oscillators using a YIG sphere. *Proc. 7th Int. Conference MOGA, Hamburg, 1968, Nachrichtentechn. Fachberichte* **35** (1968), p. 475.

7.52   Becker, R. and Bosch, B. G., Power frequency limitations of planar-type GaAs transferred electron devices. *Proc. 4th Int. Symp. on Gallium Arsenide and Related Compounds, 1970*, p. 163 (London and Bristol, The Institute of Physics, 1971).

7.53a   Becker, R. J., Late diode entries enliven solid-state source race. *Microwaves* **11** (6), June (1972), p. 34.

7.53b   Omori, M., Gunn diodes and sources. *Microwave J.* **17** (6), June (1974), p. 57.

7.54   Knight, S. and Copeland, J. A., Theoretical predictions and experimental results on LSA devices. *NEREM Record* (1967), p. 26.

7.55   Narayan, S. Y. and Gobat, A. R., High-bias-voltage operation of GaAs transferred electron oscillators. *Electron. Lett.* **4** (1968), p. 504.

7.56a   Murakami, H., Sekido, K., Ayaki, K. and Maruyama, M., GaAs epitaxial Gunn effect oscillators. *NEC Res. Dev. (Japan)* No. 10 (Oct. 1967), p. 49.

7.56b   Maruyama, M. and Watanabe, H., Gallium arsenide microwave devices. *NEC Res. Dev. (Japan)* No. 17 (April 1970), p. 1.

7.57   Edridge, A. L., Myers, F. A., Davidson, B. J. and Bass, J. C., Pulsed J band (12·4–18 GHz) Gunn-effect oscillators. *Electron. Lett.* **5** (1969), p. 103.

7.58 Califano, F. P., High-efficiency X-band Gunn oscillators. *Proc. IEEE* **57** (1969), p. 251.

7.59 Turner, I. R. and Ramachandran, T. B., Sheet Gunn oscillators. *IEEE Int. Solid-State Circuits Conference, Philadelphia, Penn., 1968;* Digest of Technical Papers, p. 80.

7.60 Parkes, E. P., Taylor, B. C. and Colliver, D. J., The performance of planar Gunn oscillators in X band. *IEEE Trans. Electron Dev.* **ED–18** (1971), p. 840.

7.61 German Federal Patent (DBP) No. 1,791,235 (inventors: B. G. Bosch and H. Pollmann; priority: June 12, 1965).

7.62 Staiman, D., Breese, M. E. and Patton, W. T., New technique for combining solid-state sources. *IEEE J. Solid-State Circuits* **SC–3** (1968), p. 238.

7.63 Hines, M. E., Network integration approaches for multiple diode high power microwave generation. *GMTT Int. Symposium, 1968*, Digest of Technical Papers, p. 46 IEEE No. 68C38.

7.64 Mitsui, S., CW Gunn diodes in composite structure. *IEEE Trans. Microwave Theory and Techniques* **MTT–17** (1969), p. 1158.

7.65 Hirayama, H. and Uchida, T., Gunn diode oscillator develops 2W cw at 12·8 GHz. *Microwaves* **10** (7), July (1971), p. 12.

7.66 Yu, S. P., Shaver, P. J., Tantraporn, W., Direct series operation of Gunn effect diodes with above critical n_0L products. *Proc. IEEE* **56** (1968), p. 2068.

7.67 Steele, M. C., Califano, F. P. and Larrabee, R. D., High-efficiency series operation of Gunn devices. *Electron. Lett.* **5** (1969), p. 81.

7.68 Baugham, K. M. and Myers, F. A., Multiple series operation of Gunn effect oscillators. *Electron. Lett.* **5** (1969), p. 371.

7.69 Carroll, J. E., Series operation of Gunn diodes for high r.f. power. *Electron. Lett.* **3** (1967), p. 455.

7.70 Boronski, S., Parallel-fed C.W. Gunn oscillators cascaded in X band waveguide for higher microwave power. *Electron. Lett.* **4** (1968), p. 185.

7.71 Kuno, H. J., Reynolds, J. F. and Berson, B. E., Push-pull operation of transferred-electron oscillators. *Electron. Lett.* **5** (1969), p. 178.

7.72a Wilson, W. E., Domain capacitance tuning of Gunn oscillators. *Proc. IEEE* **57** (1969), p. 1688.

7.72b Wilson, W. E., Pulsed LSA and Trapatt sources for microwave systems. *Microwave J.* **14** (8), August (1971), p. 33.

7.73 Shyam, M., C.W. operation of L.S.A. oscillators in R band. *Electron. Lett.* **6** (1970), p. 315.

7.74a Large, D., Octave band varactor-tuned Gunn diode sources. *Microwave J.* **13** (10), Oct. (1970), p. 49.

7.74b Kawakami, K. N., Optimize Gunn circuits for wideband varactor tuning. *Microwaves* **11** (12), Dec. (1972), p. 35.

7.75 Zieger, D., Frequency modulation of a Gunn-effect oscillator by magnetic tuning. *Electron. Lett.* **3** (1967), p. 324.

7.76 Rosenbaum, F. J. and Tsai, W. C., Gunn effect swept frequency oscillator. *Proc. IEEE* **56** (1968), p. 2164.

7.77 Hanson, D. C., YIG-tuned transferred-electron oscillator using thin film microcircuits. *IEEE Int. Solid-State Circuits Conference, Philadelphia, Pa., 1969;* Digest of Technical Papers, p. 122.

7.78 Gilbert, K. D., Dynamic tuning characteristics of YIG devices. *Microwave J.* **13** (6), June (1970), p. 36.

7.79 Tsai, W. C. and Rosenbaum, F. J., Amplitude and frequency modulation of a waveguide cavity CW Gunn oscillator. *IEEE Trans. Microwave Theory and Techniques* **MTT–18** (1970), p. 877.

7.80 Albrecht, P., The modulation of Gunn oscillators in the quenched-domain mode (in German). *Nachrichtentechn. Z.* **24** (1971), p. 516.

7.81   King, G. and Wasse, M. P., Frequency modulation of Gunn effect oscillators. *IEEE Trans. Electron Dev.* **ED–14** (1967), p. 717.

7.82   Faulkner, E. A. and Meade, M. L., Frequency-modulation sensitivity of Gunn oscillators. *Electron. Lett.* **5** (1969), p. 217.

7.83   De Sa, B. A. E. and Hobson, G. S., Thermal effects in the bias circuit frequency modulation of Gunn oscillators. *IEEE Trans. Electron Dev.* **ED–18** (1971), p. 557.

7.84   Freeman, K. R. and Hobson, G. S., The Vf_T relation of CW Gunn effect devices. *IEEE Trans. Electron Dev.* **ED–19** (1972), p. 62.

7.85a   Martin, B. and Hobson, G. S., High-speed phase and amplitude modulation of Gunn oscillators. *Electron. Lett.* **6** (1970), p. 244.

7.85b   Hobson, G. S., Kocabiyikoglu, Z. U. and Martin, B., High speed phase and amplitude modulation of Gunn oscillators. *Proc. 8th Int. Conference MOGA, Amsterdam, 1970*, p. 6/1 (Deventer, Kluwer, 1970).

7.85c   Martin, B. and Hobson, G. S., Angle modulation of frequency-locked Gunn oscillators. *Electron. Lett.* **7** (1971), p. 399.

7.86   Sugiura, T. and Sugimoto, S., A high-speed wide-band X-band FM deviator with a Gunn-effect diode. *Proc. IEEE* **57** (1969), p. 91.

7.87   Kuru, I., Frequency modulation for the Gunn oscillator. *Proc. IEEE* **53** (1965), p. 1642.

7.88   Wilson, P. G. and Minakovic, B., Development of an FM pulsed Gunn oscillator at X band. *IEEE Trans. Electron Dev.* **ED–18** (1971), p. 450.

7.89   Vane, A. B. and Dunn, V. E., A digitally tuned Gunn effect microstrip oscillator. *Proc. IEEE* **58** (1970), p. 171.

7.90a   Patel, B. C. and Hobson, G. S., A simple circuit for FSK modulation of Gunn oscillators. *Proc. IEEE* **60** (1972), p. 253.

7.90b   Magarshack, J., Gunn oscillator as a frequency memory device. *GMTT International Microwave Symposium, IEEE, 1968;* Digest of Technical Papers, p. 77.

7.91   Slater, J. C., *Microwave Electronics*, p. 205 (Princeton, N.J., D. van Nostrand Co., 1950).

7.92   Adler, R., A study of locking phenomena in oscillators. *Proc. IRE* **34** (1946), p. 351.

7.93   Khandelwal, D. D., On injection-locking figure of merit for avalanche, Gunn and other oscillators. *IEEE Trans. Microwave Theory and Techniques* **MTT–18** (1970), p. 989.

7.94a   Jochen, P., Equivalent circuit for injection-locked negative-resistance oscillators. *Electron. Lett.* **6** (1970), p. 61.

7.94b   Jochen, P., Injection phase locking of IMPATT and Gunn oscillators. *Microwave J.* **14** (2), Feb. (1971), p. 40.

7.95   Hakki, B. W., Beccone, J. P. and Plauski, S. E., Phase-locked GaAs CW microwave oscillators. *IEEE Trans. Electron Dev.* **ED–13** (1966), p. 197.

7.96   Gelbwachs, J. and Mao, S., Phase locking of pulsed Gunn oscillators. *Proc. IEEE* **54** (1966) p. 1591.

7.97   Judd, S. V., A simple repeater for frequency modulated signals using transferred electron oscillators. *Electron. Lett.* **4** (1968), p. 33.

7.98   Pollmann, H. and Bosch, B. G., Injection locking of Gunn oscillators (in German). *Nachrichtentechn. Z.* **22** (1969), p. 174.

7.99   Stickler, J. J., Injection locking of Gunn oscillators with feedback stabilization. *Proc. IEEE* **57** (1969), p. 1772.

7.100   Perlman, B. S. and Walsh, T. E., Criterion for non-reciprocal locking of bilateral microwave oscillators. *IEEE Trans. Microwave Theory and Techniques* **MTT–18** (1970), p. 507.

7.101a Bosch, B. G. and Pollmann, H., Investigation of the influence of external high-frequency signals on Gunn elements (in German). European Meeting Semiconductor Device Research, Bad Nauheim, April 1967, *Verhandl. Dtsche Phys. Gesellsch.* **2** (1967), p. 54 (abstract).

7.101b Pollmann, H., Influence of external high frequency signals on Gunn oscillators (in German). Dr.-Ing. Thesis, Technical University of Aachen, Germany, 1970.

7.102a Holliday, H. R., The effect of operating parameters and oscillator characteristics upon the phase angle between a locked X-band Gunn oscillator and its locking signal. *IEEE Trans. Electron Dev.* **ED–17** (1970), p. 527.

7.102b Bott, I. B. and Holliday, H. R., Effects of changes in operating conditions on the phase of a frequency locked Gunn oscillator in X-band. *Electron. Lett.* **6** (1970), p. 206.

7.103 Pollmann, H. and Bosch, B. G., Frequency division with power gain in Gunn oscillators. *Electron. Lett.* **3** (1967), p. 513.

7.104a Quine, J. P., Injection phase-locking characteristics of LSA-mode transferred-electron oscillators. *Proc. IEEE* **57** (1969), p. 715.

7.104b Curtice, W. R. and Quine, J. P., Comments on "Injection phase-locking characteristics of LSA-mode transferred electron oscillators". *Proc. IEEE* **58** (1970), p. 138.

7.104c Quine, J. P., Younger, C. and Pence, I. W., Bandwidth limitations of LSA and TRAPATT-mode oscillators. *IEEE Int. Solid-State Circuits Conference, Philadelphia, Pa., 1972;* Digest of Technical Papers, p. 255.

7.104d Hines, M. E., Negative-resistance diode power amplification. *IEEE Trans. Electron Dev.* **ED–17** (1970), p. 1.

7.105 Oltman, H. G. and Nonnemaker, C. H., Subharmonically injection phase-locked Gunn oscillator experiments. *IEE Trans. Microwave Theory and Techniques* **MTT–17** (1969), p. 728.

7.106a Hines, M. E., X band power amplication using Gunn effect diodes. *Proc. IEEE* **56** (1968), p. 1590.

7.106b Hines, M. E. and Buntschuh, C., Broad band power amplification with Gunn effect diodes. *IEEE Int. Solid-State Circuits Conference, Philadelphia, Pa., 1969;* Digest of Technical Papers, p. 28. Also *IEEE J. Solid-State Circuits* **SC–4** (1969), p. 370.

7.106c Hanson, D. C., Integrated X-band power amplifier utilizing Gunn and IMPATT diodes. *IEEE Int. Solid-State Circuits Conference, Philadelphia, Pa., 1972;* Digest of Technical Papers, p. 38.

7.107 Frey, W., Admittance of a phase-locked Gunn oscillator at high input powers. *Electron. Lett.* **5** (1969), p. 672.

7.108a Hashizume, N. and Kataoka, S., Transferred-electron negative-resistance amplifier. *Electron. Lett.* **6** (1970), p. 34 and 387 (erratum).

7.108b Kennedy, Jr, W. K. and Rossiter, E. L., L.S.A. microstrip phase-locked power amplifier. *Electron. Lett.* **6** (1970), p. 852.

7.109 U.K. Patent No. 1,120,550 (inventor: B. G. Bosch, priority Nov. 20, 1965).

7.110 Pollmann, H. and Bosch, B. G., Injection priming of pulsed Gunn oscillators. *IEEE Trans. Electron Dev.* **ED–14** (1967), p. 609.

7.111 Harrison, R., Elimination of delay times in X band Gunn-effect oscillators using r.f. injection. *Electron. Lett.* **5** (1969), p. 503.

7.112 Tsvirko, Yu. A. and Ivanchenko, I. A., Mode tuning and hysteresis behaviour of Gunn-effect cavity controlled generator. *Electron. Lett.* **6** (1970), p. 9.

7.113a Colliver, D. J., Morgan, J. R. and Taylor, B. C., Performance of InP 3-level oscillators in K and Q bands. *Electron. Lett.* **7** (1971), p. 50.

7.113b Colliver, D. J., Gray, K. W. and Yoyce, B. D., High efficiency microwave generation in InP. *Electron. Lett.* **8** (1972), p. 11.

7.114 White, P. M. and Gibbons, G., High-efficiency C.W. operations of 'anomalous' indium phosphide microwave oscillators. *Electron. Lett.* **8** (1972), p. 166.

7.115a Ashley, J. R., Searles, C. B. and Palka, F. M., The measurement of oscillator noise at microwave frequencies. *IEEE Trans. Microwave Theory and Techniques* **MTT–16** (1968), p. 753.

7.115b Sam, K. H., The measurement of near-carrier noise in microwave amplifiers. *IEEE Trans. Microwave Theory and Techniques* **MTT–16** (1968), p. 761.

7.116 Draysey, D. W., Court, W. P. N. and Bott, I. B., Noise performance of Gunn microwave generators in X and J band. *Electron. Lett.* **2** (1966), p. 125.

7.117 Josenhans, J., Noise spectra of Read diode and Gunn oscillators. *Proc. IEEE* **54** (1966), p. 1478.

7.118 Kodali, V. P., A.M. and F.M. noise characteristics of solid-state microwave oscillators. *Electron. Lett.* **4** (1968), p. 147.

7.119 Gilden, M., Buntschuh, C., Ramachandran, T. B. and Collinet, J. C., Avalanche and Gunn diode oscillators. WESCON Technical Papers, 1968, part 1, section 2/3, p. 1.

7.120 Sweet, A. A. and MacKenzie, L. A., The FM noise of a CW Gunn oscillator. *Proc. IEEE* **58** (1970), p. 822.

7.121a Hildsen, F. J. and Pyrah, E. D., Improvements in the power and frequency performance of Gunn devices. *Proc. Eur. Microwave Conference, London, 1969,* p. 207 (IEE Conference Publication No. 58).

7.121b Goldwasser, R. E., Berenz, J., Lee, C. A. and Dalman, G. C., IMPATT and Gunn oscillator noise. *Proc. Int. Conference MOGA, Amsterdam, 1970,* p. 12/19 (Deventer, Kluwer, 1970).

7.122 Herbst, H. and Ataman, A., Thermal modulation and FM noise of Gunn oscillators. *Arch. Elektronik Übertragungstechnik* **26** (1972), p. 359.

7.123 Ashley, J. R. and Searles, C. B., Microwave oscillator noise reduction by a transmission stabilizing cavity. *IEEE Trans. Microwave Theory and Techniques* **MTT–16** (1968), p. 743.

7.124 Schlosser, W. O., Noise in mutually synchronized oscillators. *IEEE Trans. Microwave Theory and Techniques* **MTT–16** (1968), p. 732.

7.125 Hines, M. E., Collinet, J.-C.R. and Ondria, J. G., FM noise suppression of an injection phase-locked oscillator. *IEEE Trans. Microwave Theory and Techniques* **MTT–16** (1968), p. 738.

7.126 Ashley, J. R. and Palka, F. M., Noise properties and stabilization of Gunn and avalanche diode oscillators and amplifiers. *G-MTT International Microwave Symposium, 1970;* Digest of Technical Papers, p. 161, (New York, IEEE, 1970).

7.127 Saito, T., Takagi, T. and Mano, K., Noise effect in oscillators using multiple active devices connected in series or in parallel. *Proc. IEEE* **60** (1972), p. 126.

7.128 Müller, C., Suppression of frequency modulation noise of Gunn oscillators with an external resonator (in German). *Frequenz* **23** (1969), p. 364.

7.129 Ito, Y., Komizo, H. and Sasagawa, S., Cavity stabilized X-band Gunn oscillator. *IEEE Trans. Microwave Theory and Techniques* **MTT–18** (1970), p. 890.

7.130 Clarke, J., A simple stabilized microwave source. *IEEE Trans. Instrumentation and Measurement* **IM–21** (1972), p. 83.

7.131 Day, Jr, W. R., MIC diode oscillator stabilized by a dielectric resonator. *Proc. 3rd Biennial Cornell Electrical Engineering Conference "High Frequency Generation and Amplification: Devices and Application", 1971,* p. 257 (Ithaca, N.Y., School of Electrical Engineering of Cornell University, 1971).

7.132 Sugiura, T. and Sugimoto, S., FM noise reduction of Gunn-effect oscillators by injection locking. *Proc. IEEE* **57** (1969), p. 77.

7.133 Joos, J. W., A high-stability phase-locked oscillator. *Microwave J.* **13** (9), Sept. (1970), p. 36.

7.134 Leeson, D. B., Short term stable microwave sources. *Microwave J.* **13** (6), June (1970), p. 59.

7.135 Hobson, G. S. and Warner, F. L., An automatic equipment for recording the frequency variation of X-band oscillators with temperature over the range −40 to +70°C with particular reference to Gunn oscillators. *Radio Electron. Engr* **39** (1970), p. 316.

7.136 Kocabiyikoglu, Z. U., Hobson, G. S. and De Sa, B. A. E., Relationship of the starting delay time and the frequency-temperature relation of X-band Gunn oscillators. *Electron. Lett.* **7** (1971), p. 550.

7.137a Bird, J., Bolton, R. M. G., Edridge, A. L., De Sa, B. A. E. and Hobson, G. S., Gunn diodes with improved frequency stability/temperature variations. *Electron. Lett.* **7** (1971), p. 299.

7.137b Hobson, G. S. and De Sa, B. A. E., Variation of frequency with ambient temperature of Gunn oscillators. *Proc. 1971 Eur. Microwave Conference, Stockholm*, Vol. 1, p. A4/3:1 (Stockholm, The Royal Swedish Academy of Engineering Sciences, 1971).

7.137c Edridge, A. L., Bird, J., Bolton, R. M. G. and Geraghty, S. R., Comparison of Gunn diode frequency/temperature measurements with a computer simulation. *Proc. 1971 Eur. Microwave Conference, Stockholm*, Vol. 1, p. A4/4:1 (Stockholm, The Royal Swedish Academy of Engineering Sciences, 1971).

7.138 Kohiyama, K. and Monma, K., A new type of frequency stabilized Gunn oscillator. *Proc. IEEE* **59** (1971), p. 1532.

7.139 Cleverley, M. E. and Norbury, J. R., Technique for improving the frequency/temperature characteristic of 3 cm-band Gunn oscillators. *Electron. Lett.* **5** (1969), p. 449.

7.140 Kooi, P. S. and Walsh, D., Novel technique for improving the frequency stability of Gunn oscillators. *Electron. Lett.* **6** (1970), p. 85.

7.141 Thim, H. W., Gunn amplifiers. In: *Solid State Devices*, p. 87 (London and Bristol, The Institute of Physics, Conf. Ser. no. 12, 1971).

7.142 Thim, H. W. and Barber, M. R., Microwave amplification in a GaAs bulk semiconductor. *IEEE Trans. Electron Dev.* **ED–13** (1966), p. 110.

7.143 Foyt, A. G. and Quist, T. M., Bulk GaAs microwave amplifiers. *IEEE Trans. Electron Dev.* **ED–13** (1966), p. 199.

7.144 Hayes, R. E., Saturation power in GaAs amplifiers. *IEEE Trans. Electron Dev.* **ED–15** (1968), p. 183.

7.145 Magarshack, J. and Mircea, A., Wideband CW amplification in X band with Gunn diodes. *IEEE Int. Solid-State Circuits Conference, Philadelphia, Pa., 1970;* Digest of Technical Papers, p. 134.

7.146 Walsh, T. E., Perlman, B. S. and Enstrom, R. E., Stabilized supercritical transferred electron amplifiers. *IEEE J. Solid-State Circuits* **SC–4** (1969), p. 374.

7.147 Perlman, B. S., CW microwave amplification from circuit stabilized epitaxial GaAs transferred electron devices. *IEEE International Solid-State Circuits Conference, Philadelphia, Pa., 1970;* Digest of Technical Papers, p. 136.
Also: *IEEE J. Solid-State Circuits* **SC–5** (1970), p. 331.

7.148 Perlman, B. S., Microwave amplification using transferred-electron devices in prototype filter equalization networks. *RCA Rev.* **32** (1971), p. 3.

7.149 Perlman, B. S. and Upadhyayula, C. L., Transferred electron amps challenge the TWT. *Microwaves* **9** (12), Dec. (1970), p. 59.

7.150a Perlman, B. S., Upadhyayula, C. L. and Sienkanowicz, W. W., Microwave properties and applications of negative conductance TE devices. *Proc. IEEE* **59** (1971), p. 1229.

7.150b Upadhyayula, C. L. and Perlman, B. S., Design and performance of transferred electron amplifiers using distributed equalizer networks. *IEEE Int. Solid-State Circuits Conference, Philadelphia, Pa., 1972;* Digest of Technical Papers, p. 40.

7.151 Sene, A., A wideband CW waveguide Gunn effect amplifier. *IEEE Trans. Microwave Theory and Techniques* MTT-20 (1972), p. 645.

7.152 Jeppsson, B. I. and Jeppesen, P., On the GaAs supercritical TEA. Paper presented at the *2nd European Solid-State Device Research Conference (ESSDERC), Lancaster, 1970.*

7.153 Monroe, J. W. and Kennedy, W. K., Amplifiers go solid state at X- and Ku-band. *Micro-Wave J.* **14** (12), Dec. (1971), p. 28.

7.154a Sweet, A. A., Waveguide cavity Gunn amplifiers. *Microwave J.* **15** (2), Feb. (1972), p. 41.

7.154b Sweet, A. A. and Collinet, J. C., Multistage Gunn amplifiers for FM-CW systems. *IEEE Int. Solid-State Circuits Conference, Philadelphia, Pa., 1972;* Digest of Technical Papers, p. 42.

7.155 Thim, H. W. and Lehner, H. H., Linear millimeter wave amplification with GaAs wafers. *Proc. IEEE* **55** (1967), p. 718.

7.156 Robson, P. N., Kino, G. S. and Fay, B., Two-port microwave amplification in long samples of GaAs. *IEEE Trans. Electron Dev.* **ED–14** (1967), p. 612.

7.157 Fay, B., A two-port unilateral GaAs amplifier. Ph.D. Thesis, Stanford University, Calif., USA, 1970.

7.158 Kumabe, K., Kanbe, H. and Ohara, S., Mechanism of coupling between space charge waves and microwaves in a bulk GaAs travelling wave amplifier. *Proc. 1st Conference Solid-State Devices, Tokyo, 1969;* Suppl. *J. Japan. Soc. Appl. Phys.* **39** (1970), p. 39.

7.159 Koyama, J., Ohara, S., Kawazura, K. and Kumabe, K., Bulk GaAs travelling-wave amplifier. *Rev. Electrical Commun. Lab.* (*Japan*) **17** (1969), p. 1102.

7.160 Kanbe, H., Kumabe, K. and Nii, R., High power GaAs travelling-wave amplifier. *Rev. Electrical Commun. Lab.* (*Japan*) **19** (1971), p. 917; also: *J. Jap. Soc. Appl. Phys.* **40**, Supplement, (1971), p. 144 (*Proc. 2nd Conference on Solid-State Devices, Tokyo, 1970*).

7.161 News: GaAs amplifier reaches 20 GHz. *Microwaves* **10** (4) April (1971), p. 10.

7.162 Frey, W., Becker, R., Engelmann, R. W. H. and Keller, K., CW operation of GaAs travelling-wave amplifiers for X-band frequencies. *Arch. Elektronik Übertragungstechnik* **27** (1973), p. 245.

7.163 Frey, W., Optimum r.f. power transport in nd-limited GaAs travelling-wave amplifiers. *Electron. Lett.* **9** (1973), p. 12.

7.164 Cawsey, D., V.H.F. and U.H.F. Gunn-effect oscillators. *Electron. Lett.* **3** (1967), p. 550.

7.165a Jaskolski, S. V. and Ishii, T. K., Low frequency oscillations generated by an n-type bulk effect GaAs device. *Proc. National Electronics Conference*, 1967, vol. 23, p. 342.

7.165b Jaskolski, S. V. and Ishii, T. K., Simultaneous low-frequency relaxation and high-frequency microwave oscillation of a bulk GaAs c.w. oscillator. *Electron. Lett.* **3** (1967), p. 12.

7.166 Brunt, G. A., Low-frequency negative resistance of X band Gunn diodes. *Electron. Lett.* **5** (1969), p. 151.

7.167 Olfs, P., An 'oscillating Gunn amplifier' with E_{010}-resonator. *Proc. 8th Int. Conference MOGA, Amsterdam, 1970*, p. 6/21 (Deventer, Kluwer, 1970).

7.168 Hakki, B. W., GaAs post threshold microwave amplifier, mixer and oscillator. *Proc. IEEE* **54** (1966), p. 299.

7.169 Lazarus, M. J., Bullimore, E. D. and Novak, S., A sensitive millimeter wave self-oscillating Gunn diode mixer. *Proc. IEEE* **59** (1971), p. 812.

7.170 Albrecht, P. and Bechteler, M., Noise figure and conversion loss of self-excited Gunn-diode mixers. *Electron. Lett.* **6** (1970), p. 321.

7.171 Nagano, S. and Akaiwa, Y., Behaviour of a Gunn diode oscillator with a moving reflector as a self-excited mixer and a load variation detector. *IEEE Trans. Microwave Theory and Techniques* **MTT–19** (1971), p. 906.

7.172 Sugimoto, S., Up-conversion with Gunn-effect diode. *Proc. IEEE* **55** (1967), p. 1520.

7.173 Kohiyama, K. and Shiota, H., A new type of highly frequency-stabilized self-oscillating converter using transferred-electron diodes. *Proc. IEEE* **60** (1972), p. 739.

7.174 German Federal Patent (DBP) No. 1,940,902 (inventor: B. G. Bosch; priority: Aug. 12, 1969).

7.175 Kooi, P. S. and Walsh, D., Harmonic generation using Gunn diodes. *Electron. Lett.* **5** (1969), p. 159.

7.176 Pollmann, H. and Bosch, B. G., Continuously variable microwave attenuation with Gunn-effect samples. *Electron. Lett.* **4** (1968), p. 317.

7.177 Sugimoto, S. and Sugiura, T., Microwave switching with Gunn-effect diodes. *Proc. IEEE* **56** (1968), p. 371.

7.178 Sterzer, F., Amplitude modulation of microwave signals using transferred-electron devices. *Proc. IEEE* **57** (1969), p. 86.

7.179 U.K. Patent No. 1,111,187 (inventor: B. G. Bosch; priority: Nov. 26, 1965).

7.180 Bach Anderson, J. and Majborn, B., Semiconductor rod in waveguide—field distribution for positive and negative conductivity. *IEEE Trans. Microwave Theory and Techniques* **MTT–16** (1968), p. 194.

7.181 Lewin, L., Amplifying properties of bulk negative resistance material. *Electron. Lett.* **4** (1968), p. 145.

7.182 Chawla, B. R. and Coleman Jr, D. J., Critical conductivity-length product for electromagnetic instability in negative-differential-conductivity media. *Electron. Lett.* **5** (1969), p. 31.

7.183 Baynham, A. C. and Colliver, D. J., New mode of microwave emission from GaAs. *Electron. Lett.* **6** (1970), p. 498.

7.184 Baynham, A. C., Wave propagation in negative differential conductivity media: n-Ge. *IBM J. Res. Dev.* **13** (1969), p. 568.

7.185 Baynham, A. C., Emission of TEM waves generated within an n-type Ge cavity. *Electron. Lett.* **6** (1970), p. 306.

7.186 Yokoo, K., Ono, S. and Shibata, Y., The electronically tunable Gunn-diode oscillator. *IEEE Trans. Electron Dev.* **ED–16** (1969), p. 494.

7.187a Rode, D. L., Axial electromagnetic modes and self-resonance in LSA diodes. *J. Appl Phys.* **41** (1970), p. 2402.

7.187b Rode, D. L., Self-resonant LSA oscillator diode. *Proc. IEEE* **57** (1969), p. 1216.

7.187c Rode, D. L., Dielectric-loaded self-resonant LSA diode. *IEEE Trans. Electron Dev.* **ED–17** (1970), p. 47.

7.188 Sasiela, R. J. and Berger, H., Impedance matching to self-resonant diodes. *IEEE Trans. Electron Dev.* **ED–17** (1970), p. 942.

7.189 U.K. Patent No. 1,123,145 (inventors: B. G. Bosch and H. Pollmann; priority: Dec. 21, 1965).

7.190 Levinstein, M. E., Nasledov, D. N. and Shur, M. S., Magnetic field influence on the Gunn effect. *Phys. Stat. Sol.* **33** (1969), p. 897.

7.191a Hayakawa, H., Ishiguro, T., Takada, S., Mikoshiba, N. and Kikuchi, M., Generation of high-frequency ultrasonic waves by Gunn effect. *J. Appl. Phys.* **41** (1970), p. 4755.

7.191b Hayakawa, H., Takada, S., Ishiguro, T. and Mikoshiba, N. Generation of surface waves by Gunn oscillator. *J. Japan. Soc. Appl. Phys.* **40**, Supplement (1971), p. 143 (*Proc. 2nd Conference Solid-State Devices, Tokyo, 1970*).

7.192 Petzinger, K. G., Hahn, Jr, A. E. and Matzelle, A., CW three-terminal GaAs oscillator. *IEEE Trans. Electron Dev.* **ED–14** (1967), p. 403.

7.193a Califano, F. P., Negative-resistance amplifiers using three-terminal Gunn devices. *Electron. Lett.* **4** (1968), p. 568.

7.193b Califano, F. P., Frequency modulation of three-terminal Gunn devices by optical means. *IEEE Trans. Electron Dev.* **ED–16** (1969), p. 149.

7.194 Clarke, G. M., Edridge, A. L., Griffith, I. and McGeehan, J. P., The electronic tuning effects of a control electrode on transverse Gunn oscillators. *Proc. 1971 Eur. Microwave Conference, Stockholm*, Vol. 1, p. A3/3:1 (Stockholm, The Royal Swedish Academy of Engineering Sciences; 1971).

7.195 Frey, J. and Bowers, R., What's ahead for microwaves. *IEEE Spectrum* **9** (March 1972), p. 41.

7.196 Berson, B. E., Transferred electron devices. Paper presented at *Eur. Microwave Conference, Stockholm, 1971*.

7.197 Rosenbaum, F. J., Gunn device applications. *Proc. IEEE Int. Convention, New York, 1971*, p. 522 (IEEE Cat. No. 71C8-IEEE).

7.198a Arnold, R. D., Bichara, M. R. E., Eberle, J. W. and Repert, L. M., Microwave integrated circuits applications to radar systems. *Microwave J.* **11** (7) (July 1968), p. 45.

7.198b King, G. and Heeks, J. S., Bulk effect modules pave way to sophisticated uses. *Electronics* **42** (Feb. 3, 1969), p. 94.

7.199 Anonymous, The mini-radar. *Microwave J.* **13** (2) (Feb. 1970), p. 20E.

7.200 Higgins, V. J. and Baranowski, J. J., The utility and performance of avalanche transit time diode and transferred electron oscillators in microwave systems. *Microwave J.* **13** (7) (July 1970), p. 37.

7.201 Nagano, S., Ueno, H., Kondo, H. and Murakami, H., Self-excited microwave mixer with a Gunn diode and its application to Doppler radar. *Electronics Commun. Japan* **52** (1969), p. 112.

7.202 Pauker, V. and Magarshack, J., Investigation into the oscillator-detector Gunn diode used for small doppler radar. *Proc. 1971 Eur. Microwave Conference, Stockholm*, Vol. 1, p. A4/5:1 (Stockholm, The Royal Swedish Academy of Engineering Sciences, 1971).

7.203 Anonymous, Low-cost radars—Doppler to stop a car and to catch a thief. *Microwaves* **9** (9) (Sept. 1970), p. 60.

7.204 Davis, R., Is radar-aided braking close to reality? *Microwaves* **9** (3) (March 1970), p. 12.

7.205 Anonymous, Papers on civil-radar systems support 1971 theme at GMTT. *Microwaves* **10** (6) (June 1971), p. 10.

7.206 Cornbleet, S., Solid-state activities in the U.K., Part II. *Microwave J.* **13** (5) (May 1970), p. 22E.

7.207 Anonymous, Doppler module priced for consumer applications. *Microwaves* **10** (1) (Jan. 1971), p. 16.

7.208 Keller, R. E., LSA transmitters. *Microwave J.* **14** (7) (June 1971), p. 50.

7.209 Eastman, L. F., Camp, Jr, W. O. and Bravman, J. S., LSA—new peaks in microwave power. *Microwaves* **10** (2) (Feb. 1971), p. 42.

7.210 Magarshack, J., Gunn oscillator used as a phased array aerial element. *Electron. Lett.* **3** (1967), p. 556.

7.211 Anonymous, Phased array uses bulk-diode oscillator. *Microwaves* **10** (4) (April 1971), p. 10.

7.212 Wasse, M. P. and Denison, E., An array of pulsed X-band microstrip Gunn diode transmitters with temperature stabilization. *IEEE Trans. Microwave Theory and Techniques* **MTT-19** (1971), p. 616.

7.213 Tveit, A., Gunn and Impatt diode applications in radio link equipment. *Proc. 1971 Eur. Microwave Conference, Stockholm,* Vol. 2, p. C12/1:1 (Stockholm, The Royal Swedish Academy of Engineering Sciences, 1971).

7.214 Kohiyama, K. and Monma, K., A new type of a solid-state 11 GHz band FM transmitter combining a Gunn diode and an IMPATT diode. *Proc. IEEE* **57** (1969), p. 1232.

7.215 Getsinger, W. J., Paramps beyond X-band. *Microwave Journal* **13** (11) (Nov. 1970), p. 49.

7.216 Okean, H. C., Allen, C. M., Sard, E. W. and Weingart, H., Integrated parametric amplifier module with self-contained solid-state pump source. *IEEE Trans. Microwave Theory and Techniques* **MTT-19** (1971), p. 491.

7.217 Cornbleet, S., U.K. report—energy sources. *Microwave J.* **12** (10) (Oct. 1969), p. 96E.

7.218 Lazarus, M. J., Novak, S. and Bullimore, E. D., New millimetre-wave receivers using self-oscillating Gunn-diode mixers. *Microwave J.* **14** (7) (July 1971), p. 43.

7.219 Anonymous, Waveguide mixer has built-in oscillator. *Microwaves* **10** (7) (July 1971), p. 60; also: LO/mixer unit. *Microwave J.* **14** (7) (July 1971), p. 46B.

7.220 Bass, J. C., Edolls, D. V. and Knibb, T. F., Microwave biased photodetector system with an integral Gunn-effect oscillator. *Electron. Lett.* **4** (1968), p. 429.

7.221 Walsh, T. E. and Sun, C., A packaged system of a solid state microwave-biased photoconductive detector for 10·6 *μm*. *Proc. IEEE* **58** (1970), p. 1732.

7.222 Cornbleet, S., U.K. survey of microwave solid state devices. *Microwave J.* **12** (2) (Feb. 1969), p. 32.

7.223 Anonymous, Gunn oscillator used in Japanese observatory. *Microwaves* **10** (8) (Aug. 1971), p. 25.

7.224 Guétin, P. and Boccon-Gibod, D., Franz Keldysh effect with Gunn domains in bulk GaAs. *Appl. Phys. Lett.* **13** (1968), p. 161.

7.225 Guétin, P., Interaction between a light beam and a Gunn oscillator near the fundamental edge of GaAs. *J. Appl. Phys.* **40** (1969), p. 4114.

8
Pulse Generation and Processing

8.1 INTRODUCTION

After the discovery of microwave current oscillations in n-type GaAs and n-type InP by Gunn in 1963, attention initially focused on the generation and amplification of sinusoidal electro-magnetic microwaves with the help of this new effect. Only later on, in about 1967/68, it was widely realized [8.1–8.11, 5.19] that the ultra-fast growth and decay rates of the Gunn space-charge dipole domains, with their associated relatively large current changes, could with great advantage be put to use for pulse generation and processing at sub-nanosecond speed and with micro-wave repetition rates, by employing circuits which basically consist of a Gunn element in series with an ohmic load resistor (Fig. 8.1). Since then, a whole range of logic modules and devices performing complex electronic functions, for intended incorporation in fast information processing or pulse-communications systems, have been proposed and partly verified experimentally. Further proposals for employing suitably-designed Gunn devices in these fields of electronics are, no doubt, still to come.

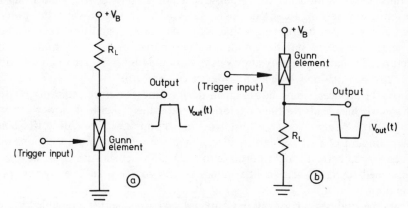

Fig. 8.1 Basic circuits for pulse applications composed of biased Gunn element and ohmic load resistor R_L. (Symbolic indication of external domain triggering applies to triggered mode of operation.)

323

Either the self-oscillation, or the triggered, dipole-domain (Gunn) mode in a TE semiconductor are essentially used. The TE devices made use of in these applications are therefore invariably *Gunn* devices in the strict sense. The potentialities of the dipole domain modes with regard to pulse generation and processing were pointed out by Gunn himself already in one of his basic patents [3.30]. Whereas analyses and experiments have so far been based on the growth or propagation of domains *parallel* to the electric bias field, a recent proposal [8.12] relies on the still faster spreading of a high-field dipole domain *transverse* to the carrier drift (cf. Section 3.3.2). A further recent addition of importance to the field is the realization of bi-stable switching in super-critically doped GaAs devices [4.48, 4.86, 8.13, 8.55, 8.56], the mechanism of which is not yet fully understood (p. 107).

As we will discuss in the following, a single bulk TE device can be designed to perform advanced electronic functions, like the generation of complex waveforms, which would require a whole range of components if they were to be realized by using conventional techniques. It has been found [8.26] that the number of components in threshold-logic [8.77, 8.83] sub-systems, for instance, is reduced to about a quarter by using TE devices instead of transistorized logic circuits. Perhaps more important, the speed at which GaAs bulk devices can operate is at least one order of magnitude higher than what is achievable with present-day bipolar or field-effect transistor logic for which a basic frequency limit is given by the product of junction, or gate-channel, capacitance times characteristic impedance of the interconnecting line. As described in Section 6.1, TE GaAs can conveniently be grown by epitaxy on a semi-insulating GaAs substrate with sufficiently high electric isolation. This is a definite advantage over silicon. In Si monolithic-integrated circuits the isolation must be provided by special means, generally by diffused-in troughs which introduce relatively large time constants owing to their capacitance and thus adversely influence the response speed of the circuit. In comparison to the also extremely fast tunnel-diode logic, GaAs TE devices possess the potential advantage of higher operational voltages and currents. In addition, they are better to integrate, easier to stabilize and are more suitable for realizing multi-electrode structures. The latter property allows us to obtain a higher degree of complexity of the electronic function which a single device can perform.

GaAs TE devices, being hot-electron devices and therefore requiring relatively high electric bias fields for their operation, exhibit the rather high power density of $p_0 = 10^7$–10^8 W cm^{-3} as it was discussed in Section 6.3.1; see eqn (6.18). Since a certain amount of package density is usually a pre-requisite in applications of the kind envisaged here, it is of importance to ascertain whether bulk GaAs devices can be designed to have such small volumes as to provide a sufficiently low power dissipation.

From the analyses carried out in Chapter 4 we know that for a reliable formation of dipole domains two criteria must be fulfilled, namely the doping-length product (n_0L) and the doping-thickness product (n_0d) of the device must exceed certain critical values. According to relation (4.51b) we have $(n_0L)_{\text{crit}} \approx 10^{11}$ cm^{-2}; to assure the formation of *completely* mature domains, a value $n_0L \approx 50$ $(n_0L)_{crit} = 5 \cdot \times 10^{12}$ cm^{-2} is more in line, cf. [8.39]. The other criterion, involving the n_0d product, is given by inequality (4.66b). Here, for domain operation, the lowest-order mode ($m = 0$) has to be considered. Assuming, for instance, a coplanar-contact sheet-type GaAs device with a semi-insulating GaAs substrate (Fig. 6.6f) and $d/L \ll 1$ for the ratio of the geometrical dimensions, we have $(n_0d)_{crit} \approx 0.135$ $(n_0L)_{crit}$. From the expression (6.18) for the power, $P_0 = p_0AL = p_0wdL$, and the two above criteria for the formation of mature domains, it can be shown that for achieving a low value of P_0 the doping level n_0 should be chosen as high as possible and the geometrical dimensions L and d as low as possible. With a highest sensible value for the doping of $n_0 \approx 10^{16}$ cm^{-3} (Section 6.1.2), we obtain a minimum length L of the device in the order of a few microns and a minimum thickness d in the submicron range, the latter thus approaching a Debye length as the ultimate dimensional limit. The Debye length also determines the lower limit on the second transverse dimension, the width w.

A device designed to have such lower-limit dimensions would exhibit the desired very low power dissipation as usage of eqn (6.18) will show, however its resistance turns out to be exceedingly high, apart from the fact that sub-micron dimensions are not particularly convenient to manufacture because of technological difficulties. Taking, then, for the minimum dimensions the more realistic values of, say, $L = 10\,\mu$m, $d = 1\,\mu$m and $w = 10\,\mu$m, we obtain for a bias field of $E_{h0} = 4$ kV cm^{-1}, together with $n_0 = 5 \times 10^{15}$ cm^{-3} and a low-field mobility $\mu_1 = 6\,250$ cm^2 (Vs)$^{-1}$, a dissipation of $P_0 \approx 6$ mW, a d.c. current $I_0 \approx 1.5$ mA, and a low-field bulk resistance $R_1 \approx 3$ kΩ. For certain applications such a resistance value will still be too high and must be traded against dissipation power.

In conclusion, we thus may expect practical Gunn-domain pulse devices made from GaAs to have lower power dissipation levels in the order of a few 10 mW. Employing the thermal analysis of Section 6.3 shows that for thermal reasons about 50 elements of this dissipation can maximally be accommodated per square milli-metre area on semi-insulating GaAs substrate. These power levels are still high compared to those of, for example, the complementary MOS-device switch which exhibits a dissipation power of a few nW and requires a switching power of roughly 1 mW. However taking the power-delay product and the power-speed ratio as useful figures-of-merit, properly designed GaAs TE devices should reach attractively low values of about 1 pJ for both parameters (cf. Fig. 8.8a), [8.21–8.23]. This low figure results from minimum response and delay times as low as a few 10 ps (Section 8.2.1) and compares favourably with what is obtainable from transistor logic which presently reaches switching times down to 0.5 ns. In addition one has to take into account the advantageous feature that signal-processing sub-systems utilizing Gunn devices only require a considerably smaller number of components than do com-parable transistor-circuit realizations. Functional Gunn devices thus appear to be attractive certainly for more special, smaller-capacity digital systems where speed is at a premium and somewhat higher power levels can be tolerated, like in ultra-fast digital instrumentation apparatus or in the pulse regeneration and processing part of a time-multiplexed binary communication system of large channel capacity. The application of Gunn devices to such systems is, at the present time, still in the research and development stage; their commercial introduction has yet to be awaited. Even so, problems will be encountered and will first have to be solved in the field of the passive, interconnecting microstrip circuitry which, so far, has been designed hardly for broadband Gigabit digital, but rather for the relatively narrow-band conventional microwave, applications.

Anticipating the results of this chapter, the main merits and capabilities of functional Gunn devices which employ the particularly advantageous form of Schottky-gate field-effect triggering (Section 8.2.2.3) can be summarized as follows [8.23, 8.26]:

Merits of practical Schottky-gate functional Gunn devices

(1) Sufficient trigger capability ($>$100 mA/V)
(2) Sufficient unidirectivity ($\approx$30 dB isolation)
(3) High input impedance (input capacitance $<$0·05 pF)
(4) High trigger sensitivity (trigger voltage of 50–100 mV)
(5) High speed (30–100 ps) and excellent power delay product (1–10 pJ)
(6) High fan-out
(7) Simplicity of construction and of use in circuits
(8) Suitability for integrated circuits.

Basic functions and properties of functional Gunn devices

(1) Threshold action
(2) Wave-shaping action
(3) Existence of refractory period (domain transit time)
(4) Constant signal propagation velocity
(5) Unidirectional signal propagation
(6) Excitatory and inhibit actions
(7) Summing action
(8) Memory action.

8.2 BASIC PROPERTIES

In this section a description is given of the most important fundamental properties and mechanisms employed in the realm of digital Gunn devices before discussing, in the following sections, specific devices and circuitry for performing particular functions in the fields of pulse generation and regeneration, as well as combinatorial and sequential logic.

8.2.1 Response Times
The initial phase of temporal growth of a charge disturbance (deviation from

neutrality) in an n.d.c. semiconducting crystal is determined by the dielectric relaxation time $|\tau_R| = \varepsilon/|\sigma'| = \varepsilon/en_0|\mu'|$ (cf. Section 3.2). We showed in Section 3.2 that $|\tau_R| \approx 1{\cdot}3$ ps for the typical parameters, exhibited by a GaAs specimen biased in the n.d.c. range, of $n_0 = 5 \times 10^{15}\,\mathrm{cm}^{-3}$, $\varepsilon = 12{\cdot}5\,\varepsilon_0$ and $\mu' = -2\,000\,\mathrm{cm}^2$ $(\mathrm{Vs})^{-1}$. The 'small-signal time constant' τ_R applies as long as we still have sinusoidal space-charge waves propagating along the sample (Section 4.3).

Charge growth in an n.d.c. sample eventually leads to the formation of travelling large-signal space-charge dipole domains, provided the doping-length, and the doping-thickness, products are super-critical as discussed in detail in Sections 4.4.1 and 4.4.2, respectively. The domain formation process [3.19–3.23, 3.60–3.62, 5.26, 8.15–8.17, 8.20] is characterized by the charging of the domain capacitance $C_D(t)$ through the low-field sample resistance R_1 and the ohmic load resistance R_L. For the limiting case of zero load resistance we define a simplified large-signal charging time constant [8.14] in analogy to eqn (4.57) or (5.3) by

$$\tau_D = R_1 \bar{C}_D = \tau_{R1} L/\bar{b} \tag{8.1a}$$

where the bar denotes average values during the charging process. Assuming $E_h \approx E_p$ we notice from Fig. 3.13 that for the domain excess potential in the steady-state case one has $\Phi_{D\infty} \approx 2V_0/3$. For our transient charging case we take $\bar{\Phi}_D \approx V_0/3 \approx E_p L/3$ as a sensible average value for the domain potential during growth. Inserting this value into eqn (5.4a) the average domain width becomes

$$\bar{b} \approx (2\tau_{R1}\mu_1 E_p L/3)^{1/2}. \tag{8.2}$$

This yields for the large-signal charging time constant according to eqn (8.1a)

$$\tau_D \approx (3\tau_{R1} L/2\mu_1 E_p)^{1/2} \tag{8.1b}$$

and, with $\mu_1 E_p \approx 2\,v_D$ (cf. Fig. 3.15a),

$$\tau_D \approx (3\tau_{R1}\tau/4)^{1/2} \tag{8.1c}$$

where $(\tau = L/\bar{v}_a \approx L/v_D)$. An ohmic load resistor R_L can conveniently be taken into account by substituting for L an effective sample length

$$L_{\mathrm{eff}} = L(1 + R_L/R_1) \tag{8.3}$$

which is obtained from $E_1 L = IR_1$ and the definition

$$E_1 L_{\mathrm{eff}} = I(R_1 + R_L).$$

As eqn (8.1c) indicates, the time constant for dipole-domain build-up is proportional to the geometric mean value of the dielectric relaxation time τ_{R1} and the domain transit time τ, i.e. it is proportional to the square root of the sample length L and also, because of eqn (8.3), to that of the load resistance R_L. With the numerical values of the above-quoted example, together with $v_D = 10^7\,\mathrm{cm\,s}^{-1}$ and, say, $L = 30\,\mu\mathrm{m}$, one obtains $\tau_D \approx 17$ ps as a typical value for the zero-load case.

The domain charge is proportional to the domain field amplitude $E_2 - E_1$.

From Section 3.4.4. we know that the field E_1 outside of the domain decreases monotonically during the domain formation phase [Fig. 3.18; convert to time dependence by employing eqn (3.26)]. Thus τ_D is also the time constant governing the drop of E_1 [8.16–8.18]. In most practical cases, however, the time response of the *total device current* density (Figs 3.18, 8.2) of eqn (3.16),

$$J_t(t) = \varepsilon \frac{dE_1}{dt} + en_0 v_1; \quad v_1 \equiv v\,[\,E^1(t)\,],$$

will be of interest rather than that of E_1 itself. During domain build-up, when the sample bias is $E_h > E_p$, the *drift*-current component $I = en_0 v_1 A$ first rises with time before it drops, since the maximum of the $v(E)$ curve has to be passed through. This initial rise of I is less obvious in the *total* current because of the displacement current having a counteracting effect. Also due to the displacement current, the total current drops appreciably only when domain build-up (drop of E_1) has well advanced.

Furthermore we are interested in the current response when the domain is being extinguished at the anode, i.e. when its accumulation layer disappears in the anode contact. An approximate expression for the domain extinction time can be obtained if we assume that the domain shape remains unaltered during the dissolution process (as in Section 3.4.4; cf. [8.16]) and, secondly, that the accumulation-layer velocity v_a keeps constant at the steady-state domain velocity v_D. Then the domain extinction time is simply determined by b_∞ / v_D where b_∞ denotes the width of the steady-state domain. The domain width is given by eqn (5.4a) whereas $v_D = v_{1\infty}$ can be obtained from eqn (3.28b) together with eqns (3.23) and (3.24), see Figs 3.9–3.11. During domain extinction the low field E_1 and the total sample current rise with an appreciable displacement current adding to the drift current, cf. [8.16], see Fig. 3.18.

The assumption of an unaltered accumulation-layer velocity during domain extinction, i.e. $v_a = v_D = $ const., is not entirely correct. In Section 3.4.4.3 we have seen that during the extinction process the excess potential of the rest domain *grows* (Fig. 3.19). This causes a certain reduction in accumulation-layer velocity v_a according to eqn (3.26) [3.39b, 8.25] and consequently a prolongation of the extinction time. In the actual case with diffusion the velocity of the trailing domain edge (Fig. 3.14) drops monotonously until domain dissolution has completed.

The total-current waveforms measured on two GaAs samples of length $L = 380\,\mu$m and $L = 35\,\mu$m are reproduced in Fig. 8.2 where the other relevant sample parameters are given in the caption. In both cases the samples were biased slightly below the threshold voltage V_T ($\mu' > 0$) and then triggered by superimposing on the bias a trigger voltage pulse which raised the sample bias beyond threshold ($\mu' < 0$) ('bias triggering'; see Section 8.2.2.2).

Calculated values of pulse delay time t_d, fall time t_f, and rise time t_r as a function of n_0L and L are given in Figs 8.4–8.6 for GaAs elements operated with zero load*. These characteristical time constants are defined in Fig. 8.3 where the time $t = 0$ corresponds to the instant at which the threshold electric field has been reached.

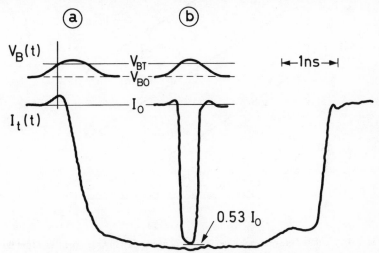

Fig. 8.2 Response of total current $I_t(t)$ measured (a) on bulk GaAs sample of $n_0L = 2{\cdot}8 \times 10^{13}$ cm^{-2}, $L = 380$ μm, $R_1 = 87$ Ω, $R_L = 50{\cdot}5$ Ω (after Sugeta *et al.* [8.17]); and (b) on co-planar GaAs sample of $n_0L = 1{\cdot}25 \times 10^{13}$ cm^{-2}, $L = \mu$m, $d = 10$ μm, $R_1 = 280$ Ω, $R_L = 100$ Ω (after Mause [8.24]). In both cases a domain was launched by superimposing a trigger pulse on the bias voltage $V_B(t)$.

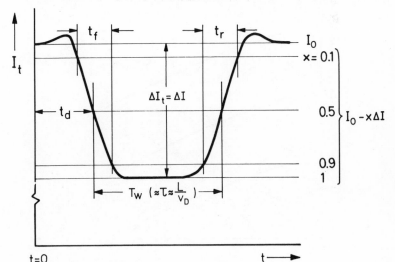

Fig. 8.3 Schematic representation of current response $I_t(t)$ defining delay time t_d, fall time t_f, and rise time t_r. T_w denotes pulse width.

*For a remark on pulse jitter see Section 8.4.1.

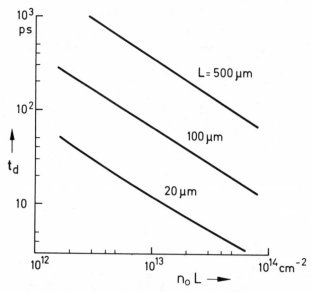

Fig. 8.4 Calculated current-pulse delay time t_d of n-type GaAs element as a function of doping-length product n_0L and of length L for zero load resistance R_L. After Heinle [8.25].

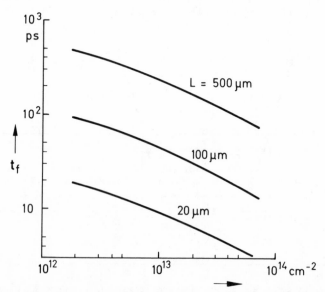

Fig. 8.5 Calculated current fall time t_f of n-type GaAs element as a function of doping-length product n_0L (abscissa) and of length L for zero load resistance R_L. After Heinle [8.25].

330

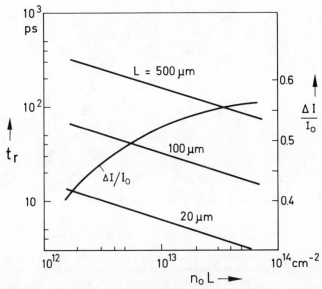

Fig. 8.6 Calculated current rise time t_r and normalized current drop ΔI_t of n-type GaAs element as a function of doping-length product $n_0 L$ for zero load resistance. After Heinle [8.25].

Figure 8.6 shows, furthermore, the dependence on $n_0 L$ of the drop ΔI_t in total current where ΔI_t is determined by the *drift* current drop ΔI according to the static/dynamic $I(V)$ characteristic (Fig. 3.15) if we assume that the steady state has been reached. The samples were considered to be biased at $E_{h0} = 3 \cdot 1$ kV cm^{-1} ($\mu' > 0$) and triggered at the instant $t = 0$ by superimposing on the bias field a step pulse of height $\Delta E_A = 0 \cdot 6$ kV cm^{-1} and of duration $t_w = \tau/2 = L/2v_D$. The calculation was based on the one-dimensional model discussed in Section 3.4.4, however a notch in the doping density, as sketched in Fig. 8.7, was included for serving as the

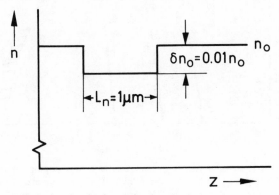

Fig. 8.7 Notch in doping density for nucleating dipole domain assumed in calculations to Figs 8.4–8.6.

331

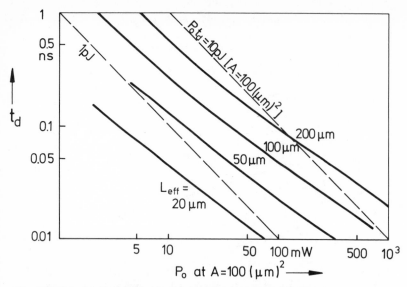

Fig. 8.8a Calculated dependence of pulse delay time t_d on dissipated device power P_0 (power-delay product), with effective device length L_{eff}, according to eqn (8.3), as parameter. After Sugeta *et al.* [8.26b].

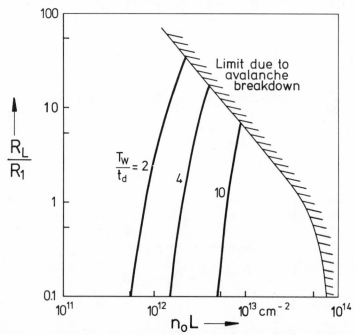

Fig. 8.8b Calculated accessible values for load resistor R_L and doping-length product $n_0 L$, with ratio of pulse width T_W to pulse delay time t_d as parameter. After Sugeta *et al.* [8.26b].

domain nucleation centre. Figures 8.8a and b relate the delay time t_d to the dissipated power P_0 and the load resistance R_L, respectively, illustrating the important fact that a trade-off exists between these quantities. There are practical limits on the tolerable values of R_L and device length L due to avalanche breakdown in the domain ($E_{bd} \approx 150 \, \text{kV cm}^{-1}$; see Fig. 6.25) as indicated by Fig. 8.8b.

Comparing the measured response as given in Fig. 8.2 with the calculated parameters of Figs 8.4–8.6 shows—not surprisingly—larger values in the experiment, by a factor of two to three. For example, from the waveform (a) of Fig. 8.2 one finds $t_d \approx 0.4$ ns, $t_f \approx 0.53$ ns, and $t_r \approx 0.37$ ns. For determining the corresponding theoretical values one has to use $L_{eff} \approx 600 \, \mu\text{m}$ according to eqn (8.3) for the relevant length L in Figs 8.4–8.6 since the load resistance is not negligible. In this case the calculation yields $t_d \approx 0.2$ ns, $t_f \approx 0.2$ ns, and $t_r \approx 0.15$ ns. For a check on curve (b) one must introduce $L_{eff} \approx 48 \, \mu\text{m}$. The discrepancy between theory and experiment may partly be due to the assumption of an ideal triangular domain shape (Fig. 3.7; diffusion-free case) and, in addition, must probably to a large extent be attributed to the detrimental influence of parasitic circuit elements [8.21]. Particularly the capacitive loading of the Gunn sample caused by the interconnecting leads and by bonding pads is an unwanted effect.

The delay time t_d depends to some degree on the triggering method and conditions (cf. Section 8.2.2); consequently the values of Fig. 8.4 can only be taken as a rough guide when comparing with results obtained for triggering mechanisms other than bias triggering. In practical samples current-drop delay may be caused by domain nucleation starting from a small-sized, laterally-limited inhomogeneity (notch) as discussed in Section 3.3.2. In that case the nucleation first spreads laterally (Fig. 3.4) before the growing domain begins to travel towards the anode contact. The calculated effect of the extension of a doping notch on pulse delay is shown in Fig. 8.9 for a particular example. The curves obtained for the total current I_t normalized to the threshold current I_T indicate a lateral spreading velocity of approximately $10^8 \, \text{cm s}^{-1}$.

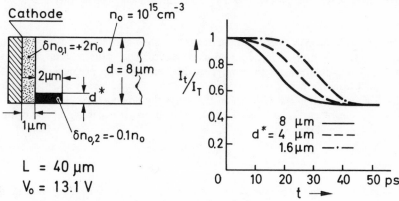

Fig. 8.9 Calculated effect of lateral extension d^* of doping notch ($\delta n_{o,2}$) on current-pulse delay in bulk n-type GaAs element. After Yanai *et al.* [3.67a].

It has been proposed [8.12] to make use of these extremely high spreading velocities (10^8–10^9 cm s^{-1} [3.67–3.70]) for realizing devices which perform particularly fast logic functions. For example, in an H-shaped GaAs sample a travelling domain set up in one arm can be made to trigger, by lateral spreading across the H-junction bar, a domain in the second arm where a field-effect gate electrode is provided at the junction to control this triggering.

There are two further effects, neglected so far, which cause pulse delay [7.52, 8.82] and may appear particularly in coplanar devices of the widely-used kind shown in Fig. 6.6f. In these devices the thickness d of the active TE layer generally is small and dielectric loading is provided at least by the semi-insulating substrate. As a consequence, domain formation time and, thus, drop in device current may appreciably be prolonged according to eqns (4.57) and (4.60).

The second effect can become significant in coplanar devices if they are mounted, in the common way, on metal heat sinks and the height of the semi-insulating substrate is relatively low. Such devices constitute a strip line for electro-magnetic wave propagation, with the epitaxially-grown active TE layer and the metal heat-sink acting as the two conductors, and the semi-insulating semiconductor substrate in between forming the dielectric layer. This line can be represented by the familiar equivalent circuit shown in Fig. 8.10. An essential feature of this line is a reduced propagation velocity of (quasi-TEM) electro-magnetic

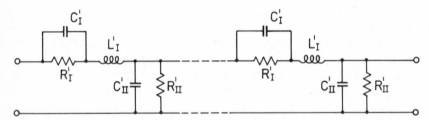

Fig. 8.10 Equivalent circuit of stripline formed by co-planar TE device with semi-insulating substrate, mounted on metal heat sink. Symbols represent series and shunt components of resistance, capacitance and inductance per unit length. After Becker *et al.* [8.82].

waves, caused by the properties of the dielectric and by the finiteness of the electrical conductivity in the active TE layer. If a dipole domain nucleates near the cathode contact the drift current at this point begins to drop. This drift-current drop propagates along the device toward the anode with the *reduced* phase velocity; consequently a corresponding time delay is experienced in the response of the externally flowing total current.

The propagation of an instantaneous drift-current drop ΔI occurring at the cathode ($z = 0$) toward the anode ($z = L$) is described by [7.52]

$$I(z,t) = I_0 - \Delta I \left[1 - \mathrm{erf} \left(\frac{z^2 R'_{\mathrm{I}} C'_{\mathrm{II}}}{4t} \right)^{1/2} \right]$$

with I_0 denoting the original current level. If we define the delay time $t_d^\star$ of the external total current by the time necessary for the drift current at the anode to drop by an amount of $0.9\ \Delta I$, the argument of the error function, erf, in the above equation is approximately unity. This yields

$$t_d^* \approx \frac{L^2}{4} R_I' C_{II}' = \frac{L^2}{4} \frac{\varepsilon}{d_I d_{II} e n_0 \mu_0}$$

where ε is the permittivity of the semiconductor material, d_I the thickness of the active TE layer, and d_{II} that of the semi-insulating substrate. As an example, for a coplanar GaAs device with the parameters $L = 50 \ \mu m$, $d_I = 5 \ \mu m$, $d_{II} = 200 \ \mu m$, $n_0 = 3 \times 10^{15} \ cm^{-3}$ one obtains an additional delay time $t_d^* \approx 23$ ps, i.e. a value which may not always be negligibly small.

In the extreme case the phase velocity of the wave may be reduced below the dipole-domain velocity, and domain formation then is completely inhibited since the domain would overtake its induced drift-current drop. The particular condition of equal velocities for domain and wave can be shown [7.52, 8.82] to lead to a doping-thickness criterion for stability against dipole-domain formation, similar to that derived in Section 4.4.2.2 on different arguments. The discussed additional time delay in current drop is practically avoided if the heat sink below the thin semi-insulating semiconductor substrate is not a metal but an insulating material of high thermal conductivity like BeO, or if the active TE layer is deposited directly on such an electrically insulating but thermally well-conducting material (hetero-epitaxy).

8.2.2 Pulse Triggering

8.2.2.1 *General Remarks*

Current-pulse triggering, i.e. dipole-domain triggering, by an external signal (cf. Fig. 8.1) is a basic feature of functional TE devices for signal-processing applications. We now investigate the different methods available for domain triggering and the requirements which the triggering pulse must fulfil. First a short general description is given of the main triggering mechanisms. This is followed in Section 8.2.2.2 by a more detailed treatment of the simplest triggering method, namely bias triggering, on which principal domain triggering properties of TE elements are studied. A further analysis is in Section 8.2.2.3 then given of triggering via an additional Schottky-barrier contact (rectifying metal-semiconductor contact) which turns our device into a three-terminal structure. It is found that Schottky-gate triggering offers the best performance.

The following main methods for domain triggering can be employed:

(1) *Bias triggering* by applying a positive voltage pulse to the anode, or a negative pulse to the cathode, of a subcritically-biased TE element [3.30, 3.31].

(2) Use of a *third ohmic contact* [3.30, 8.4, 8.27, 8.28] near the cathode contact as a gate electrode for receiving the trigger pulse.

(3) Use of an *isolated (capacitively-coupled) gate contact* [3.30, 8.29, 8.30] as the third terminal.

(4) Use, instead, of a *Schottky-barrier gate contact* [8.31, 8.32].

(5) Involving *light irradiation* onto a photoconducting TE element [8.33].

335

Bias triggering is depicted in Fig. 8.11 (ignore here the gate electrode). Initially the sample is subcritically biased at an anode voltage V_0 of $V_S^\star < V_0 < V_T$, i.e. the working point is on the static branch of the $I(V)$ characteristic, Figs 3.15 and 8.13. If now a positive voltage step of height $\Delta V_A \geq V_T - V_0$ is applied to the anode, a domain will be formed at the cathode since the threshold voltage V_T is reached (Fig. 8.11). As a result of this, the device current drops and remains at its low value during the domain transit time $\tau \approx L/v_D$. Even if the trigger pulse has ended after domain formation, the domain continues to exist since the initial anode voltage was chosen to exceed the domain sustaining voltage $V_S^\star$. Examples of experimentally-obtained waveforms of trigger pulse and generated current pulse had been shown in Fig. 8.2. Instead of applying a *positive* voltage pulse to the *anode*, the threshold electric field E_T inside the sample can equally be reached by applying a *negative* voltage pulse to the *cathode*. The obtained current pulses are readily transformed to voltage pulses by providing an ohmic load resistor. Bias triggering suffers from the disadvantage that the device employed is a two-terminal device where the transfer function is the same for both directions, i.e. it offers no separation between input and output.

Obtaining unidirectivity of signal flow in a circuit containing two-terminal TE devices affords adding circuit elements with non-reciprocal properties. In a particularly simple

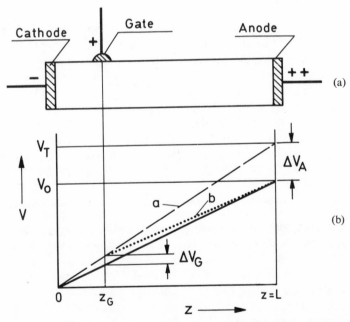

Fig. 8.11 Schematic representation of domain triggering (a) by raising of terminal (anode) voltage by ΔV_A ('bias triggering') and (b) by applying positive voltage pulse ΔV_G to ohmic gate contact.

way this may be achieved by triggering the TE devices via external gate diodes [5.19], for example Schottky diodes [8.34]. In this case the unidirectivity obtained depends on the ratio of the sum of diode bulk resistance and depletion-layer resistance to bulk resistance.

By providing a *third ohmic contact* near the cathode as a gate electrode (Fig. 8.11) we obtain a three-terminal device which principally offers greater ease in devising circuits. The Gunn domain is launched at the cathode by applying a positive trigger voltage pulse between gate and cathode, thereby increasing the current flowing in that portion and raising the electric field to or beyond the threshold value. From Fig. 8.11, which shows the case for constant terminal voltage, it is noted that a gate voltage-pulse amplitude of

$$\Delta V_{\mathrm{G}} = \frac{z_{\mathrm{G}}}{L} \Delta V_{\mathrm{A}} \qquad (8.4)$$

is required for triggering. This implies an improvement in trigger sensitivity over bias triggering which depends on the position z_{G} of the gate electrode and on the length L of the TE element. In practice, however, the sensitivity is found not to be higher than that of the two-terminal device, probably because of the voltage drop owing to the gate spreading resistance and to the space-charge accumulation underneath the ohmic gate contact [8.23b, 8.35]. The gate contact should be of small area and of high resistance in order to minimize interference with the domain passing by. Although here we have a three-terminal device, the isolation between input (gate) and output (anode) is not particularly high because of the galvanic connection between these two terminals. A higher degree of freedom from unwanted reverse signal transfer should be obtainable in a proposed special Y-shaped device configuration in which the three ohmic contacts are positioned at the end of the Y-arms [8.36, 8.42b]. Reflection insensitivity is then obtained for correctly chosen bias voltages and load impedances.

Galvanic separation between the gate contact near the cathode and the other two electrodes results if a *capacitively-coupled third electrode* is used. This electrode may be realized by evaporating metal onto a previously applied isolation layer of, for example, SiO_2. Domain triggering can be achieved either by applying a positive pulse or a negative pulse to the gate. In the first case the displacement current through the contact capacitance, due to the trigger pulse, adds to the drift current in the cathode-gate portion. In order to raise the device field to E_{T} and, thus, trigger a domain near the *cathode*, the drift current I_0 in that portion must increase simply by

$$\Delta I_{\mathrm{A}} \approx I_{\mathrm{T}} - I_0 \approx \frac{\Delta V_{\mathrm{A}}}{R_1} = \frac{\Delta V_{\mathrm{G}} L}{R_1 z_{\mathrm{G}}} \qquad (8.5)$$

where R_1 denotes the low-field sample resistance and ΔV_{A} indicates, in Fig. 8.11, the amount by which the anode voltage is below the trigger level. Since, however, the actual gate trigger voltage is increased owing to the finite voltage drop across the

337

contact capacitance, this triggering method offers in practice no improvement apart from the isolated input [8.23b, 8.35].

By applying a negative voltage pulse to the isolated gate contact, domain nucleation below the *gate zone* can be achieved by depletion-type field-effect action [8.37, 8.38]. The depletion layer which if formed underneath the negatively-biased gate causes a narrowing of the conducting channel and, hence, an increase of the electric field within it (analogous to the case of Fig. 8.12). A TE device containing such an isolated gate is in principle, of course, a MIS-FET device.

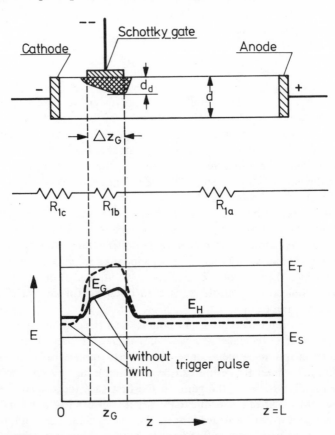

Fig. 8.12 Schematic representation of TE element with Schottky gate for domain triggering by field-effect action and d.c. equivalent circuit. Spatial dependence of electric field *E* in 'channel' without and with trigger pulse applied to gate. Adapted from Heime [8.53].

Depletion-type field-effect control of domain nucleation can equally be obtained by using a *Schottky-barrier gate contact*. This contact has the definite advantage over the isolated gate of simpler technological manufacture. When a Schottky gate

338

is employed there are basically three possible triggering methods [8.23b, 8.35]:

(1) Forward bias at the Schottky barrier and positive trigger pulse (same operation as that of the ohmic trigger contact);

(2) reverse or zero bias and positive trigger pulse; } (same operation as that of and finally } the isolated trigger contact)

(3) reverse or zero bias and negative trigger pulse. }

The last case is that of the field-effect action (MES-FET device). The principal structure of the device capable of this type of triggering is shown in Fig. 8.12. The depth d_a of the depletion layer below the gate increases with increasing reverse bias on the Schottky barrier as in corresponding FET structures. The depth is not constant over its length owing to the potential drop caused by the bias voltage between cathode and anode. A domain is nucleated in the channel below the depletion layer if there the field reaches the threshold value E_T at sufficiently high gate bias values. Schottky-gate field-effect triggering (Section 8.2.2.3) has major advantages as already listed above at the end of Section 8.1. This triggering method is, however, restricted to *negative* triggering pulses at the gate whereas in digital circuitry of some sophistication it may be required to process pulses of either polarity. Triggering which involves a depletion layer below a gate electrode is sometimes called 'notch triggering'.

Depletion-type field-effect action in TE devices can in principle, of course, also be achieved by employing a *pn-junction gate* on the device [8.8b]. But for technological reasons and because of the higher switching times for pn-junctions, Schottky-barrier gates are preferable for notch triggering. On the other hand, pn-junction control electrodes on GaAs Gunn elements have successfully been used to vary the frequency of transit-time domain oscillations over up to more than one octave by changing the junction potential [7.192–7.194, 8.40] as briefly mentioned in Section 7.3. The underlying effect is most probably a variation of transit length and point of domain nucleation.

Illumination of a TE element, made from a photoconducting semiconducting material like GaAs, raises the carrier density by generation of electron-hole pairs and thus causes an increase in sample current. By projecting a thin line shadow onto a sample illuminated by (white) light, a local zone of increased resistivity can be created which serves as the domain nucleation notch [8.33a], cf. [8.51]. A domain then nucleates at the shadow and travels to the anode. Frequency tuning of domain transit-time oscillations by moving the shadow line along the sample was obtained over a frequency range of 5:1, cf. [8.50]. Optical triggering of domains should in principle also be achievable by employing a photosensitive device as the load in series to the TE sample [8.41, 8.42]. Varying the resistance of the photosensitive device by light causes the terminal voltage across the TE sample to change. In order that the high-speed properties of the TE element may be utilized, the photoconducting device must possess a sufficiently fast recombination of the excess carriers produced, a requirement which will be difficult to fulfil.

339

For illumination the TE sample or the photosensitive load, monochromatic light, e.g. laser light, having a wavelength lower than that corresponding to the absorption edge of the semiconductor can also be used. By focusing the beam of a He-Ne laser onto a point immediately near the cathode of a *super*-critically-biased GaAs element, the nucleation of domains was inhibited owing to the lowering of the electric field by the increased conductivity at the nucleation site [8.33b]. When, on the other hand, the same element was biased *sub*critically and was then illuminated at a point remote from the cathode, the locally-created electron-hole pairs reduced the overall sample resistance. This led to the launching of domains as a result of the current being raised to an above-threshold level.

In another experiment, the formation of domains at the cathode of a GaAs TE element could be speeded up by approximately 150 ps compared to the case without light irradiation by focusing the laser beam onto a spot at the lateral edge of the cathode contact [8.43]. It is believed that the details of stable domain launch are in this case controlled by localized charge distributions near the lateral edge of a growing three-dimensional dipole domain.

We will now discuss bias triggering and Schottky-gate triggering, as the two most important triggering methods, in more detail.

8.2.2.2 *Bias Triggering*

The process of bias triggering, as depicted above in Fig. 8.11 (without gate electrode), is illustrated in a different way in Fig. 8.13 on the static and dynamic branches of the

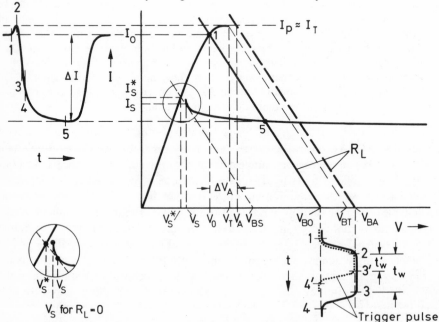

Fig. 8.13 Illustration of bias triggering on drift-current (I)/voltage (V) characteristic using load-line concept.

$I(V)$ characteristic of the TE element and the load line given by the assumed ohmic load resistor R_L. The applied bias voltage pulse $V_{B0} \rightarrow V_{BA}$ raises the initially subcritical sample voltage V_0 $(> V_S^\star)$ by an amount ΔV_A to a value V_A above the threshold $V_T = E_T L$, thereby inducing domain nucleation and generation of the negative pulse in sample drift current I as indicated at the left of Fig. 8.13. In the following investigation we make the simplifying assumption of $I_t \approx I$, i.e. of neglecting the static capacitance C_h (cf. Fig. 3.21) that is in parallel to the resistance of the part of the sample outside of the domain. For this case of neglecting C_h it has been shown [3.66] that the device resistance outside of the domain can well be approximated by the sample low-field resistance $R_1 = G_1^{-1}$ as done in Fig. 3.21. Actually these resistances differ since the domain-outside field E_1 reaches values above the velocity-peak field E_p during the domain-formation process.

(a) *Trigger Pulse Length.* The triggering characteristics of a sample can conveniently be studied in the plane formed by drawing the domain-field amplitude $(E_2 - E_1)$ against low field E_1 [3.66, 8.6] or in the plane of domain potential Φ_D against low field E_1 [8.17, 8.35] shown in Fig. 3.13 with the device line included. We choose the here more practical representation of domain excess potential Φ_D against drift current I [3.66] (Fig. 8.14). Owing to the neglect of the static sample capacitance C_h we are able to include the influence of the load resistor R_L by employing a combined 'device-plus-load line' $R_1 + R_L$. Figure 8.14 shows such lines for a subcritical bias (0) and for the case of a trigger pulse applied (A). The line is determined by the appropriate extension of the voltage balance eqn (3.23), viz.

$$V_B = E_B L = E_h L + I R_L \approx I(R_1 + R_L) + \Phi_D. \tag{8.6}$$

If trigger pulses are applied to the device according to Fig. 8.13, the locus of Φ_D against I moves along trajectories as indicated in Fig. 8.14 (arrows). The time variation of the process is determined by eqn (3.25) which gives the growth rate of the domain potential. If the dynamic working point (Φ_D, I) is positioned at the right-hand side of the steady-state domain characteristic $\Phi_{D\infty}(I)$, a nucleated domain *grows* since there $f(E_1, E_2)$ of eqn (3.25) is negative. On the other hand, a domain decays if the working point lies to the left of the characteristic. In our particular examples of Figs 8.13 and 8.14, it is assumed that the longer trigger pulse (t_w) leads to the build-up of a steady-state domain, with the working point passing along $1 \rightarrow 2 \rightarrow 3 \rightarrow 4$ at the corresponding time instants and finally reaching the steady-state point W at 5. Contrary, applying the narrow trigger pulse $(t_w', $ dotted curves) does not lead to the formation of a mature domain. At the end of the pulse the operating point moves here to 4 which means that the sample must revert to the starting condition 1. There exists a critical length $t_w = t_{crit}$ of the trigger pulse for initiating steady-state dipole domains. This critical length is determined by the position of the separatrix S. The trigger pulse length is sufficient if the working point has passed S before the trigger terminates, cf. [3.66].

In our simplifying assumption of an ideal rectangular trigger pulse and the neglect of the static sample capacitance C_h, the dynamic working point moves on two device-plus-load

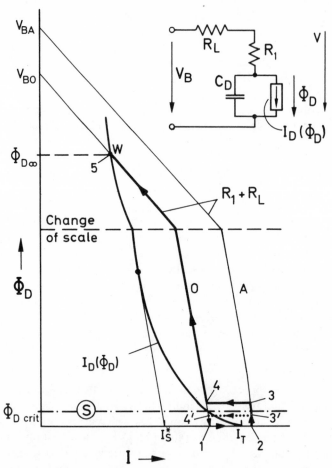

Fig. 8.14 Bias triggering process shown in domain-potential (Φ_D)/drift-current (I) plane for two triggering pulses of different width (see Fig. 8.13). Solid trajectory indicates formation of steady-state domain, whereas dotted trajectory shows inability of trigger pulse to initiate domain.

lines corresponding to trigger 'on' and 'off'. However trajectories outside of these lines are being traced if C_h is included [3.66, 8.6] and parasitic circuit elements are taken into account [8.48] and also if the trigger pulse amplitude is not constant [3.66].

An analytic expression for the critical trigger pulse length t_{crit} can be derived by integrating eqn (3.25) which yields

$$t_{crit} = - \int_{\Phi_{D,n}}^{\Phi_{D,crit}} \frac{d\Phi_D}{f(E_1, E_2)} \tag{8.7a}$$

where $\Phi_{D,n}$ is the excess potential of the initial nucleation domain. Since the original terminal voltage V_0 is generally close to the threshold V_T we can restrict ourselves to small starting domains (points 3 or 3' in Fig. 8.14) and thus find for $f(E_1, E_2)$, by expanding eqn (3.21) with respect to $(E_2 - E_1)$,

$$f(E_1, E_2) \approx \mu'_A (E_2 - E_1)^2.$$

Here $\mu'_A \equiv \mu'(E_A)$ is the differential mobility at the trigger level $E_h = E_A = V_A/L$ (see Fig. 8.13). The term $(E_2 - E_1)^2$ can be eliminated by using eqn (3.24). Thus eqn (8.7a) becomes

$$t_{\text{crit}} \approx \frac{\varepsilon}{2\,en_0(-\mu'_A)} \int_{\Phi_{D,n}}^{\Phi_{D,\text{crit}}} \frac{d\Phi_D}{\Phi_D} =$$

$$= \frac{\varepsilon}{2\,en_0(-\mu'_A)} \ln \frac{\Phi_{D,\text{crit}}}{\Phi_{D,n}} \tag{8.7b}$$

showing a dependence on the ratio of potential of the critical domain to that of the initial domain. The nucleation centre is supposed to have a spatial doping dependence like that of Fig. 8.7, i.e. to be a notch of length L_n and depth $\delta n_0/n_0$. This causes the initial domain to have an excess potential

$$\Phi_{D,n} = L_n E_{h0}(\delta n_0/n_0)$$

and thus, to be exact, the trace $1 \to 2$ in Fig. 8.14 does not lie on the abscissa but above it by an amount of $\Phi_{D,n}$. For determining $\Phi_{D,\text{crit}}$ the intersection point of the steady-state characteristic $I_D(\Phi_D)$ and the device-plus-load line has to be found (Fig. 8.14). This is accomplished by making use of the steady-state condition $f(E_1, E_2) = 0$ in evaluating eqn (3.24). Approximating $v(E)$ near its peak by a parabola, one eventually obtains

$$\Phi_{D,\text{crit}} \approx \frac{9\varepsilon}{2\,en_0} (E_p - E_1)^2.$$

If we restrict ourselves to the case $R_L = 0^*$, we additionally have, according to the, device-line eqn (3.23), $E_1 = E_{h0} - \Phi_{D,\text{crit}}/L$. The nucleating domains are small enough to warrant neglecting the term $\Phi_{D,\text{crit}}/L$ compared to E_{h0}. The critical trigger-pulse width therefore becomes, cf. [8.35]

$$t_{\text{crit}} \approx \frac{|\tau_{RA}|}{2} \ln \frac{9\varepsilon(E_p - E_{h0})^2}{2\,en_0 L_n E_{h0}(\delta n_0/n_0)} \tag{8.8}$$

where $\tau_{RA} = \varepsilon/en_0\mu'_A$ is the dielectric relaxation time at the working point with trigger applied.

Equation (8.8) indicates that t_{crit} is a sensitive function of the nucleation-centre parameters. It is also evident that t_{crit} depends on the trigger-pulse amplitude since this quantity determines μ'_A (for mobility values see Fig. 4.2).

---

*The dependence on R_L should be only small and possibly can be neglected in most practical cases.

Normalized values of t_{crit} as a function of the sample bias E_{ho} for the case of zero load resistance are shown in Fig. 8.15 with carrier concentration n_0 as parameter. The evaluation was based on the doping notch of Fig. 8.7 and on quantities determined from the $v(E)$ characteristic as given by Fig. 2.8 or eqn (3.2). As an example of an absolute value, one obtains $t_{\mathrm{crit}} = 13\cdot73$ ps for $n_0 = 10^{15}$ cm^{-3}, $E_{\mathrm{ho}} = 3\cdot1$ kV cm^{-1} and a trigger field amplitude $\Delta E_{\mathrm{A}} = 400$ V cm^{-1}. Measured values of t_{crit} are higher by a factor 2–10 [8.35, 8.49]; this discrepancy may be explainable particularly by the somewhat arbitrary assumptions regarding the nucleation centre.

(b) *Trigger Sensitivity*. Besides having in the sample a nucleation centre (doping notch) which causes a local field increase being constant in time, there are always

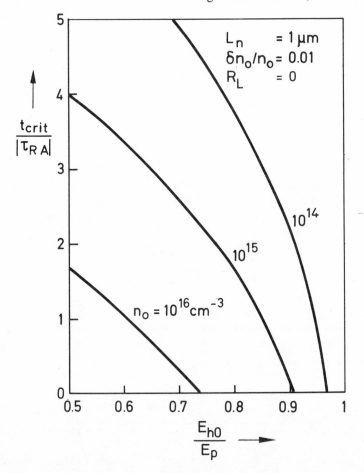

Fig. 8.15 Calculated dependence of normalized minimum trigger-pulse width t_{crit} on normalized bias field E_{ho} with net-doping concentration n_0 as parameter. Normalizing quantities are the peak field E_{p} and the dielectric relaxation time τ_{RA} at the working point with trigger applied. After Heinle [8.25].

344

superimposed random electric-field variations. These field variations are primarily caused [5.89, 6.77] by random electron-density fluctuations as a result of diffusion noise, i.e. of the 'generalized thermal (Johnson) noise' introduced in Section 5.3. The random field variations may give rise to the nucleation of a domain within a finite cube limited in its three dimensions to a number of Debye lengths L_{Db}, say 15 L_{Db}*. The sensitivity of the bias trigger process is, consequently, to first order determined by the smallest allowable difference between the threshold field E_T and the terminal field E_{h0} [8.44, 6.77b]. The bias must not exceed the value beyond which the probability of triggering unwanted pulses due to spontaneous field variations becomes too high, cf. [8.45].

Thermal noise has a Gaussian distribution [8.46, 8.47]. It is assumed that also the generalized thermal noise, represented according to eqn (5.60) by a noise temperature T_N, remains Gaussian in its distribution so that the probability of field fluctuations much greater than their r.m.s. value is small. The r.m.s. value of the fluctuations in domain excess potential in the nucleation cube is

$$(\delta\Phi_{D,N})_{rms} = \left(\frac{k_B T_N}{C_{D,N}}\right)^{1/2} \tag{8.9}$$

and that of the electric-field fluctuations in the cube, as obtained by dividing by the cube side length 15 L_{Db},

$$(\delta E_N)_{rms} = \left[\frac{k_B T_N}{\varepsilon(15 L_{Db})^3}\right]^{1/2} \tag{8.10}$$

where $C_{D,N} = \varepsilon 15 L_{Db}$ denotes the capacitance of the nucleation cube. For the limit trigger-field value we can, therefore, roughly assume

$$(\Delta E_A)_{min} = (\delta E_N)_{rms}. \tag{8.11}$$

With $n_0 = 5 \times 10^{15}$ cm^{-3} and a temperature $T_{e,1} = 680$ $K \approx T_N$ of the light electrons just below threshold [2.26] (Fig. 2.5); one finds $(\Delta E_A)_{min} = 58.5$ V cm^{-1}.

A useful characteristic quantity for describing the trigger process of a Gunn element (being a current source) is the maximum current gain factor [8.21b, 8.23a, c]

$$\alpha_{max} = \frac{\Delta I_t}{(\Delta I_{tA})_{min}} \approx \frac{\Delta I}{(\Delta I_A)_{min}} = \frac{E_T - E_{1\infty}}{(\Delta E_A)_{min}} \tag{8.12}$$

which is the amplitude ratio between triggered current pulse ΔI and minimum triggering current pulse $(\Delta I_A)_{min}$. A sample of high $n_0 L$ operated with sufficiently low R_L delivers according to Fig. 8.6 a current pulse $\Delta I \approx 0.55 I_0$; furthermore we have $E_T - E_{1\infty} \approx 2.8$ kV cm^{-1}. With the above value $(\Delta E_A)_{min} = 58.5$ V cm^{-1} one then obtains a gain $\alpha_{max} \approx 48$. Maximum current gains of $\alpha_{max} \approx 10$–40 have been

---

*The size of the nucleation volume is not yet precisely known. An estimate of the volume to $(15 L_{Db})^3$ [6.77b] can only be considered as being preliminary since the lateral propagation of the carrier-density fluctuations had been neglected.

measured [8.21b, 8.23a, c]. Figure 8.16 shows how α_{max} first rises with increasing load resistance R_L (cf. Fig. 8.13) but then drops sharply when a parasitic capacitance C_{sh} is present. In this case the terminal voltage cannot drop instantaneously after domain extinction because of the time constant $R_L C_{sh}$, and successive domain triggering may occur (see Section 8.4.3.1). To ensure the desired single-domain triggering one is forced to lower the device bias current I_0 which necessitates an increase in the amplitude of the triggering current pulse $(\Delta I_A)_{min}$.

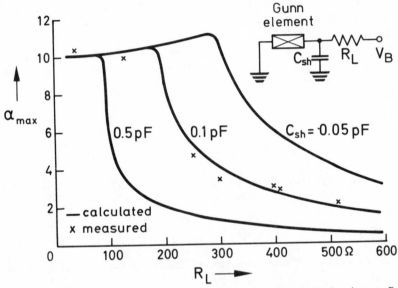

Fig. 8.16 Dependence of maximum current gain α_{max} on ohmic load resistance R_L with parasitic capacitance C_{sh} as parameter. $R_1 = 100\,\Omega$, $n_0 = 10^{15}\,\text{cm}^{-3}$, $L = 100\,\mu\text{m}$, static sample capacitance $C_h = 0.015\,\text{pF}$. After Sugeta *et al.* [8.21b].

(c) *Fan-out.* In pulse-processing circuits, a number of devices is usually connected in series where each device is triggered by the preceding one. Furthermore, many applications require that a device has to successfully trigger an array of parallel-lying succeeding devices (Fig. 8.17). Successful triggering means in our case of TE devices that the TE element, acting as a pulse-current generator, must give rise to a triggering electric-field pulse of an amplitude exceeding $(\Delta E_A)_{min}$ of eqn (8.11) inside each of the m immediately following, identical devices. (Disregard the polarity problem here if imagining TE elements in the circuit of Fig. 8.17.) The maximum possible value of m, for successful triggering by one device under worst-case conditions, is termed the 'fan-out' of that device. The magnitude of m for a TE device [8.21b, 8.23c, 8.26, 8.44, 8.52] depends on the ratio of low-field resistance R_1 to load resistance R_L and on α_{max}. A calculation yields [8.21b]

$$m = \frac{R_1(1 - \zeta_G) + R_{La}}{R_1 + R_{La} + R_{Lc}}\left(\alpha_{max} - \frac{R_{Lc} + \zeta_G R_1}{R_{La}}\right) \tag{8.13}$$

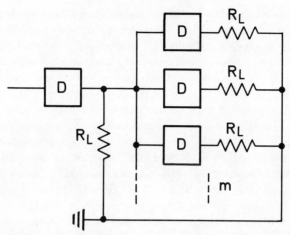

Fig. 8.17 Schematic representation of a device D connected to m further devices of the same kind, all having identical load resistors R_L. After Hartnagel [8.44].

for the arrangement shown in the inset of Fig. 8.18 with $R_L = R_{La} + R_{Lc}$. The evaluation of eqn (8.13) is given in Fig. 8.18 for a current gain $\alpha_{max} = 10$. The case of $\zeta_G = 1$ and $R_{Lc} = 0$ is the case of bias triggering. For $\zeta_G < 1$ it is assumed that

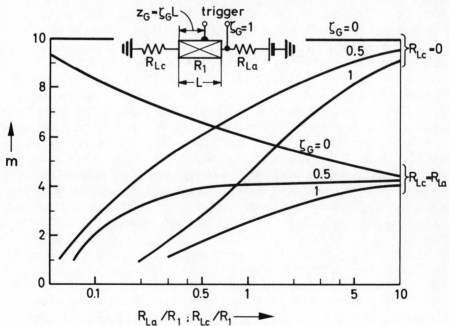

Fig. 8.18 Dependence of fan-out m on ratio of load resistance R_L to low-field sample resistance R_1 with distance between cathode and ohmic-contact trigger gate as parameter according to eqn (8.13), with $\alpha_{max} = 10$. After Yanai and Sugeta [8.52].

347

there is an ideal ohmic contact acting as trigger gate. For ohmic-gate triggering the possible fan-out m is larger than for bias triggering owing to the (theoretically at least) increased triggering sensivity (cf. Section 8.2.2.1). For bias triggering, m grows with increasing resistance ratio R_L/R_1. There exist, however, restrictions on the magnitude of R_L. It can only be increased to the value at which the maximum domain field reaches the avalanche breakdown value (Fig. 8.8b), cf. [8.21b]. Also the current-pulse fall time t_f must remain smaller than half the domain transit time τ, cf. [8.44]. Even more drastic limitations may result from demands, set by the intended applications, on the absolute values of the pulse response times (Section 8.2.1).

During experiments with bias-triggered GaAs devices, fan-outs of up to three have been obtained so far [8.34]. This points to not yet optimally-designed devices and circuits. Somewhat higher fan-outs are expected when using improved gate diodes as unidirectional elements for connecting the individual TE devices [8.34]. Schottky-gate triggered devices, on the other hand, should exhibit substantially higher fan-outs since they possess a relatively high input impedance [8.23c, 8.26]. For a Schottky-gate capacitance of 0·1 pF as a typical value and assuming channel, and gate, bias levels of 5 % below threshold, fan-outs of $m \geq 10$ are expected to be readily obtainable [8.24, 8.34].

8.2.2.3 *Schottky-Gate Triggering*

Schottky-gate triggering in the field-effect mode has been introduced above in Section 8.2.2.1 (Fig. 8.12). Self-biasing of the gate electrode is normally obtained because of the potential drop in the device between the cathode and the position of the gate. Domain nucleation occurs below the depletion layer of the Schottky contact, at its anode edge, for a sufficient reduction in cross-section of the current-conducting channel by the trigger signal which must raise the electric field in that region up to E_T (notch triggering). The main advantages of Schottky-gate triggering are higher trigger sensitivity, as it will be investigated below, and an excellent input-output isolation of more than 30 dB [8.23]. Since the Schottky contact exhibits a depletion-layer capacitance, a time constant is introduced for the trigger process which adds to the time constants deduced in Section 8.2.1 regarding response times for current-pulse generation in *two*-terminal Gunn elements. However, Schottky capacitances as low as 0·05 pF can be realized, which yield, together with input-circuit resistances in the 50 Ω range, time constants of a few ps which are negligibly small for most purposes.

A geometrical representation of a Schottky-gate functional TE device is given in Fig. 8.19, and a photograph of an experimental version, as they have been success-fully operated, is reproduced in Fig. 8.20. In both cases we have coplanar-type mesa devices with a TE-active n-type GaAs layer epitaxially deposited on a semi-insulating GaAs substrate. A dumb-bell shape is chosen particularly for reasons of obtaining larger areas for bonding and of avoiding breakdown effects at the anode contact (Section 6.4). Cr-Ni, Cr-Au, Cr and Ag have proved to be satisfactory as Schottky-contact metals on GaAs [8.23, 8.53].

Schottky gate (Ag, Cr, or Cr - Ni)

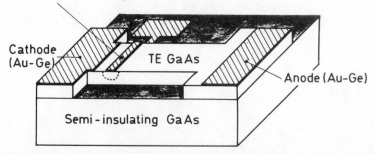

Fig. 8.19 Geometrical representation of Schottky-gate functional TE device of planar-type mesa configuration.

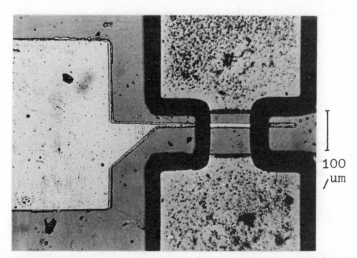

100 /um

Fig. 8.20 Micro-photograph of Schottky-gate functional TE device made from GaAs epitaxially grown on semi-insulating GaAs substrate. Au-Ge for ohmic contacts, Cr for Schottky contact. (By courtesy of K. Heime, Fernmeldetechnisches Zentralamt der Deutschen Bundespost, 1971.)

We now investigate the trigger behaviour of the Schottky-gate TE device, based on the simple model shown in Fig. 8.12. For additional simplicity we assume that the depth of the depletion layer be constant along z at the value d_a. This condition is approximately fulfilled for a small width of the gate stripe, i.e. for $\Delta z_G/L \ll 1$. First we calculate the electric fields in the channel below the gate and outside of the gate region [8.53]. This is followed by a derivation of the so-called 'trigger capability', being the amplitude ratio of output current pulse to trigger pulse voltage [8.23, 8.26]. The theoretical results will be compared with experimental findings.

349

According to Fig. 8.12 the overall d.c. device resistance R_{1G} is considered to consist of

$$R_{1G} = R_{1a} + R_{1b} + R_{1c}. \tag{8.14}$$

If we denote the resistance for the case of no Schottky gate present ($d_a = 0$) with R_1 it follows

$$R_{1a} + R_{1c} = R_1 \frac{L - \Delta z_G}{L}, \tag{8.15}$$

$$R_{1b} = R_1 \frac{\Delta z_G}{L} \frac{d}{d - d_a} \tag{8.16}$$

and

$$R_{1G} = R_1 \left(1 + \frac{\Delta z_G}{L} \frac{d_a/d}{1 - d_a/d} \right). \tag{8.17}$$

For $d_a \to 0$ and/or $\Delta z_G \to 0$ we have $R_{1G} = R_1$. The current in the device is given by

$$I_0 = \frac{V_0}{R_{1G}} = \frac{V_0}{R_1} \left[1 + \frac{\Delta z_G}{L} \frac{d_a/d}{1 - d_a/d} \right]^{-1}. \tag{8.18}$$

With the current density in the channel below the gate being

$$J_G = \frac{I_0}{(d - d_a)w}, \tag{8.19}$$

where $w = A/d$ denotes the second transverse dimension of the device (channel), we obtain for the electric field in that region

$$E_G = \frac{J_G}{\sigma_1} = \frac{V_0}{\sigma_1 A R_1} \left[1 - \frac{d_a}{d} \left(1 - \frac{\Delta z_G}{L} \right) \right]^{-1}. \tag{8.20}$$

There, σ_1 is the conductivity in the channel. Using $R_1 = L/\sigma_1 A$ and $E_{h0} = V_0/L$ for the field in the case of constant cross-sectional area, one finds

$$E_G = E_{h0} \left[1 - \frac{d_a}{d} \left(1 - \frac{\Delta z_G}{L} \right) \right]^{-1}. \tag{8.21}$$

The relation between the depletion-layer depth d_a and the voltage V_G applied across the Schottky contact is given by Schottky's theory, cf. [2.1, 8.54], as

$$d_a = [2\varepsilon(V_G + V_{df})/en_0]^{1/2}. \tag{8.22}$$

The diffusion voltage V_{df} for Schottky-contacts to n-type GaAs with $n_0 = 10^{16}$ cm^{-3} is about 0·75 eV, e.g. [2.1, 2.2].

For the electric field outside of the gate region we have

$$E_{\mathrm{H}} = J_{\mathrm{H}}/\sigma_1 = I_0/\sigma_1 A. \tag{8.23}$$

With $\sigma_1 A = L/R_1$ follows

$$E_{\mathrm{H}} = I_0 R_1/L = E_{\mathrm{h0}} \left[1 - \frac{\Delta z_{\mathrm{G}}}{L} \frac{1}{1 - d/d_{\mathrm{d}}} \right]^{-1}. \tag{8.24}$$

The voltage pair V_{G} and $V_0 = E_{\mathrm{h0}}L$ at which dipole-domain nucleation commences below the gate, can be calculated from eqns (8.21) and (8.22) if E_{G} is set equal to the Gunn threshold field E_{T}. Plots obtained in this way are shown in Fig. 8.21 for two values of channel thickness d together with an experimental result [8.53]. There is qualitative agreement between theory and experiment. The discrepancy is thought to be mainly due to the difficulty of exactly measuring the voltage V_{G} across the depletion layer. From Fig. 8.21 and eqn (8.21) it can be seen that the sensitivity with regard to domain triggering increases if:

(1) the anode voltage is increased;
(2) the width of the Schottky-gate stripe Δz_{G} and/or the channel thickness d is decreased; and

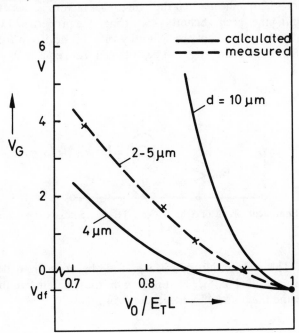

Fig. 8.21 Dependence of Schottky-gate voltage V_{G} on anode voltage V_0 for onset of $w \simeq 100 \ \mu\mathrm{m}$, $n_0 = 3 \cdot 2 \times 10^{15} \ \mathrm{cm}^{-3}$. After Heime [8.53]. (N.B. Read $-V_{\mathrm{G}}$ instead of V_{G}.)

(3) the carrier concentration n_0 is decreased, i.e. the depletion layer depth d_d is increased according to eqn (8.22).

Trigger voltage amplitudes in the 50–100 mV range should be necessary for anode voltages about 5% below threshold and channel thicknesses of a few microns.

The TE device is primarily a current generator delivering, upon domain formation, a current pulse of amplitude ΔI. It appears, therefore, to be sensible to use for describing the performance of the Schottky-gate device, parameters which are related to ΔI. Such a parameter is the 'gate-trigger capability' [8.23, 8.26]

$$\left.\begin{array}{l} \gamma_G = \dfrac{\Delta I}{(\Delta V_G)_{\min}} = \dfrac{\Delta I}{(\Delta I_A)_{\min}} \cdot \dfrac{(\Delta I_A)_{\min}}{(\Delta E_A)_{\min}} \cdot \dfrac{(\Delta E_A)_{\min}}{(\Delta V_G)_{\min}} \\[2mm] \approx \alpha_{\max} \cdot (A\bar{\sigma}') \cdot g_t \end{array}\right\} \qquad (8.25)$$

where $\bar{\sigma}'$ is a mean value taken between E_{h0} and E_T but which approximately can be set equal to σ_1, and the third factor on the r.h.s. has been termed 'gate trigger factor' g_t. The quantities of minimum triggering current $(\Delta I_A)_{\min}$, minimum triggering electric field $(\Delta E_A)_{\min}$, and maximum current gain $\alpha_{\max}$ have been introduced and defined in connection with bias triggering of the two-terminal TE device in Section 8.2.2.2. Equation (8.20) indicates that the gate trigger factor g_t is here the important quantity describing the specific Schottky-gate triggering properties. Equation (8.25) suggests an equivalent circuit of the Schottky-gate TE device in the ON state as shown in Fig. 8.22. The quantity g_t can be obtained from eqns (8.22) and (8.24) by

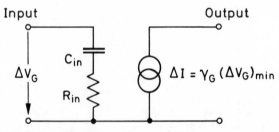

Fig. 8.22 Equivalent circuit of Schottky-gate TE device in ON state. After Yanai *et al.* [8.23a].

differentiation. The relationship between g_t and the normalized depletion-layer depth d_d is evaluated in Fig. 8.23. Parameters in the diagram are the voltage V_G across the gate and the pinch-off voltage, cf. [8.54],

$$V_{po} = ed^2 n_0 / 2\varepsilon. \qquad (8.26)$$

As an example, a device with $\alpha_{\max} = 20$, $R_1 = 500\ \Omega$, $n_0 = 10^{15}\ \mathrm{cm}^{-3}$, $d = 5\ \mu\mathrm{m}$, and $L = 100\ \mu\mathrm{m}$ yields $g_t \approx 3 \cdot 6 \times 10^2\ \mathrm{cm}^{-1}$ and thus $\gamma_G \approx 144\ \mathrm{mA\ V}^{-1}$. A typical experimental value is $\gamma_G \approx 40\ \mathrm{mA\ V}^{-1}$ [8.23a].

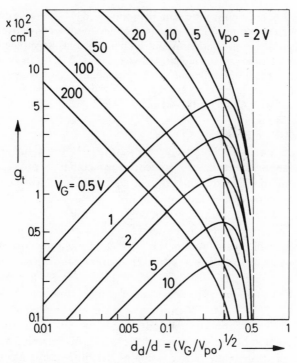

Fig. 8.23 Gate trigger factor g_t as a function of normalized depletion-layer depth d_d with voltage V_G across depletion layer and pinch-off voltage V_{po} as parameters. After Yanai *et al.* [8.23a]. [N.B. Read ($V_{df} - V_G$) for V_G]

For comparing Schottky-gate, and bias, triggering it is useful to define in analogy to eqn (8.25) a trigger-capability parameter also for bias triggering, viz.

$$\gamma_A = \frac{\Delta I}{(\Delta V_A)_{\min}} = \frac{\Delta I}{(\Delta I_A)_{\min} R_1} = \frac{\alpha_{\max}\sigma_1 A}{L}. \tag{8.27}$$

The ratio of the two capability parameters is then

$$\frac{\gamma_G}{\gamma_A} \approx \frac{(\Delta V_A)_{\min}}{(\Delta V_G)_{\min}} = Lg_t. \tag{8.28}$$

Introducing the numerical values of the above example for the device with $d = 5\,\mu\text{m}$, it follows $\gamma_G/\gamma_A \approx 3\cdot6$. Reducing the channel thickness to $d = 2\,\mu\text{m}$ yields even $\gamma_G/\gamma_A \approx 25$, i.e. the triggering amplitude at the Schottky gate can be lowered to 4% of that required at the anode (or cathode) in bias triggering.

Practical pulses obtained from TE devices naturally show jitter of the leading and trailing pulse edges due to thermal noise, bias-voltage fluctuations, etc. Particularly the thermal noise influences the instant of beginning and the duration of the domain formation process.

353

It is found [8.26b, 8.63] that pulse jitter is more pronounced in Schottky-gate Gunn devices than in two-terminal structures. The former type of device shows a relatively pronounced dependence of the domain nucleation instant on anode-bias voltage (fluctuations) if the bias conditions are adjusted to give a high trigger sensitivity [8.53]. An analysis for Schottky-gate devices, taking into account thermal noise, yielded a jitter of 1–6 ps for the leading pulse edge if typical sample parameters are chosen, whereas values lower by one or two orders of magnitude were found for two-terminal devices [8.26b 8.84].

8.2.3 Conduction-Path Profile

A whole class of functional Gunn devices, particularly those for generating fast pulses of various shapes (Section 8.3.2), relies on the fact that the propagation velocity of a mature dipole domain remains, over a wide range, almost constant against sample non-uniformities which the domain meets along its drift path. The domain acts therefore as a wall of constant current density. This important domain property arises from the special form of the basic $v(E)$ characteristic (Figs 2.8 and 2.9 for n-type GaAs) and will be analysed below. A second phenomenon of importance, relied on in this group of application, is the fact that no more than one fully-developed dipole domain can generally exist in a single Gunn device*. If a second domain nucleates, one of the two domains decays after a short time for thermo-dynamical reasons [3.3b, 3.7, 3.23, 3.50]. This property is directly related to the instability encountered with series-connected current generators since a mature domain is electrically represented by a current generator [8.57]; cf. Section 3.4.5.

Consider now a steady-state dipole domain travelling across a sample (Section 3.4.3). The instantaneous current density within the depletion layer of the propagating domain, being solely displacement ('charging') current, is in the one-dimensional limit given by eqns (3.9) and (3.15) as

$$J_{ch}(t) = \varepsilon \frac{\partial E(t)}{\partial t} = e v_D n_0. \tag{8.29}$$

Assuming a small domain width $b_\infty \ll L$, together with a not too abrupt z-dependence of doping density n_0 [3.26] and cross-sectional area A [8.61], the instantaneous total current through the sample becomes (because of current continuity)

$$I_t(t) = J_{ch}(t) A[z_a(t)] = e v_D[z_a(t)] n_0[z_a(t)] A[z_a(t)], \tag{8.30}$$

where the dipole-domain position is characterized by the accumulation-layer position $z_a(t)$.

This equation implies that the sample current depends linearly on the domain velocity, the net-doping density, and the cross-sectional area of the sample at the momentary position of the dipole domain which sweeps across the sample. We will show below that the domain velocity v_D is a quantity remaining almost constant

---

*It has been shown that two stable domains can exist simultaneously in multi-terminal Gunn devices of particular geometry [8.67].

irrespective of changes in sample properties. Consequently the current at any instant is, to first order, directly proportional to the momentary product of net-doping concentration n_0 and cross-sectional area A at the point the domain is just passing, cf. [8.1–8.3, 8.57, 8.58, 8.59–8.62, 6.127]. The product n_0A has been termed 'conduction-path profile' [8.1c, 6.88a].

Possible methods for influencing and varying the conduction path profile are [8.1c]:

(1) localized alloying or diffusion of impurity atoms along the drift path;
(2) varying the cross-sectional area of the drift path, cf. [8.2, 8.3, 8.57];
(3) illumination of parts of the drift path, cf. [8.55];
(4) localized impact ionization of deep donor levels due to high domain fields, cf. [3.32, 6.32a];
(5) penetration of electric fields from electrodes adjacent to the drift path.

A further possibility consists of influencing the 'effective conduction-path profile' by providing a suitable current shunt along part of the domain drift path. The domain, as a current generator, drives an additional current through this shunt. In one method:

(6) the shunt circuit is provided between additionally attached, and possibly switch-operated, small ohmic or Schottky-contact electrodes on the sample [8.2, 8.57].

In a different method:

(7) the sample surface is partly loaded by a distributed capacitance consisting of a dielectric layer and a metal plate [8.57, 8.63].

In method (7) the domain with its excess potential step Φ_D, when travelling underneath the metal plate, induces equal but opposite charges on the semiconductor surface at either side of the domain and on the metal. Charging and discharging of the metal plate is associated with a displacement current which adds to the sample current [8.64, 4.73]. The charging current can also be taken off an additional electrode on the metal plate. It can be shown [8.63, 8.64] that the charging current is, to first order, directly proportional to the capacity of the metal-electrode segment just above the domain. The temporal response of the current is, then, a fairly exact replica of the physical shape of the metal plate. The influence of the distributed capacitance on device behaviour is investigated in more detail in Section 8.3.2.2.

In methods (1), (2), and (7) the conduction-path profile is determined during the manufacture of the device and, therefore, the information content remains fixed to perform a specific function. Methods (3) to (6), on the other hand, permit to change the profile, and consequently the current waveform, during operation.

Methods (2), (5)–(7) pose the least technological difficulties, and they have successfully been applied, for example, for realizing fixed-waveform and controllable-waveform generators (Section 8.3.2), and coding devices or analogue-pulse converters (Section 8.4.5). These devices are preferably constructed in integrated-

355

circuit form, and the name 'domain-originated functional integrated circuits' (DOFIC) has been proposed for them [8.1c].

More sophisticated applications and extensions of the DOFIC concept, however hardly put into practice yet, include multi-terminal arrangements for high-speed time-division multiplexing (as an equivalent to a delay-line system) and two-dimensional arrays or electronic components which are sequentially scanned or actuated by travelling domains [6.106, 8.65]. For instance, applying method (6) above, a number of contacts can be attached along the sample and connected to a resistive shunt path. Whenever a domain passes underneath the shunt path some additional current is forced by the domain to flow through the shunt, increasing the total anode current of the sample. In this way the domain is able to 'read out' the state of external switches.

Rather interesting prospects arise from the interaction between domains and light. By making use of method (3) above, a DOFIC solid-state camera can be envisaged, and by employing the recombination radiation which may be produced in method (4) a DOFIC display is potentially feasible [8.1c, 8.66]. Both these opto-electronic DOFIC devices would have a considerably simpler structure than solid state imaging and display devices of the mosaic type which necessarily require complex circuits for addressing the mosaic.

Many practical systems possess a degree of complexity which would necessitate the use of one or even a few rather long DOFIC Gunn devices to realize and accommodate all the particular functions required. Such device lengths can become inconvenient, not the least for thermal reasons (cf. Section 6.3). The design problem may then often be solved easier by employing a larger number of individual short Gunn devices which are operated or addressed sequentially cf. [8.14, 8.24; Section 8.4.5] instead of using one or a few long DOFIC devices carrying the required large number of terminals or complex information content along it. As an example, a fast time-division multiplexing circuit has been proposed [8.24] which employs several monostable AND gates containing one short uniform Gunn element each as the essential component (cf. Sections 8.4.1, 8.4.2), rather than using the one long multi-terminal Gunn device as mentioned above for this application [6.106, 8.65]. Input 1 of the AND gates receives a sequentially delayed clock pulse (continuous pulse train) whereas the binary information to be time-multiplexed is applied to the input terminals 2, delivering the multiplexed information as a stream of short Gunn pulses across a common load resistor of the gates.

Now it remains to prove the approximate constancy of the domain velocity v_D on which the principle of the conduction-path profile is based. Consider a sample consisting of two sections (a) and (b) with differing doping $n_{0,a} > n_{0,b}$, see inset to Fig. 8.24. A domain is conveniently described by the characteristical eqn (3.24) which in our case of stable propagation reads

$$\Phi_{D\infty,i} = \frac{\varepsilon}{2en_{0,i}} (E_{2\infty,i} - E_{1\infty,i})^2; \quad i = \text{a,b} \tag{8.31}$$

with the subscript to be taken for that section in which the domain is just travelling.

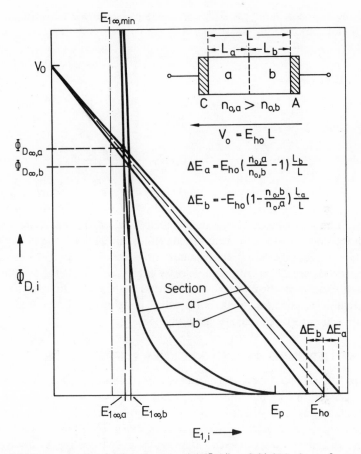

Fig. 8.24 Demonstration in domain-potential (Φ_D)/low-field (E_1) plane of approximate invariance of steady-state low field $E_{1\infty}$ ($\propto$ domain velocity v_D) against changes in net-doping concentration n_0; $i = $ a, b.

The two characteristics relevant to our case, as obtained from eqn (8.31), are schematically shown in Fig. 8.24 (cf. Fig. 3.13). For small values of $E_{1\infty}$ the curves tend to merge, asymptotically approaching $E_{1\infty,\text{min}}$. This is a consequence of the specific relationship between $E_{1\infty}$ and $E_{2\infty}$ (Butcher's equal-areas rule) resulting from the basic $v(E)$ curve. The domain velocity v_D is determined by $E_{1\infty}$ according to eqn (3.25). The appropriate value of $E_{1\infty}$ can be obtained from the intersection between domain characteristic and the device line (working point W) as we know from Section 3.4.4.2, cf. Fig. 3.13. For our device with a doping step, the device-line eqn (3.23) of the one-dimensional case must be modified and reads, cf. [8.60],

$$E_{h0}L = E_{1,a}L_a + E_{1,b}L_b + \Phi_{D,i}. \tag{8.32}$$

357

By using the expression for the drift-current density outside the domain both in sections (a) and (b), viz.

$$J = en_{0,a}\mu_a E_{1,a} = en_{0,b}\mu_b E_{1\ b}$$

it follows with $\mu_a \approx \mu_b$

$$n_{0,a}E_{1,a} \approx n_{0,b}E_{1,b}$$

and thus eqn (8.32) becomes

$$E_{h0}L = (n_{0,b}L_a + n_{0,a}L_b)\frac{n_{0,i}}{n_{0,a}\ n_{0,b}}\ E_{1,i} + \Phi_{D,i}. \qquad (8.33)$$

The straight device lines for our two differently-doped sections are schematically inserted in Fig. 8.24. It is seen that $E_{1\infty,a}$ and $E_{1\infty,b}$ as given by the intersections, differ only little. Accordingly, the domain velocity v_D changes (rises) hardly when the domain is passing from section (a) to section (b) of our device.

If the doping concentration is uniform but rather we have device sections differing in cross-sectional area A, the domain velocity is the same while the domain travels within each section of constant cross-section. Here, the one-dimensional domain dynamics can be applied for each section. Only in regions of abrupt area transition the domain will experience velocity changes [8.68].

8.3 PULSE GENERATION AND REGENERATION

In Sections 8.3 and 8.4 we describe particular devices and circuits in which the basic concepts of functional Gunn-domain electronic as outlined above are put to use for specific applications. The intention here is not to provide an exhaustive coverage of all the proposals and realizations that have appeared so far. Rather, some examples are chosen and investigated in each category, thought to be representative and to show the underlying principles particularly well.

8.3.1 Switching and Pulse Regeneration

8.3.1.1 *Introductory Remarks*
An electronic device capable of externally-triggered switching from a high-current to a low-current state, and vice versa, constitutes a basic element of numerous circuits for logic operations, memory functions, pulse regeneration, etc. The tunnel diode, allowing fast switching from a high-current/low-voltage condition to a stable opposite operational state, is an example of this class of device.

In a TE device with its markedly higher current and voltage levels than those of the tunnel diode, switching is obtained when the current of the device drops and the voltage across it increases as a result of dipole-domain formation (Fig. 3.18). This process was illustrated in Fig. 8.13 (cf. Fig. 8.2) for a resistively-loaded and sub-critically biased device of supercritical n_0L product, exposed to a short bias-trigger pulse. As we know, in the general case the current reverts to its pre-trigger value after the domain transit time $\tau \approx L/v_\mathrm{D}$. This means that the low-current state is available only during this time interval. This type of switching is therefore *monostable*.

Suggested by the form of the static $v(E)$ characteristic (Fig. 2.8) one might in analogy to the LSA mode of operation imagine a possible *bistable* switching process by triggering the device to a homogeneous high-field state with associated low current density [3.66, 8.6]. This would require that the homogeneous field increase takes place faster than the domain formation and, furthermore, that the $v(E)$ curve possesses, above the peak field, a second range with positive differential mobility, if only of small magnitude, to provide a stable working point (cf. Fig. 2.8). However, even in this case the high-field state is stable only for a limited time. Since, normally, we have a zero-field boundary condition at the cathode (injection of electrons) and thus necessarily a disturbed field distribution immediately in front of it, the (quasi-) homogeneous field state switches back to the original sub-critical low-field state after a time of the order of the transit time L/v of the injected electrons. Although particular doping profiles or layered structures at the cathode for field shaping may provide a solution here, a bistable homogeneous-field switching has not been realized so far. On the other hand, a true bistable switching process of a different nature, albeit not yet fully understood, has been discovered more lately [4.48, 4.86-8.13]. It will be described below after having first dealt with the more conventional monostable domain switching and its applications.

8.3.1.2 *Monostable Operation*

The simplest circuit for domain switching consists of the basic series arrangement of a super-critically-doped Gunn element and an ohmic load resistor R_L as indicated in Fig. 8.25a (cf. Fig. 8.1.) Here, we have bias triggering (Section 8.2.2.2),

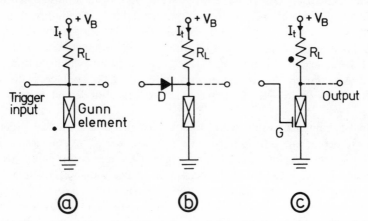

Fig. 8.25 Basic switching circuit (a), and extended versions to provide undirectivity by employing external gating diode D (b) or Schottky gate G (c).

and the initial drop to a lower total current I_t can be taken off across R_L or the Gunn element as a corresponding voltage change (cf. Fig. 8.1). Theoretical and practical response times, including delay and fall times, have been discussed in Section 8.2.1, cf. Figs 8.2–8.5; there we found theoretical current fall times t_f in the 10–100 ps range with t_f being approximately proportional to the square root of R_L. More superior circuits which provide undirectivity by using an additional external gate diode [8.34] or a Schottky-contact gate (Section 8.2.2) are shown in Figs 8.25b and c. As already mentioned, the use of both the external gate diode and the Schottky gate reduces the switching speed, but this deterioration should be negligible for most practical purposes [8.24, 8.23b]. Since the domain triggering process is rather sensitive with regard to the form of the trigger pulse (Section 8.2.2), it has been proposed to use the domain switching process for pulse discrimination [3.66, 8.6]. In experiments current changes of up to 1 A into a 50 Ω load have typically been achieved, cf. [5.19, 8.42a]; there being, however, a trade-off between achievable power and speed (Fig. 8.8a). Examples of the current response in the case of bias triggering have been given in Fig. 8.2. Oscillograms of the response obtained in preliminary experiments with Schottky-gate triggering are reproduced in Fig. 8.27b.

In the case of monostable domain switching discussed here the current switches back to the original state when the domain reaches the anode so that an approximately rectangular pulse is produced in response to a weak triggering signal of short, however sufficiently long, duration. As an important application of TE devices, this property can be used for pulse amplification, cf. [1.27, 5.19, 8.8a] with the additional advantages of possibly improving the pulse edges [8.8a] and of a limiting action [8.70]. The subcritically-biased Gunn element with provision for domain triggering is thus an excellent device for regenerating nanosecond and subnanosecond pulses, cf. [8.23b, c, 8.42b] as already demonstrated by Fig. 8.2b, and applications as regenerators or repeaters [8.23, 8.26] in pulse communication systems have been proposed [8.42a, 8.71]. The basic series-type circuits of such a pulse regenerator are those shown in Fig. 8.26, with an output terminal added. The unidirectional circuit of Fig. 8.25c obtained by employing a Schottky gate is here particularly attractive. Circuits (a) and (b) require a positive input signal whereas circuit (c) must be triggered by a negative signal. In all three cases a positive-going output pulse is obtained. A negative pulse, on the other hand, can be taken off across the load if it is put at the other side of the Gunn element between cathode and ground (see Fig. 8.1b).

The pulse gain achievable with subcritically-biased Gunn-effect regenerators has already been dealt with in Sections 8.2.2.2 and 8.2.2.3 in connection with the sensitivity of the mechanisms of bias triggering and Schottky-gate triggering, respectively. There, in particular the maximum current gain α_{max} (cf. Fig. 8.16) for bias triggering and its modification by the gate triggering factor g_t in the case of Schottky-gate triggering (Fig. 8.23) had been investigated. It was found that the minimum trigger-field value $(\Delta E_A)_{min}$, determined primarily by diffusion noise, sets the ultimate upper limit for the achievable gain. One can show [8.44] that the

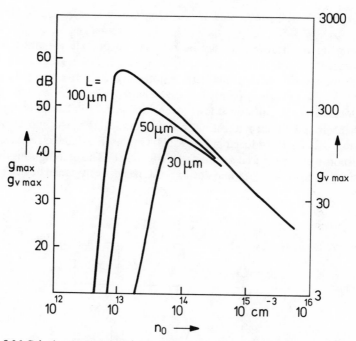

Fig. 8.26 Calculated maximum pulse gain g_{max} (pulse voltage gain $g_{v\ max}$) of bias-triggered GaAs element against net-doping concentration n_0 for various sample lengths L. After Hartnagel [8.44].

maximum pulse voltage gain, being the ratio of maximum output pulse voltage to applied input voltage, is

$$g_{v\ max} = \frac{E_T}{2(\Delta E_A)_{min}\left[\dfrac{L}{R_{L\ max}\,A\,\mu_1\,en_0} + 1\right]} \qquad (8.34)$$

for bias-trigger operation, with $R_{L\ max}$ denoting the maximum value of the load resistance allowable for full domain growth during the transit time. An evaluation of eqn (8.34) for GaAs devices is given in Fig. 8.26 as a function of sample doping n_0 and length L. Maximum gain is obtained for large values of L and not too high doping levels. Experimentally, pulse gains of from 13·5 dB ($L = 200\ \mu\text{m}$, $n_0 = 5 \times 10^{14}\ \text{cm}^{-3}$, $R_1 = 100\ \Omega$, $R_L = 50\ \Omega$, $\tau_f \approx 100\ \text{ps}$) to about 30 dB ($L = 500\ \mu\text{m}$, $n_0 = 5 \times 10^{14}\ \text{cm}^{-3}$, $R_1 = 100\ \Omega$, very large R_L) have been measured [5.19, 8.42b, 8.69]. For devices employing Schottky-gate triggering the maximum gain is simply given by $g_{v\ max} \approx \gamma_G R_L$ where R_L denotes the load resistor and γ_G the gate trigger capability factor according to eqn (8.25). Experimentally, Schottky-gate GaAs devices showed pulse gains near 14 dB for input-voltage amplitudes of 0·2 V

($L = 100$ μm, $w = 100$ μm, $d = 2$–5 μm, $\Delta z_G/L = 0\cdot1$, $n_0 = 3\cdot2 \times 10^{15}$ cm^{-3}) [8.53].

For particular applications it may be required to operate a number of pulse amplifiers/regenerators in series or in parallel. Figure 8.27a shows the circuit diagram of *series*-connected Gunn elements with Schottky gate ('active line'), [8.26]. The measured response at various points in such a chain of GaAs devices with the load resistors located at the cathode side is given in Fig. 8.27b. The photograph of a 4-element integrated-circuit realization, on semi-insulating GaAs substrate, of the circuit of Fig. 8.27a is reproduced in Fig. 8.28a. On this integrated circuit a cathode-current response was measured as shown by Fig. 8.28b. The oscillograms displayed were obtained from device no. 1 and device no. 4, respectively, when device no. 1 was

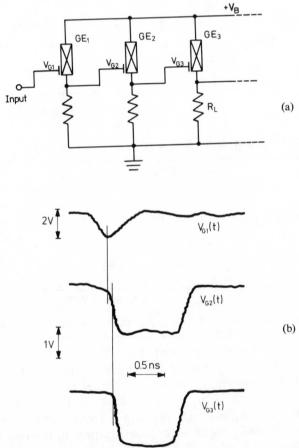

(a)

(b)

Fig. 8.27 (a) 'Active line' made up of series-connected Gunn elements with Schottky gate. (b) Temporal response of gate voltages obtained from practical circuit. After Sugeta and Yanai [8.23b, 8.26b].

362

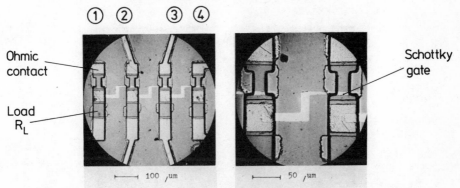

Fig. 8.28a Photograph of series-connected Schottky-gate GaAs Gunn devices ('active line') as integrated circuit. Au-Ge for ohmic contacts; Cr for Schottky contacts. By courtesy of K. Mause, Fernmeldetechnisches Zentralamt der Deutschen Bundespost; also [8.39].

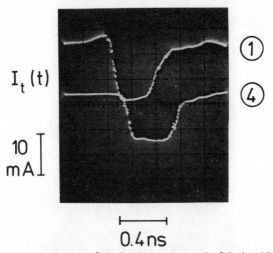

Fig. 8.28b Temporal response of total cathode current I_t of devices No. 1 and No. 4 in circuit of Fig. 8.28a when first device was bias-triggered. By courtesy of K. Mause [8.39].

bias-triggered. The overall pulse delay of approximately 160 ps indicates a delay per stage of about 40 ps.

A corresponding integrated GaAs circuit, however, with *parallel* feed of three Schottky-gate Gunn devices, is shown in Fig. 8.29a. The associated cathode-current responses in Fig. 8.29b, obtained if device no. 1 was bias-triggered, do not exhibit time delay, at least within the measuring accuracy which amounted to 20 ps. Also reproduced in Fig. 8.29b is the current response of the first device without it being loaded by the three gates. One notices that the 'predominantly capacitive' loading by the gates causes pulse broadening.

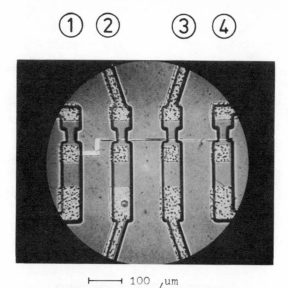

⊢————⊣ 100 ₋um

Fig. 8.29a Photograph of integrated-circuit version of three Schottky-gate GaAs Gunn devices being parallel fed from preceding two-terminal Gunn device. Au-Ge for ohmic contact; Cr for Schottky contacts. By courtesy of K. Mause [8.39].

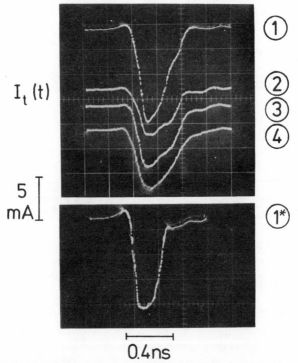

$I_t(t)$

①
②
③
④

5 mA

①*

0.4ns

Fig. 8.29b Temporal response of total cathode currents I_t of devices Nos 1–4 in circuit of Fig. 8.29a when device No. 1 was bias-triggered. Trace 1* shows response of device No. 1 when not loaded by the Schottky-gate devices. By courtesy of K. Mause [8.39].

It has been proposed to trigger semiconductor laser diodes by the current pulse generated in Gunn elements of high n_0L product [8.42b, 8.72]. Suitably-designed coplanar-type GaAs Gunn elements theoretically deliver under continuous pulse-train operation currents with a d.c. component of up to 800 mA and a pulse-signal amplitude of 700 mA [8.72]. These current values exceed the threshold current of modern laser diodes, particularly of the heterojunction laser of advanced geometry. If, now, such a Gunn element and a laser diode are operated in parallel and the system is supplied with a constant-current bias, then a Gunn domain which is made to nucleate will force more current through the laser diode resulting in lasing action during the period of domain transit. The main advantage of using a Gunn element for switching the laser in this way appears to be the possibility (a) of electronic processing of the pulse signals as described in the course of this chapter, and (b) of switching the relatively high currents at high speed. Current switching of semiconductor lasers has also been proposed by using Gunn elements operating in the bistable switching mode discussed below (Section 8.3.1.3) [8.78]. In preliminary experiments rise times and fall times down to 200 ps and 400 ps, respectively, have been achieved for switching between lasing levels of GaAs double heterojunction lasers.

8.3.1.3 *Bistable Operation*
In logic and other circuitry devices capable of bistable switching are very frequently employed. More recently it could be shown [8.13] that a TE device which satisfies particular requirements also possesses bistable switching properties, at sub-nanosecond speed.

Short sandwich-type GaAs elements of super-critically doping-length product and operated in a constant-voltage circuit show switching to a stable lower-current state although the simplified small-signal theory (Section 4.4.1.2) predicts current instability for such devices. When the sample bias voltage is increased above the critical value $V_T = E_T L$ the current of the device drops to the lower stable value. The result of a preliminary experiment is shown in Fig. 8.30, indicating a current decrease from 0·81 A to 0·63 A with an associated voltage increase from 3·5 V (V_T) to 7 V for the case of a 25 Ω load. By shortly lowering the sample bias below V_T the device could then be switched back to the original high-current/low-voltage state. Switching times down to 100 ps have been obtained so far [8.13b, c]. The original quasi-ohmic high-current state is associated with a fairly uniform electric-field distribution along the sample. When the bias is increased to the threshold V_T a high-field domain develops near the cathode and the current drops. This domain travels then to the anode in the usual way (Section 3.4.4). However, as the characteristical feature of this type of operation [8.13, 4.48, 4.86], it remains *stationary* there (see inset to Fig. 8.30), thus preventing the formation of a new domain at the cathode and keeping the current at the low value. The switching time is determined by the domain formation time, which indicates minimum switching times of the order of 10 ps for *dipole* domains developing, according to Section 8.2.1 (see Fig. 8.5). If only *accumulation* layers could be made to form, switching times of well less than 10 ps may be expected [8.13a]. A practical bistable TE switch of the kind discussed here should preferably be fitted with set and reset electrodes which might possibly be formed by Schottky

365

gates. A device of this configuration would constitute a true operational equivalent to a flip-flop circuit, yet with the property of yielding considerably higher switching speeds than present-day transistors permit.

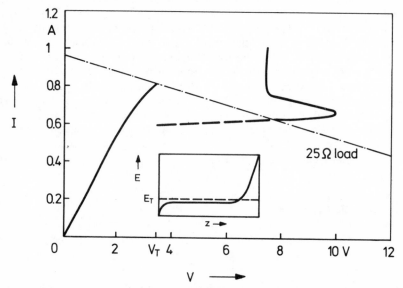

Fig. 8.30 Static current I against terminal voltage V for a stable GaAs Gunn element ($n_0 = 3 \times 10^{15}$ cm^{-3}, $L = 10\,\mu$m). Onset of avalanche breakdown for $V > 10$ V. Inset shows spatial electric-field distribution in low-current state. After Thim [8.13a].

The nature of the observed bistable switching phenomenon involving a stationary high-field layer at the anode is not yet completely understood (cf. last paragraph in small-print part at end of Section 4.4.1.2). One proposal for an explanation [4.48] requires that the charge readjustment in the accumulation layer of the domain approaching the anode takes place quicker than the movement of the layer into the anode contact (velocity v_a). From the discussion of the transient dipole-domain behaviour in Section 3.4.4.3 we know that the domain potential Φ_D diminishes when the domain passes into the anode. This leads, under the constant-voltage condition assumed there, to an increase in outside field E_1 (drift current) and consequently to a recharging of the rest domain (see Fig. 3.19). This effect may eventually prevent E_1 in the up-stream part of the sample from reaching again the threshold value, which means that no new domain can form at the cathode. Rather, the process settles in a stationary state with the accumulation layer of the dipole domain being trapped at the anode due to its particular geometry. Also, there seems to be evidence that the dipole determined by diffusion. A condition describing the stationary situation is obtained [4.48] if the time for the charge readjustment is taken as the (negative) dielectric relaxation time τ_R and the layer width is assumed to be a few (e.g. three) Debye lengths L_{Db}, thus with eqns (3.3a) and (3.3b) yielding [cf. eqn (4.54)]

$$\frac{\varepsilon}{e|\mu'|n_0} \lesssim \frac{3}{v_a}\left[\frac{\varepsilon D}{e|\mu'|n_0}\right]^{1/2}. \tag{8.35}$$

For n-type GaAs one obtains from expression (8.35) a required doping concentration $n_0 \gtrsim 5 \times 10^{14}$ cm^{-3}. Further conditions to be fulfilled are [8.13] a doping-length product greater than the critical value for domain formation ($\gtrsim 5 \times 10^{11}$ cm^{-2} for GaAs, see Section 4.4.1.2), an amount of doping fluctuations below a certain critical value ($\lesssim 8\%$ for GaAs and InP) and, finally, a $v(E)$ characteristic which does not show a steeply rising high-field branch. On the other hand, as already briefly mentioned in the last paragraph of Section 4.4.1.2, it was concluded from *small-signal* calculations [4.47a, 4.86] that the bistable switching mechanism could be based on the appearance of an absolute (temporal) instability. At sufficiently high doping levels the absolute instability is obtained for purely imaginary frequencies from the dispersion relation. Because of the boundary conditions involved, the found exponential rise in amplitude leads to a stationary state. The boundary conditions imply a higher electric field at the anode because of the diffusion tail. There, then, is the nucleation centre for the absolute instability. One can show [4.86] that the transition to the stable state occurs with or without the intermediary formation of a travelling dipole domain.

A kind of bistable switching was experimentally also obtained in thin co-planar epitaxial GaAs devices of a type similar to that shown in Fig. 6.6f but with the active n-layer extending right underneath the metal contacts and, as an essential feature, with an isolated metal shield covering the semiconductor surface region in front of the anode [8.55, 8.56]. Current changes of even up to 60% could be measured. In this case the underlying mechanism may again be different since the device exhibited a doping-thickness product ($n_0 d \approx 7 \times 10^{11}$ cm^{-2}) near the critical value (see Section 4.4.2.2) and, probably most important, the co-planar device investigated possessed high d.c. electric field regions at the cathode and the anode due to its particular geometry. Also, there seems to be evidence that the dipole domain launched when the switching voltage is applied, does not settle in a completely stationary state but rather is merely slowed down due to trapping effects [8.58].

8.3.2 Pulse Generators

8.3.2.1 *Simple Pulse-Train Generators*
The simplest possible form of pulse-train generator results if a TE element of super-critical doping-length and doping-width products is biased into the dipole-domain transit-time mode and provided with an ohmic load resistor. In this case a train of voltage spikes, of the kind shown in Fig. 3.20, can be taken off across the load resistor, with the pulse repetition frequency f_r being equal to the domain transit-time frequency $f_\tau \approx v_D/L$ (see Section 3.4.4.3). The repetition frequency is, thus, determined by the sample length L and may easily be made to lie in the microwave frequency range for a steady-state domain velocity $v_D \approx 10^7$ cm s^{-1} for GaAs (see

Fig. 3.9). Because of the slight dependence of v_D on the bias voltage, a bias-depending frequency variation is obtained over a restricted range [see Fig. 3.10 and equation (3.29)]. In the example of Fig. 3.20 (pulsed bias) the repetition frequency is only $f_r \approx 0.455$ GHz for the particular value $L = 220 \ \mu$m. For sufficiently short sample lengths (as suited for C.W. bias operation) the domain transit-time becomes comparable to the domain formation and extinction times and the generated pulse waveform approaches a train of sine half-waves. Pulse widths of a few 10 to some 100 ps are normally obtained (see Section 8.2.1 on response times). Experimental peak pulse powers for the case of intermitted bias operation (up to 1 μs duration) reach some 10 W with associated voltage amplitudes of some 10 V. To obtain voltage spikes of particularly high amplitude, the basic pulse generator circuit is preferably extended by inserting an inductor between the low-impedance bias source and a parallel combination of Gunn element and ohmic load resistor [8.73]. As in similar tunnel-diode circuits, the series inductor causes the domain voltage to switch to a high value while maintaining almost a constant supply current. A train of 1 ns pulses having amplitudes exceeding 200 V (ten times the Gunn threshold voltage in the particular example) have been obtained across a 3 000 Ω load in this way [8.73].

According to what has been said above on the repetition frequency, a pulse-train generator with low repetition rate (100 MHz and lower) requires Gunn elements with lengths that are prohibitive for thermal reasons (see Section 6.3). However generators for pulse trains with repetition rates of up to an order of magnitude lower than the reciprocal domain transit time can be constructed using short C.W.-operated devices by inserting them into a relaxation mode circuit containing, again, an inductor [5.19, 8.74]*. A possible configuration of such a circuit, with the inductor in parallel to a load resistor $R_L \gg R_1$ (R_1 denotes low-field resistance of device), is shown in Fig. 8.31 together with the generated voltage waveform which

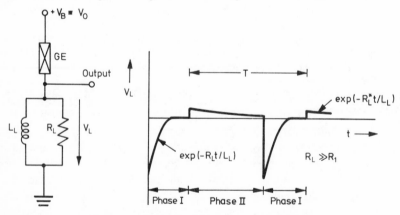

Fig. 8.31 Circuit diagram of relaxation-type pulse-train oscillator and voltage waveform obtained across load resistor. After Fisher [5.19].

*Such a configuration can be regarded as operation in the delayed dipole-domain mode (cf. p. 130).

can be determined from a simplified network transient analysis. For this analysis the oscillation cycle is divided into two linear phases governed by straight-line static and dynamic current-voltage characteristics (cf. Fig. 3.15b), respectively. The formation time of a domain is sufficiently short for the inductor L_L to represent an open circuit during the formation period, and a (negative-going) pulse is obtained as the useful output (phase I). After the short time L_L/R_L has passed, the inductor, however, acts as a short circuit so that the voltage across R_L becomes zero. It remains there until the domain reaches the anode and a (positive-going) voltage step is produced. Now the stored energy in L_L is removed via $R_L^\star$, being the parallel combination of R_1 and R_L (phase II). For $R_L \gg R_1$, this is a slow process, having the effect of delaying the nucleation of a second domain. A finite voltage during phase II is necessary for circuit operation but it should be kept as low in value as possible. For the time duration of phase II the simplified analysis yields [5.19]

$$t_1 - t_0 = \frac{L_L}{R_L^\star} \ln \frac{\Delta I(1 - R_L^\star/R_L)}{\left[\dfrac{V_0}{R_1} - I_T\right]} \tag{8.36}$$

where I_T is the threshold current (end point of static characteristic). Equation (8.36) indicates that the relaxation oscillator exhibits voltage tunability because of the dependence of $t_1 - t_0$ on sample bias voltage V_0.

A pulse generator delivering approximate square waves can be constructed [8.75] by connecting a short section of open-ended transmission line to the load resistor as shown in Fig. 8.32, case (a). The Gunn element is supplied with a constant-

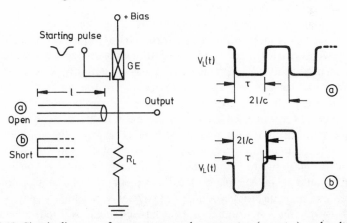

Fig. 8.32 Circuit diagram of square-wave pulse generator (repeater) and voltage waveform obtained across load resistor. Adapted from Boccon-Gibod [8.75], and from Sugeta *et al.* [8.23].

voltage bias somewhat below the threshold V_T but above the sustaining value $V_S^\star$ and a 'starting pulse' is provided at the Schottky trigger gate to initiate a first domain. The leading edge of the generated first Gunn pulse will be reflected at the open end of the transmission line with the same polarity and arrive back at the cathode side of the device with a time delay determined by the line length *l*. Assuming the relation

$$\tau < 2l/c \tag{8.37}$$

between domain transit time τ and the time delay on the line ($c =$ phase velocity of wave on line), the returning pulse then causes a second domain to be launched, and so on. This process of domain formation, and the associated pulse generation, therefore repeats after the specific delay time. (Circuits of this kind are called mono-polar 'pulse-repeater' circuits.) The pulse-to-space ratio is determined by the values of the device length *L* and transmission-line length *l*. The repetition frequency can be altered by changing *l*. The upper frequency limit is given by the inherent oscillation frequency f_τ of the Gunn element. If the transmission line is provided with a *short* circuit at its end according to case (b) in Fig. 8.32, the reflected wave possesses *opposite* polarity with respect to the forward wave and a (single-event) bipolar pulse repeater is obtained [8.23].

The generator of Fig. 8.32 has the inconvenience that the pulse-train obtained is superimposed on a d.c. level as a consequence of the shape of the fundamental $v(E)$ characteristic of TE semiconductors (see Figs 3.15, 3.18). The d.c. component can be eliminated by connecting in parallel with the load a transmission-line section, cf. [8.75], which is terminated by a short-circuit or an inductor. This line shorts out the d.c. component but has a high impedance for the fundamental square-wave frequency and a number of its higher harmonics if the length of the line is chosen appropriately.

8.3.2.2 *Complex-Waveform Generators*

The pulse generators considered so far employ Gunn elements with supercritical parameters regarding doping level and geometrical dimensions, however these parameters were—or, at least, could be—uniform across the sample length. Now, we have seen in Section 8.2.3 that the current of a TE sample in which a steady-state dipole domain propagates is directly proportional to the conduction-path profile (doping concentration times cross-sectional area) at the point the domain is just passing, a fact that can be used for generating pulse trains of more or less complex waveform (sawtooth, staircase, etc.) at extremely fast rates. By employing modern sophisticated technological processes like ion implantation or high-resolution electron-beam masking for photo-etching, almost arbitrarily complex current waveforms should be producible by locally influencing in an appropriate way the doping of the sample, its cross-section, or the '*effective* conduction-path profile' (for example by providing a distributed outside capacitance), as listed in Section 8.2.3. Yet, to demonstrate the principle, we shall in the following restrict our description to a few examples of functional generators of this kind which possess only moderate complexity. These examples comprise two generators the conduction-path profiles of which are fixed once for all during their manufacture, and one case in which it can still be altered during operation.

A sawtooth current waveform is obtained from a linearly tapered device made from a uniformity-doped TE material [8.3, 8.57] as shown in Fig. 8.34*. In the usual way the waveform can be taken off an ohmic load resistor as a proportional voltage. The measured waveform of such a tapered device at a super-critical bias is reproduced in Fig. 8.45. A similar behaviour is obtained with an annular device geometry requiring concentric electrodes [8.77]. In the devices discussed here the cross-section changes along the sample, cf. [8.76], but the doping concentration is constant. The opposite situation would in principle, of course, provide the same result. In Fig. 8.33 the bias voltage and the cross-section at the anode have been

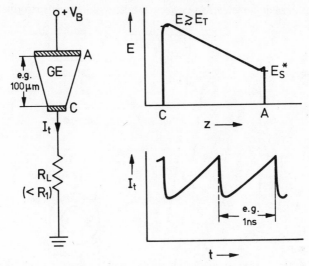

Fig. 8.33 Tapered Gunn-element (GE) sawtooth generator. Device shape, electrical circuit, spatial dependence of electric field E in device just before domain formation, and total-current waveform $I_t(t)$ obtained.

$$I_t(t)$$

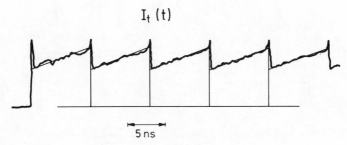

Fig. 8.34 Sawtooth current waveform experimentally obtained from tapered GaAs Gunn element. After Shoji [8.57].

---

*A somewhat more elaborate analysis indicates that the obtained waveform rather displays a slight upward concavity for a linear taper [8.61].

371

chosen such that the electric field, which decreases linearly along the sample, reaches the domain sustaining value $E_s^\star$ just at the anode. In this way no overshoot pip occurs and the maximum possible modulation ratio of the current is obtained. By varying the bias voltage, the point at which the field has dropped to $E_s^\star$ and an arriving domain has dissolved, can be shifted across the sample, thus allowing a control of the transit-time frequency [6.126, 6.127, 8.59, 8.79]. In this manner the repetition frequency of the simple spike-train (now: saw-tooth) generator of Section 8.3.2.1, operating into a non-resonant load, can be made voltage-tunable over a wide range. The same frequency tuning effect is obtained in a sample of *uniform* cross-sectional area if an electron-concentration gradient is introduced by main-

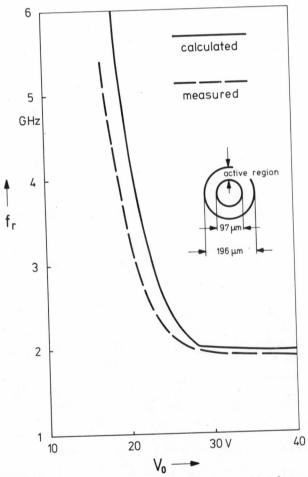

Fig. 8.35 Calculated and measured dependence of pulse repetition frequency f_r on terminal voltage V_0 for annular-shaped GaAs Gunn element. After Newton and Bew [8.79].

taining a temperature gradient [8.80]. Examples of the calculated and measured bias dependence of the frequency are shown in Fig. 8.35 for an annular coplanar-type GaAs device. Extending the case to a double-tapered device, in which the cross-section first increases from the cathode side to the middle and then decreases again towards the anode, an abrupt jump to a lower frequency occurs when the bias voltage exceeds a certain value [8.57].

Employing the concept of shaping the 'effective conduction-path profile' (Section 8.2.3) a functional generator can be constructed by partly covering a Gunn element with a metal plate of suitable geometry but isolated from the semiconductor by a high-resistivity (ohmic) layer. As the useful output, either the device cathode current is employed which is altered by the capacitive loading (Figs 8.36a and b), or the

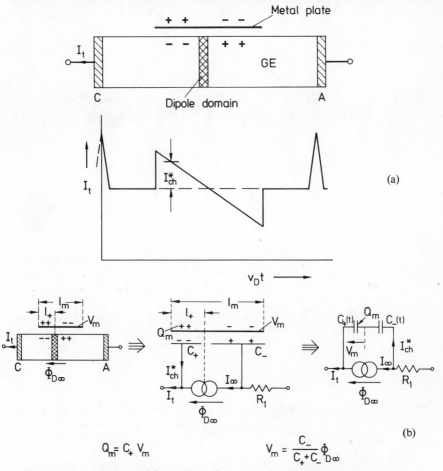

$$Q_m = C_+ V_m \qquad\qquad V_m = \frac{C_-}{C_+ + C_-}\Phi_{D\infty}$$

Fig. 8.36 Gunn device with isolated metal plate. (a) Configuration, and time dependence of charging component I_{ch}^* in total device current I_t (after Hofmann [8.64]). (b) Derivation of equivalent circuit.

charging current is taken off the capacitive electrode itself (Figs 8.37a and b). In the first case, depicted in Fig. 8.36a for an isolated metal electrode of length l_m and constant width w_m, the total device current I_t contains a component due to the charging of the plate when the dipole domain passes underneath it. Figure 8.36b interprets the derivation of an appropriate equivalent circuit of the configuration. The charging current $I_{ch}^{\star}$ is determined in the following way, cf. [1.33, 4.73, 8.64].

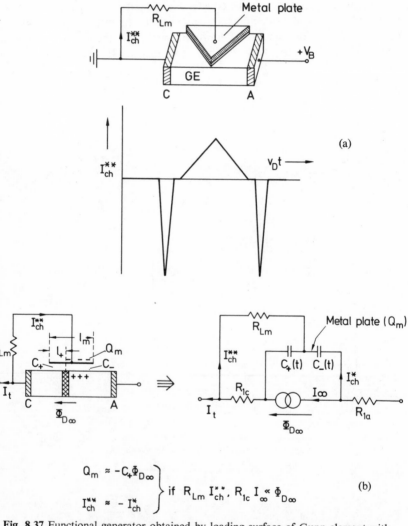

$$Q_m \approx -C_+\Phi_{D\infty}$$
$$I_{ch}^{\star\star} \approx -I_{ch}^{\star}$$
$$\left. \right\} \text{ if } R_{Lm} I_{ch}^{\star\star}, R_{1c} I_\infty \ll \Phi_{D\infty}$$

(b)

Fig. 8.37 Functional generator obtained by loading surface of Gunn element with isolated metal plate of suitable geometry. (a) Configuration, and time dependence of external charging current $I_{ch}^{\star\star}$ (adapted from Shoji [8.57, 8.63]). (b) Derivation of equivalent circuit.

For the positive charge on the metal plate we can write (see Fig. 8.36b)

$$Q_m(t) = C'w_m l_+ V_m \tag{8.38a}$$

where C' denotes the capacitance per unit area and l_+ the length of the positively charged part of the plate. The voltage V_m induced on the plate is

$$V_m = \Phi_{D\infty} (1 - l_+/l_m)$$

where $\Phi_{D\infty}$ denotes the domain excess potential. Thus one obtains

$$Q_m(t) = C'w_m \Phi_{D\infty} l_+ \frac{l_m - l_+}{l_m}. \tag{8.38b}$$

If we assume $l_+ = v_D t$ with a constant domain velocity v_D, the charging current $I_{ch}^\star = dQ_m/dt$ becomes in case of a constant plate width w_m

$$I_{ch}^\star = C'w_m \frac{d}{dt}\left(\Phi_{D\infty} l_+ \frac{l_m - l_+}{l_m}\right) =$$

$$= \frac{C'w_m \Phi_{D\infty} v_D}{l_m} (l_m - 2l_+) + \frac{C'w_m}{l_m} l_+ (l_m - l_+)\frac{d\Phi_{D\infty}}{dt}. \tag{8.39}$$

The first term on the r.h.s. of this equation is represented in Fig. 8.36a. The second term has a finite value only if $\Phi_{D\infty}$ changes with time. For a first estimate we can ignore a time dependence*. Experiments have proved that it is indeed possible to realize device configurations almost free of a reaction on $\Phi_{D\infty}$. The charging current $I_{ch}^\star$ adds to the total current of the unloaded device. By a suitable arrangement of the capacitive loading, the total device current can, for example, be made to approach a sine waveform giving an increased r.f. power output (higher efficiency; Section 5.2) in oscillator applications [4.73].

In the second case (Fig. 8.37a and b) the waveform of the charging current $I_{ch}^{\star\star}$ taken off the metal plate is a rather exact replica of the metal-electrode shape. For an estimate of this current, cf. [1.33, 8.63], we assume again that the metal electrode exhibits a *constant* potential $\Phi_{D\infty}$ (requiring $R_{Lm}I_{ch}^{\star\star}$, $R_{1c}I_\infty \ll \Phi_{D\infty}$) and that the charging current is considerably smaller than the cathode current of the device to make reaction effects negligible. If the domain travels underneath the plate for a distance dz the negative charge on it decreases by an amount (see Fig. 8.37b)

$$dQ_m = C'w_m(z)\Phi_{D\infty}\,dz.$$

The externally flowing charging current is then

$$I_{ch}^{\star\star} = \frac{dQ_m}{dt} = C'\Phi_{D\infty}w_m(z)\frac{dz}{dt}. \tag{8.40a}$$

---

*The time dependence is negligible if $C'w_m l_m \ll C_D$ (domain capacitance). Otherwise the charges induced on the plate react on the domain charge and, thus, on $\Phi_{D\infty}$.

With the relation $z = v_\mathrm{D}t$ one eventually obtains

$$I_\mathrm{ch}^{\star\star}(t) = C'w_\mathrm{m}(z)v_\mathrm{D}\Phi_{\mathrm{D}\infty} = \text{const. } w_\mathrm{m}(z). \qquad (8.40\mathrm{b})$$

The two additional current spikes shown in Fig. 8.37a are caused by the domain-formation, and the dissolution, processes and can easily be clipped.

Further examples of functional generators which rely on producing a particular 'effective conduction-path profile' and permit a control of the current waveform during operation are demonstrated in Figs 8.38 and 8.40*. In these cases the device itself consists of uniformly-doped and uniformly-shaped TE material but with

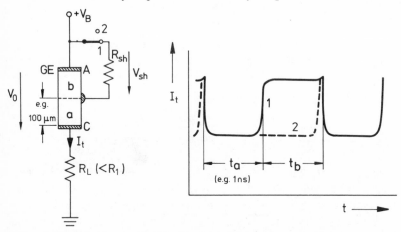

Fig. 8.38 Simple controlled-waveform functional generator showing circuit and obtained current waveform $I_\mathrm{t}(t)$. Adapted from Shoji [8.57].

additional ohmic contacts attached for providing suitable current shunt paths over part of the sample [8.57].

Figure 8.38 shows a rather simple example for explaining the principle involved. A small ohmic contact is provided at the middle of the Gunn element GE, dividing it into the two parts (a) and (b), and an ohmic shunt resistor R_sh is connected, via a switch, between the middle contact and the anode contact. Assuming now that a dipole domain has been launched at the cathode due to a sufficiently high bias voltage and that it is just traversing part (a) of the sample, an electrical equivalent circuit can be drawn as shown in Fig. 8.39a. During this time interval (t_a) the steady-state device current is given by [eqns (3.27), (3.29)]

$$I_\mathrm{t}^{(\mathrm{a})} = I_\infty = en_0v_\mathrm{D}A$$

i.e. it is solely determined by the steady-state domain dynamics (Section 3.4.4) and is

---

*It may well be that eventually the most important class of controlled-waveform generators will rather rely on influencing the conduction path profile (carrier concentration) by light irradiation onto the TE sample (see Section 8.2.3.).

practically not influenced by the presence of the shunt resistor as long as the domain can be considered as a true d.c. current generator (current saturation range of $\Phi_{D\infty}$). An entirely different situation results subsequently when the domain has passed into part (b) of the device (Fig. 8.39b). Then the current flowing within part (b) is still equal to $I_t^{(a)}$ above, however, the external current becomes

$$I_t^{(b)} = I_\infty + \Delta I_t \tag{8.41}$$

since the domain as a current generator is equivalent to an infinitely high resistance, and the bias V_B now drives a substantial current through the shunt R_{sh}. From the circuit of Fig. 8.39b one obtains

$$\Delta I_t = \frac{V_B - I_\infty(R_L + R_{1a})}{R_L + R_{1a} + R_{sh}} \tag{8.42}$$

where, however, R_{sh} is to be chosen to provide $I_t^{(b)} < I_T$, cf. [8.57].

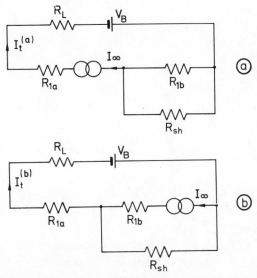

Fig. 8.39 Electrical equivalent circuit of functional generator according to Fig. 8.38. (a) For time interval while domain traverses part *a* of the device, (b) while domain in part *b*. Adapted from Shoji [8.57].

For a suitable choice of the values of R_{1a}, R_{1b} and R_{sh} an approximate rectangular current waveform may be produced as shown in Fig. 8.38. The appropriate choice of parameters yielding a total device current $I_t^{(b)}$ which remains just below the threshold value ensures that the domain is able to travel across the entire sample length without premature interference with a new domain nucleation and that, at the same time, the maximum modulation ratio is obtained. If the shunt path is interrupted by opening the switch, the spike waveform typical of the uniform device is generated (Section 8.3.2.1, Fig. 3.20). The shunt-path principle can readily be extended to construct more complex functional generators as shown in Fig. 8.40 for the case where there are three taps on the device. By appropriately operating the switch the pulse-to-space ratio of the rectangular wave obtained can be controlled as indicated.

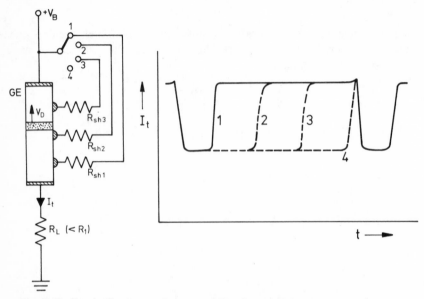

Fig. 8.40 Circuit diagram and current $I_t(t)$ of rectangular-wave generator with controllable pulse-to-space ratio.

8.4 LOGIC CIRCUITS

8.4.1 General Remarks

The existence of the distinct electric-field threshold near $3\cdot8$ kV cm^{-1} (cf. Section 3.4.4) for launching Gunn dipole domains in TE devices readily suggests to investigate a possible use of these devices for threshold-logic* applications. In the field of transistorized logic circuitry it is recognized, cf. [8.83], that binary threshold logic displays certain advantages over conventional Boolean-type logic in terms of reduced number of components and improved speed, besides possessing greater logic flexibility due to its capability of weighting and summing the inputs. (A major disadvantage, however, should not be concealed, namely a higher susceptibility with

---

*A threshold-logic gate has binary inputs and outputs like in Boolean-type gates. As the decisive difference, however, the inputs to the threshold gate may be weighted and a binary decision made as to whether the total weight is smaller or greater than the predetermined threshold level. Mathematically, threshold gates are described by using a separation-function representation, cf. [8.77, 8.83].

regard to load and bias variations). During the past years it has become clear that Gunn devices are indeed excellent contenders for certain logic applications which promise definite improvements even over threshold transistor logic with regard to speed and number of components required to perform a particular logic function.

Many binary-logic functions can, at least principally, be performed by *two-terminal* Gunn devices as put forward primarily in early proposals [1.27, 8.5, 8.7, 8.85–8.90]. As usual, a particular association of the electrical quantities of the circuit to the binary values must be defined* in these operations; in the simplest case the presence of a positive pulse corresponds to the binary number '1' and no pulse represents the number '0'.

The basic building blocks of *combinatorial* logic are the AND and the OR gates and the inverter. As a first brief example of *two-terminal* Gunn-device logic, Fig. 8.41 shows the circuit of a simple gate which contains one Gunn element besides several

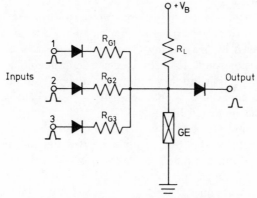

Fig. 8.41 Circuit diagram of AND/OR threshold gate employing a two-terminal Gunn device. After Sugeta *et al.* [8.26b].

gate diodes and ohmic resistors, cf. [8.24, 8.69]. The values of the resistors R_{G1} to R_{G3} and R_L have, together with the working point of the Gunn device, to be chosen according to the relative weight attributed to the individual input terminals. If the gate has, for instance, n equally weighting inputs, the threshold level can be set to lie between 1 and n. In the first case the circuit performs an OR function, yielding an output if an input pulse is applied to one, or (at the same time) more than one, input terminal. The latter case (threshold level n) is that of an AND gate which requires input pulses simultaneously at all input terminals to give an output signal. In the intermediate case of a threshold level $(n + 1)/2$ the function of MAJORITY is obtained since now an input must be present simultaneously at the majority of the input terminals to yield an output. As a further example, an exclusive-OR function can be performed by a circuit [8.69, 8.44] in which several, identical, uniform two-

---

*For possible ways of relating in Gunn-effect logic the electrical quantities to the binary numbers see [8.26, 8.88].

terminal Gunn elements operate into a common load resistor (output) between the cathodes and ground, cf. [8.91], and the input pulses are fed to the anode of the individual Gunn elements, preferably via gating diodes. The exclusive-OR gate (or comparator) provides no output if input pulses appear simultaneously at more than one Gunn element.

The further basic function required for combinatorial logic is the negation (NOT gate, inverter). In the simplest case an inverter is obtained [8.90] by employing the pulse regenerator circuit of Fig. 8.26a, but now with the bias voltage adjusted to an above-threshold value. Applying a negative-going pulse of appropriate amplitude to the anode of the Gunn element stops the continuous pulse train generation for the duration of the input pulse, thus providing the desired function. Without providing special means, no exact synchronization exists in this simple inverter between the leading edge of the input pulse and the beginning of the pulse-train interruption; rather there is an uncertainty interval which may range from zero to τ (domain transit time). On the other hand, the *end* of the input pulse is always synchronized to the *end* of the interruption period. Sometimes the polarity of a pulse must be reversed. This is easily accomplished by again using the simple regenerator circuit, but with an additional load resistor connected between cathode terminal and ground. If now a positive input pulse is applied to the anode, a negative pulse appears at the cathode; the reverse polarity conversion is obtained by triggering at the cathode and taking off the output at the anode (cf. Fig. 8.1).

Sequential logic (shift registers, counters, etc.) requires a memory unit as a basic circuit. A memory function is, for example, carried out by the circuit shown in Fig. 8.33 (version a) which naturally can also be constructed by employing a two-terminal Gunn device subjected to bias triggering. A memory loop is obtained if 2 two-terminal devices connected in series via capacitors and biased alternatively with opposite polarity, are provided with a capacitive feedback loop from the output of the second Gunn element to the input of the preceding one [8.92]. By making use of these basic units together with Gunn-element pulse generators and the just mentioned gate circuits, high-speed shift registers [8.69, 8.85, 8.93] have been proposed and partly verified.

For realizing certain logic functions (e.g. comparator) it may be of advantage to employ Gunn elements which possess several isolated ohmic-contact cathode electrodes of small area instead of a simple large one [3.30, 8.5b, 8.69, 8.89].

A number of further memory applications of TE devices, all of which principally yield very low switching or cycle times, have been proposed, cf. [8.19] and Section 8.4.3.1. A memory loop with circulating information is also obtained by providing a suitable feedback path with time delay on the exclusive-OR gate described above [8.85], or on a device fitted with a capacitive third electrode and triggered by the pulse to be stored [8.8b, 8.10]. In a further proposal, for dynamic information storage [8.4, 8.11a], having a resemblance to the memory circuit of Fig. 8.33a, a Gunn element is connected to a resonator and biased slightly below threshold. The pulse signal to be stored is applied to the resonator and the bias-triggers a dipole domain in the Gunn device. The ringing of the resonator as a result of the induced current step subsequently causes the threshold voltage repetitively to be reached and exceeded (again exciting the resonator), so that the oscillation continues until an erase pulse of opposite polarity is applied*. A memory effect can further be obtained by creating localized impact ionization in a Gunn device, leading to a short increase in device current, if the voltage pulse to be memorized is superimposed on the (super-critical) bias voltage of the sample while a dipole domain is in transit (cf. Section 8.2.3) [6.32, 8.1]. Because of the only limited lifetime of the excess charge carriers created by the impact ionization, the storage times (repetitive appearance of current pip) are limited to the order of 1μs, i.e. to a certain number of domain transits. A similar memory effect of limited storage time is obtained by irridating light pulses on super-critically biased n-type photo-conductive GaAs [8.94]. Without light the device is in an n_0L-stable state. A short light pulse of suitable wavelength increases the electron density beyond the critical value for nL instability, thus creating current oscillations. These oscillations terminate only after the lifetime of the photoconductive electrons. A re-set of the memory is possible with a short light pulse of appropriate longer wavelength to quench the photoconductivity. An extremely suitable device for static information storage would, no doubt, be the super-critically-doped Gunn element with stationary high-field domain at the anode, providing bistable current switching (Section 8.3.1.3). However, the further development has here to be awaited before the practical importance of this device can be judged. A fixed-information storage cell can be constructed by applying the principle of shaping the conduction-path profile of Gunn devices (Section 8.2.3). The conduction-path profile of the device is laid out according to the information to be stored, and a dipole domain scanning along the sample reads out this information by a sample current proportional to the value of the conduction-path profile at the point which the domain is just passing (see Sect. 8.2.3).

Gunn-effect logic using two-terminal devices with additional but not d.c.-isolated electrodes suffers from the lack of unidirectivity which poses a problem particularly in circuits of some complexity. Difficulties, like reduction of operational speed, may also be encountered by the necessity to provide special means for separating the d.c. bias, and the signal, paths. Decoupling can be obtained to a certain degree by employing gating diodes in between the various Gunn elements as shown, for example, in Fig. 8.41. However, as we have discussed in Section 8.2.2.3, the best way for achieving undirectional operation is to employ Gunn devices fitted with an additional Schottky-contact gate near the cathode for triggering the dipole domains. Since it is possible to make the width of the gate stripe very small (≈ 0.5 μm) the input gate capacitance obtained can be extremely low. There is a difference in the design of Schottky-gate Gunn devices for use as pulse regenerators (Section 8.3.1.2) and for logic applications, cf. [8.26b]. In the former case the value of the load resistor is determined by the characteristic impedance of the employed transmission line and, consequently, is relatively low (50–75 Ω). This requires large values of the gate trigger capability factor γ_G (Section 8.2.2.3) in order to obtain sufficiently high signal amplitudes and gain. In logic circuits, on the other hand, more optimum values for the load resistor of Schottky-gate devices can be chosen (≈ 200 Ω), though still keeping a low-power delay product, and lower values of γ_G suffice.

---

*See Fig. 5.15, dashed part of d.d. curve.

The more specific description of Gunn-logic circuitry and devices in the following sections is, therefore, restricted to only considering the *three-terminal* Schottky-gate TE elements as the probably most important class of Gunn device for practical logic applications. First, basic circuits for combinatorial logic (gates) and then for sequential logic will be dealt with below. This is followed by the description of a full-adder circuit as an example of a logic sub-system. In the final section, eventually, we leave the field of pure digital electronics and the restriction to Schottky-gate devices. There we will show how Gunn elements can be employed for the hybrid operation of analogue-to-digital conversion, an operation required before the digital processing of an analogue signal or quantity can commence.

The main advantages of Gunn-device logic compared to other methods of realization are the ultrafast speed of operation (low power-delay product) and the reduced number of required components which also means a reduced number of parasitic circuit elements. These properties have been investigated and discussed in preceding sections. Yet, before a final assessment can be made on the eventual role and importance of Gunn-logic circuitry, quite a number of factors and properties will still have to be investigated and compared with those of existing transistor logic. These factors certainly include flexibility, reliability, permissible circuit tolerances [8.26b], noise margin [8.26b], temperature behaviour, safety against instabilities, pulse jitter, fan-in and fan-out capabilities. Some brief information on fan-out behaviour and on pulse jitter has been given in Sections 8.2.2.2 and 8.2.2.3, respectively. Time eventually will show what impact Gunn functional devices are really able to make in the field of digital signal processing. Not the last, a further maturation of GaAs technology and the specific development of microstrip circuitry for Gigabit *digital* systems must be awaited before a final judgement can be made.

8.4.2 Combinatorial Threshold Logic
In the following we investigate logic circuits which employ Schottky-gate Gunn devices. These Schottky-gate devices require a negative-going voltage pulse at the gate for triggering a dipole domain (notch-triggering, Section 8.2.2.3). It is therefore appropriate to take the output (for feeding succeeding Schottky-gate devices) off across an ohmic load resistor between the cathode of the device and ground since there a negative voltage pulse is obtained as the result of domain formation and extinction (see Fig. 8.1b). Working, then, with negative pulses, at least predominantly, suggests to adopt here an agreement, cf. [8.26a, b] which assigns the presence of a negative pulse to the binary number '1' whereas the state of no pulse (or that below a certain threshold level) corresponds to the number '0'.

When considering in the following the oscillograms obtained from practical circuit realizations, it should be borne in mind that these circuits were prototypes, not yet optimally designed with regard to reaching the limits of achievable response speed and low power consumption.

8.4.2.1 *AND/OR Gate*

The fundamental AND and OR gate functions can be performed by a circuit as indicated in Fig. 8.42a. The essential feature is here the operation of several identical or similar Gunn elements into a common cathode load resistor. Such a circuit exhibits the properties of 'threshold level', 'summing', and 'weighting'. If the signal pulses applied to the inputs of the two Gunn elements in the particular example of Fig. 8.42a have a time difference larger than the dipole-domain transit time $\tau \approx L/v_\mathrm{D}$ (assumed to be equal for both elements), two separate output pulses of low amplitude are obtained across the load resistor R_{L1} as shown by the display A in Fig. 8.42b. In

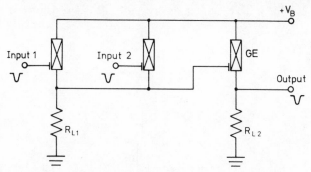

Fig. 8.42a Circuit diagram of AND/OR threshold gate employing Schottky-gate Gunn devices. After Sugeta *et al.* [8.26, 8.81].

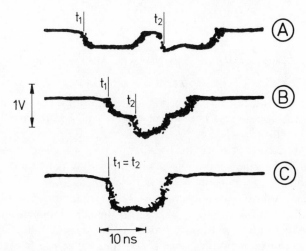

Fig. 8.42b Measured voltage waveforms across common cathode resistor R_{L1} of circuit shown in Fig. 8.42a ($R_{\mathrm{L1}} = 25\,\Omega$; device parameters: $L \simeq 110\,\mu\mathrm{m}$, $z_\mathrm{G} \simeq 20\,\mu\mathrm{m}$, $\Delta z_\mathrm{G} \simeq 10\,\mu\mathrm{m}$, $d \simeq 9\,\mu\mathrm{m}$, $w \simeq 60\,\mu\mathrm{m}$, $n_0 \simeq 2\cdot5 \times 10^{15}\,\mathrm{cm}^{-3}$, $R_1 \simeq 300\,\Omega$). Trace A: Input pulses applied have time difference $t_2 - t_1 = \Delta t > \tau\,(\simeq L/v_\mathrm{D})$. Trace B: $\Delta t < \tau$. Trace C: $\Delta t \simeq 0$. After Sugeta and Yanai [8.26].

case of the time difference being smaller than τ, the outputs are summed giving a large amplitude (displays B and C). Depending on the value of the common load resistor R_{L1} in a gate realization containing n inputs (parallel Gunn devices) and on the threshold level of the output Gunn element, the combined circuit can perform OR, AND, or MAJORITY functions [requiring a threshold 1, n, or $(n + 1)/2$, respectively]. The summed signal is weighted by changing the load resistor R_{L1} or the device current.

8.4.2.2 NOT Gate and Inhibitor

The basic circuit performing the functions of an inhibitor or a NOT gate is shown in Fig. 8.43a [8.26, 8.81]. Two similar devices and a load resistor are connected in

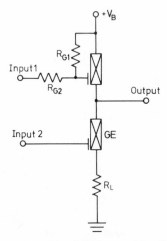

Fig. 8.43a Circuit diagram of inhibitor/NOT gate. After Sugeta and Yanai [8.26, 8.81].

series, with two ohmic resistors incorporated at the upper device to adjust the required gate bias conditions. The output signal is taken off from the common contact of the two devices.

If the time difference between the two input pulses applied to the terminals 1 and 2 is smaller than the width τ of the Gunn pulses initiated in the devices, the earlier appearing pulse inhibits the later one as demonstrated by the oscillograms in Fig. 8.43b. In case A the time difference $\Delta t = |t_1 - t_2|$ between the application of an input pulse each at the terminals 1 and 2 is larger than the domain transit time τ, so that two individual output pulses are produced. In cases B and C the inhibiting action occurs because of the only small $(<\tau)$ time difference between the input pulses. When a dipole domain has been triggered in one of the Gunn devices, the electric field within the other device is so low that a domain cannot be triggered in it by the input pulse, thus providing the inhibitor function.

The fundamental circuit of Fig. 8.43a can be employed as an inhibitor in the way

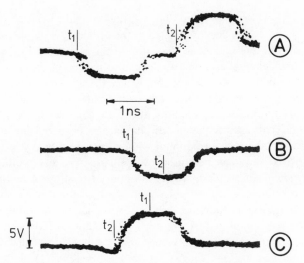

Fig. 8.43b Voltage waveforms measured at output terminal of circuit shown in Fig. 8.43a (for different time relations between input pulses), demonstrating inhibitor action ($R_{G1} \simeq 270\,\Omega$, $R_{G2} \simeq 200\,\Omega$, $R_L \simeq 50\,\Omega$, device parameters same as in Fig. 8.42). After Sugeta and Yanai [8.26, 8.81].

described, with two input-signal terminals. Alternatively, it is possible to operate the circuit as a NOT gate when input terminal 1 is used as the signal input and a clock signal (continuous pulse train) is applied to the input terminal 2. In this case an output pulse is obtained if, and only when, no input pulse appears at terminal 1.

8.4.3 Sequential Logic

8.4.3.1 *Memory and Delay Circuits*

The possibility of achieving monostable current switching in Gunn devices according to Section 8.3.1.2 suggests to realize dynamic, rather than static, memory circuits. The problem posed in this particular envisaged application is to have the current switching and re-switching, once it has been triggered in a subcritically biased Gunn sample, continue even after the termination of the trigger (set) pulse, storing in this way the set pulse in the form of an induced current oscillation. To achieve this goal the threshold voltage must be exceeded repetitively after the set pulse has initiated the current oscillation process.

Various means of triggering continuous current oscillations in Gunn devices by a simple pulse event have been briefly described in Section 8.4.1. A particularly simple, and thus attractive, basic memory circuit of this kind, employing a Schottky-gate device, is shown in Fig. 8.44a [8.26, 8.81]. It relies again on the principle of providing a suitable load circuit to achieve successive domain triggering, the load at the cathode in this case consisting of a small capacitor in parallel to an ohmic resistor. When in this circuit a dipole domain is triggered in the subcritically-biased device by the application of a set (write) pulse to the gate, the domain travels along the sample in the usual way and finally is extinguished at the anode. Yet, while and just after the domain disappears, the voltage at the cathode can rise again to the high bias level only with a time delay determined by the charging time of the capacitor in parallel to the cathode load resistor. In consequence, there exists immediately after the disappearance of a domain at the anode a time interval during which the voltage drop across the Gunn device remains above the bias level, i.e. above or at the threshold level. This means that a new domain is triggered underneath the gate contact, and that the process of domain launching is to repeat constantly.

The memorized state can be erased by applying a positive voltage pulse (re-set pulse) to the gate as indicated by the oscillographic displays A and B in Fig. 8.44b. The applied positive gate pulse reduces the depth of the domain nucleation notch underneath the gate electrode (see Fig. 8.12) to such a low value that a renewed domain formation is inhibited. A ring-type, yet somewhat more elaborate, dynamic memory unit employing a Schottky-gate device can be constructed by having the

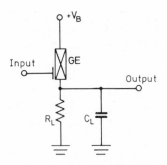

Fig. 8.44a Circuit diagram of memory circuit. After Sugeta and Yanai [8.26, 8.81].

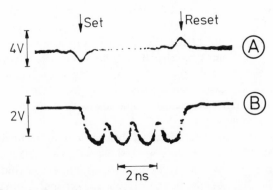

Fig. 8.44b Trace A: Set pulse (negative) and reset pulse (positive) for operating memory circuit shown in Fig. 8.44a. Trace B: Output pulse at cathode demonstrating memory action ($R_L \simeq 50\,\Omega$, $C_L \simeq 4\,\mathrm{pF}$, device parameters same as in Fig. 8.42). After Sugeta and Yanai [8.26, 8.81].

ohmic load resistor at the anode of the device (Fig. 8.1a) and providing a feedback path, possibly with some additional time delay introduced, from the anode to the gate electrode [8.26].

Reading of the stored information, without destroying it, can be accomplished by using a (Gunn device) AND gate [8.69, 8.85]. One input terminal of this gate receives the pulse train of the memory unit and the other one a clock signal. The minimum time required for recognition of a stored information bit in these dynamic Gunn effect memories is determined by the pulse repetition frequency, i.e. it is equal to the domain transit time τ. As a certain disadvantage of the memory circuits discussed, the stored information is lost when the bias supply becomes interrupted. However, this property will probably be only of minor importance for the more special applications envisaged for Gunn logic.

Sequential logic circuits require units providing pulse delay. A possible circuit realization which employs Schottky-gate Gunn devices, providing a 1-bit delay per stage, is shown in Fig. 8.45a [8.26]. The circuit consists of the cascade connection of

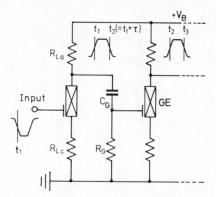

Fig. 8.45a Circuit diagram of delay circuit. After Sugeta and Yanai [8.26].

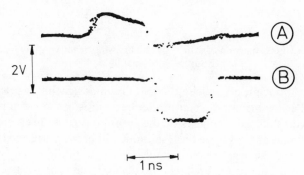

Fig. 8.45b Voltage waveforms measured in delay circuit shown in Fig. 8.45a, device parameters same as in Fig. 8.42. Trace A: Differentiated anode output pulse. Trace B: Anode pulse of following device triggered by negative-going part of this differentiated pulse. After Sugeta and Yanai [8.26].

identical devices, cf. [8.8b]. A positive output pulse is then used to trigger a domain in the succeeding device. The experienced pulse delay is approximately equal to the domain transit time $\tau \approx L/v_D$. Trace A in Fig. 8.45b is the differentiated waveform of the anode voltage and trace B the triggered output cathode voltage of the following device, demonstrating the delay action obtained in an experimental circuit.

8.4.3.2 *Shift Register*

The first proposals for realizing Gunn-logic shift registers were based on the use of purely ohmic-contact Gunn devices [8.85, 8.69] and have been partly verified experimentally [8.93]. A simpler construction of a shift register is possible by using Schottky-gate Gunn devices in a configuration shown in Fig. 8.46a. This register is made up of 1-bit units, each consisting of a cascade connection of the delay circuit and the memory circuit described in Section 8.4.3.1 above.

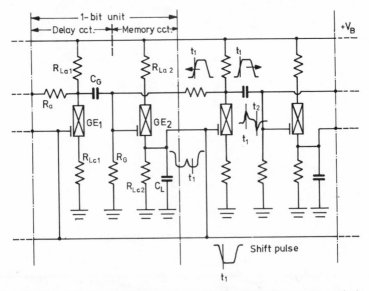

Fig. 8.46a Circuit diagram of shift register employing Schottky-gate Gunn devices. After Sugeta and Yanai [8.26, 8.81].

The working principle of the Schottky-gate Gunn-device shift register is as follows. We assume that the memory circuit of the first unit is in the memorized state and that a (negative) shift pulse appears on the appropriate line. Then the Gunn device in the delay circuit of the succeeding 1-bit unit will become triggered by the sum of the shift pulse and the negative output pulse of the first unit, and a positive pulse is obtained across the anode load $R_{La,1}$ in this delay circuit. The positive pulse appearing there is fed back via R_a to the gate of the Gunn device in the first memory circuit, erasing its memorized state. Also, the positive pulse is differentiated by C_G, R_G and applied to the gate of the device in the memory circuit of the second unit, thereby triggering this circuit into the memorized state with 1-bit time delay. Consequently, the stored information can be shifted from one shift-register unit to the next one.

The shifting operation of an experimental shift register circuit of this kind has been verified by the oscillograms reproduced in Fig. 8.46b. The first 1-bit unit was triggered by a negative set pulse initiating current oscillations in its memory circuit (trace B). However, when the applied shift pulse appears (trace A) it adds to the output pulse of the first unit (trace B), and this triggers the memory device of the second unit into continuing oscillations (trace C) while extinguishing the information in the first unit (trace B).

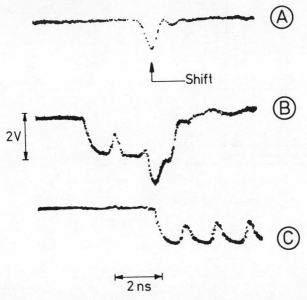

Fig. 8.46b Voltage waveforms measured in shift register shown in Fig. 8.46a. Trace A: Shift pulse. Trace B: Output pulse at cathode of first memory circuit. Trace C: Output pulse at cathode of following memory circuit. $R_{\text{La},2} \simeq 160\,\Omega$, $R_{\text{Lc},2} \simeq 50\,\Omega$, $C_{\text{L}} \simeq 4\,\text{pF}$, device parameters same as in Fig. 8.42. After Sugeta and Yanai [8.26, 8.81].

8.4.4 Full Adder as Example of Sub-System

The preceding sections have dealt with the basic modules of high-speed logic circuitry utilizing Schottky-gate Gunn devices; only the shift register described immediately above may beyond that be regarded as representing already a signal-processing sub-system. As an instructive example of a further sub-system we are here to describe the circuit of a full 1-bit adder* which particularly well demonstrates

---

*See [8.26 and 8.14, 8.69, 8.85] for further examples of signal-processing sub-systems (including a comparator, register, and binary counter) employing Schottky-gate, or bias-triggered Gunn devices, respectively.

the fact that the Schottky-gate Gunn-device logic requires a considerably reduced number of component parts compared to equivalent transistor-logic realizations, besides possessing the potential property of operating much faster than transistor logic [8.26].

First, a block diagram of the full-adder circuit using conventional Boolean-logic gates is shown in Fig. 8.47. A circuit of this type consists of seven AND gates and

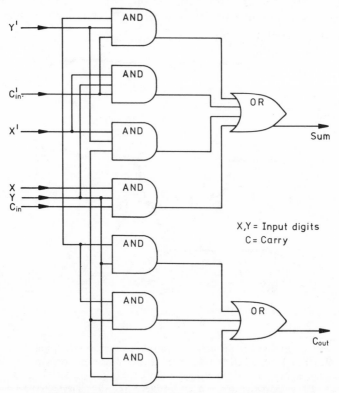

Fig. 8.47 Block diagram of Boolean-logic full adder.

two OR gates. Its realization in the form of high-speed emitter-coupled transistor-logic (ECL) technique requires approximately 40 circuit components. Figure 8.48 demonstrates the simplification obtainable if the full adder is constructed rather by using threshold gates. The number of necessary gates has reduced to two. A complete circuit diagram of this threshold-logic full adder, again employing ECL technique, is given in Fig. 8.49. The number of components is markedly lower than that of the Boolean-logic equivalent, being now about 20 to 30. Finally, Fig. 8.50 shows the circuit realization utilizing Schottky-gate Gunn elements and the corresponding truth table. An extremely simple circuit has evolved. The number of component parts has dropped again considerably, namely to nine, including two **Gunn** elements

GE*. The symbol D in the circuit of Fig. 8.50 indicates a small time delay of approximately 50 ps which must be introduced to ensure that the Gunn element GE_1 without the delay is triggered ahead of GE_2. The gate threshold voltages V_{GT}

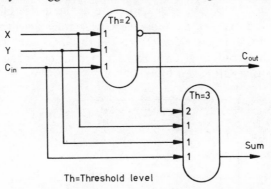

Fig. 8.48 Block diagram of threshold-logic full adder.

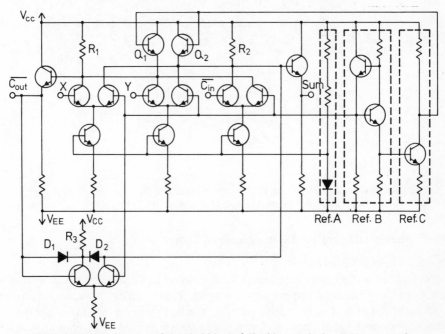

Fig. 8.49 Circuit diagram of threshold-logic full adder employing ECL gates. After Hampel and Winder [8.83].

---

*It is realized that a reduction in component parts is not necessarily an advantage by itself, particularly in the days of reliable and comparatively inexpensive large-scale integration. Other parameters must equally be taken into account, like required energy per logic operation and the power/speed ratio (cf. Section 8.1). Also it may appear somewhat unfair to compare an advanced technique (Boolean-type ECL logic) with prototype (Gunn) threshold versions.

(corresponding to threshold level Th in Fig. 8.48) of the Gunn elements must be chosen as indicated, with the factor m selected to match the d.c. bias level of the element GE_2. In this circuit, V_{G1} is the sum of the inputs at X,Y and the carry input terminal C_{in}. Because the positive output pulse at the anode of element GE_1 is adjusted to provide a level $-2m$, it inhibits the triggering of element GE_2 when the relative input level is 2. However, when the input is 3, GE_2 will be triggered. The truth table, therefore, reads as given in Fig. 8.50. On the other hand, the circuit can be used as an exclusive-OR gate if $C_{in} = 0$.

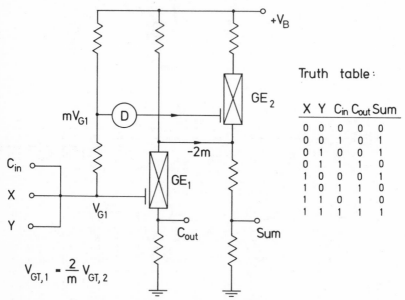

Truth table:

X	Y	C_{in}	C_{out}	Sum
0	0	0	0	0
0	0	1	0	1
0	1	0	0	1
0	1	1	1	0
1	0	0	0	1
1	0	1	1	0
1	1	0	1	0
1	1	1	1	1

Fig. 8.50 Circuit diagram and truth table of threshold-logic full adder employing Schottky-gate Gunn devices. After Sugeta and Yanai [8.26].

8.4.5 Analogue/Digital and Digital/Analogue Conversion

8.4.5.1 *General Remarks*

In various fields of electronics it is required to convert an analogue electrical quantity into an equivalent digital signal and vice versa. The necessity for such conversions arises, for example, if an analogue quantity is to be processed in a digital computer, if a digital read-out is desired in a measurement process, or in communication systems employing pulse-code modulation. In all these cases a quantization of the continuous amplitude range of the analogue quantity must be carried out, followed by a counting or coding process for the representation of the digital values, and/or possibly the reverse procedures. A number of proposals for employing Gunn devices, with their potentiality for extremely high-speed operation, in such applications have been put forward and partly verified experimentally. Most attention has up to the present been given to the basic process of converting an analogue signal (voltage) into a pulse train, where the number of the pulses in the produced sequence is proportional to the amplitude of the analogue signal. Some methods of such analogue-to-pulse conversion by utilizing Gunn devices are described below.

The obtained pulse sequence has to be processed further in a counter or an adder, depending on the particular application problem, to yield the required binary-code information, for instance in a dual-code representation. For reasons of compatibility and consistency, and for realizing high operation speed, these counters, comparators, and adders might equally be constructed using Gunn devices, in appropriate circuits [8.14, 8.26, 8.69, 8.85] made up of the basic logic units described in Sections 8.4.1–8.4.4.

For decoding, i.e. for digital-to-analogue conversion, a semi-conventional technique utilizing at least partly Gunn elements has been proposed [8.14, 8.42]. In this proposal the pulses of a sequence representing one analogue value have to be differently amplified such that the pulse amplitudes become proportional to $q + 1$ for pulses representing the digits 2^q. Consequently, the pulse describing 2^1 has twice the amplitude of that of 2^0, and so on. The weighting, cf. [8.9] is carried out with the help of Gunn devices of different cross-sectional area A, since A is proportional to the amplitudes of the Gunn pulses produced. The individual Gunn devices can be triggered by a synchronizing (clock) signal. The weighted pulses are then applied to an integrating network which yields the reproduced analogue signal.

8.4.5.2 *Analogue/Pulse Conversion*

In one class of *conventional* analogue-to-digital converter the analogue input signal is first converted into another, intermediate analogue quantity which is proportional to the input signal. This intermediate quantity may, for instance, be frequency or time. In the first case, then, an a.c. signal with a particular frequency value has to be processed into a digital information, whereas in the second case this applies to a time interval which one normally fills with counting pulses. Both tasks can be performed by relatively simple counting processes.

The basic principle of employing an intermediate analogue quantity is also found in *Gunn-effect* analogue-to-digital converters. A very simple converter of this type which delivers a pulse sequence of particular frequency is obtained by utilizing Gunn devices linearly tapered along the direction of domain propagation. The narrow side of the taper must be the cathode side as in the sample shown in Fig. 8.34. In Section 8.3.2.2 we have seen that in devices of such geometry (or equally of annular geometry, Fig. 8.36) the repetition frequency of an initiated Gunn oscillation can be varied by altering the bias voltage across the device (Fig. 8.36). This effect relies on controlling the time a dipole domain spends in the sample by shifting the point at which the electric field in the sample drops to the domain-sustaining value $E_s^\star$ and the domain collapses. Accordingly it is possible to convert the amplitude of the bias voltage, as our analogue signal, into a pulse train with a repetition frequency proportional to the analogue signal.

In a variant of this tapered converter device, a number of equi-distant notches are cut into the sample along the taper [8.1] as shown in Fig. 8.51 for a thin epitaxial layer of GaAs with the edge profile constructed by photoetching [6.88a]. The analogue voltage is added to the bias voltage, being slightly below threshold. As in the case above, the initiated dipole domain travels for a particular time (distance) across the device, where this time is determined by the level of the analogue voltage. Here, however, the obtained triangular output pulse is quantized by the geometrical ripple superimposed on the taper since the device current reproduces the conduction-path profile (Section 8.2.3, DOFIC device). The number of the quanta (pulses)

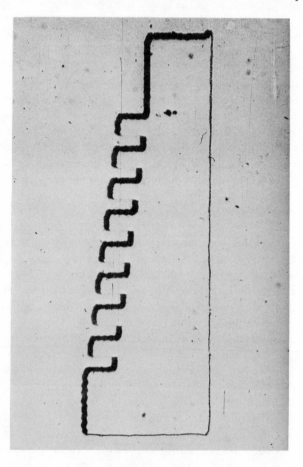

250 μm

Fig. 8.51 Analogue-to-pulse converter device fabricated in epitaxial GaAs on semi-insulating substrate. After Heeks *et al.* [6.88a].

appearing during one domain transit are a measure of the analogue-signal level and must subsequently be counted. Figure 8.52 shows the quantized output measured on a device similar to that in Fig. 8.51. An experimentally achieved quantization of up to 16 has been reported [6.88a].

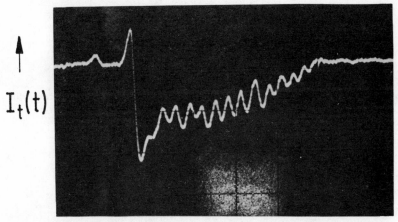

Fig. 8.52 Output current obtained from analogue-to-pulse converter similar to that shown in Fig. 8.51. After Heeks *et al.* [6.88a].

One particular type of *conventional* analogue-to-digital conversion via time as an intermediate analogue quantity is the sawtooth-voltage technique. Similar methods can be applied to *Gunn-device* converters. Gunn-device converters relying on this principle possess several advantages over the DOFIC notch-type converter described above (Fig. 8.51). For instance, they are easier to construct and are electronically controllable with regard to sensitivity and degree of quantization. A rather simple kind of analogue-to-pulse converter results, for instance, if the analogue signal in the form of a rectangular voltage pulse or step function is first differentiated in an appropriate RC circuit, the output of which is then superimposed on the bias of a subcritically operated Gunn element of 'normal' configuration, i.e. of *uniform* geometry and doping [8.18]. Depending on the height of the applied positive analogue voltage pulse, which produces a positive voltage step followed by an exponential decay behind the differentiating network (admittedly not an exact sawtooth waveform in this case), the Gunn device becomes biased above its threshold level for a certain proportional time interval. During this interval Gunn pulses are produced, with the number of these pulses, appearing in the device current, being a measure for the amplitude of the input analogue voltage. Because of the exponential output of the differentiating circuit, the number of output pulses is not linearly related to the analogue level, however.

In an improved configuration, exhibiting a true linear conversion law, a separate sawtooth generator is employed [8.42, 8.71, 8.90]. As a further advantage of such a

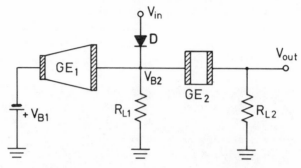

Fig. 8.53a Circuit diagram of sawtooth-type analogue-to-pulse converter containing two Gunn devices GE and gating diode D.

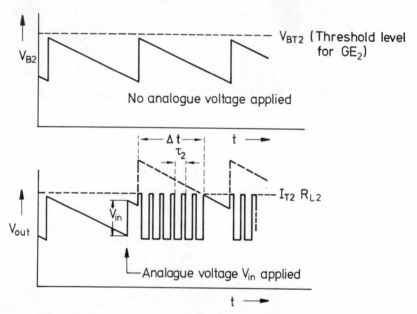

Fig. 8.53b Voltage waveforms obtained in converter circuit of Fig. 8.53a.

realization compared to the above one, the analogue voltage may now be continuous or continuously varying. For producing the sawtooth voltage a second Gunn element may preferably be used. A possible circuit configuration of a two-Gunn-element converter is shown in Fig. 8.53a. A relatively long, linearly-tapered Gunn element (GE_1) generates the sawtooth voltage (see Section 8.3.2.2). As indicated in Fig. 8.53b, the level of the sawtooth voltage and that of the bias across the short, uniform device (GE_2) are chosen so that the overall bias voltage of the second device remains just below the threshold value as long as no analogue signal is applied. An analogue voltage V_{in} appearing at the input then raises the bias above threshold, for a time interval Δt proportional to the analogue level. A corresponding number of Gunn pulses, namely approximately Δt divided by the domain transit time τ_2 of the second Gunn element, is produced during one sawtooth-voltage cycle (Fig. 8.53b). The output V_{out} is obtained across the cathode load resistor R_{L2}. The output voltage of an experimental sawtooth-type converter, however with the sawtooth not generated by a Gunn device, is reproduced in Fig. 8.54. In experiments a quantization of up to 17 has been achieved [8.90].

The maximum conversion frequency (rate) of the analogue to pulse conversion process in the circuit of Fig. 8.53a is given by

$$f_{A/P,max} \approx \frac{1}{(N_{max} - 1)\tau_2} \tag{8.43}$$

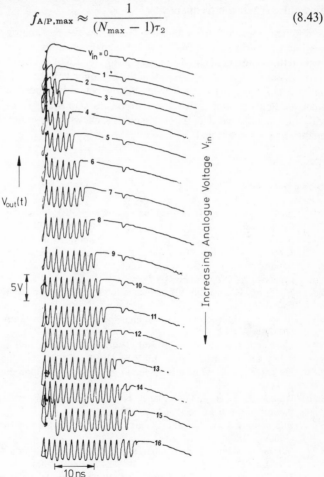

Fig. 8.54 Output voltage $V_{out}(t)$ measured in analogue-to-pulse converter circuit similar to that shown in Fig. 8.53a. (By courtesy of S. H. Izadpanah and H. L. Hartnagel, Sheffield University, 1969.)

where N_{max} denotes the maximum number of quanta (approximately given by the ratio of sawtooth duration to transit time τ_2). If, for example, $N_{max} = 20$ is required, one obtains $f_{A/P,max} \approx 0.5$ GHz with an assumed minimum value $\tau_{2,min} \approx 100$ ps for a GaAs element (see Section 8.2.1). The sawtooth voltage with a necessary length of approximately 2 ns ($L = 200~\mu m$) could possibly be generated continuously by a tapered planar-type GaAs Gunn device of advanced heat sinking. A larger degree of quantization ($N \gtrsim 20$) requires longer sawtooth durations, i.e. a greater length of the geometrically-shaped Gunn device ($L \gtrsim 200~\mu m$). Such long devices must be pulse-biased for thermal reasons. The conversion process, i.e. the quantization of a particular analogue level, is then possible only on a sampling basis at a rate of the bias-pulse repetition frequency.

A different approach to employing Gunn devices for analogue-to-pulse conversion is possible by making use of the travelling-signal detection technique [8.14]. In a proposed method the analogue signal propagates along a chain of intermediate attenuators and delay sections. After each of the attenuators the analogue voltage is applied to a Gunn device (of uniform geometry) which is biased slightly below its threshold level. If the attenuated analogue signal is sufficiently large, a dipole domain is triggered. The first N Gunn devices will, consequently, produce a domain pulse each, which together represent N quanta of the quantized analogue signal.

REFERENCES

8.1a   Sandbank, C. P., Functional integrated circuits. *Proc. Conference Advanced Electronic Techniques, July 1966* (London, United Trade Press, 1966).

8.1b   Sandbank, C. P., Domain-originated functional integrated circuits. *Int. Solid-State Circuits Conference, Philadelphia, Pa., Feb. 1967*, Digest of Technical Papers, p. 96.

8.1c   Sandbank, C. P., Synthesis of complex electronic functions by solid-state bulk effects. *Solid-State Electronics* **10** (1967), p. 369.

8.2   Shoji, M., Controlled bulk semiconductor current pulse generator. *Proc. IEEE* **55** (1967), p. 710.

8.3   Shoji, M., Bulk semiconductor high-speed current waveform generator. *Proc. IEEE* **55** (1967), p. 720.

8.4   Copeland, J. A., Hayashi, T. and Uenohara, M., Logic and memory elements using two-valley semiconductors. *Proc. IEEE* **55** (1967), p. 584.

8.5a   Hartnagel, H. L., Digital logic-circuit applications of Gunn diodes. *Proc. IEEE* **55** (1967), p. 1236.

8.5b   Izadpanah, S. H. and Hartnagel, H. L., Experimental verification of Gunn-effect comparator. *Proc. IEEE* **55** (1967), p. 1748.

8.6   Engelmann, R. W. H. and Heinle, W., Proposed Gunn-effect switch. *Electron. Lett.* **4** (1968), p. 190.

8.7   Sugeta, T., Ikoma, T. and Yanai, H., Bulk neuristor using the Gunn effect. *Proc. IEEE* **56** (1968), p. 239.

8.8a   U.K. Patent No. 1,119,020 (inventors: B. G. Bosch, H. Pollmann and H. Ebert; priority: Dec. 17, 1965).

8.8b   U.K. Patent No. 1,170,607 (inventors: B. G. Bosch, H. Pollmann and H. Ebert; priorities: Nov. 6 to Dec. 22, 1965).

8.9   U.K. Patent No. 1,121,010 (inventors: B. G. Bosch, H. Pollmann and H. Ebert; priority: Dec. 21, 1965).

8.10   Bosch, B. G. and Pollmann, H., The Gunn effect (in German). Pt I: Mechanism of the Gunn effect. *Int. Elektron. Rundschau* **20** (1966), p. 497. Pt II: Gunn elements and their applications. *Int. Elektron. Rundschau* **20** (1966), p. 590.

8.11a U.K. Patent No. 1,210,427 (inventor: B. G. Bosch; priority: Feb. 28, 1967).

8.11b Pollmann, H. and Bosch, B. G., R.f. triggering of self-sustaining Gunn oscillations. *Proc. IEEE* **56** (1969), p. 2168.

8.12   Tomizawa, K., Kawashima, M. and Kataoka, S., New logic functional device using transverse spreading of a high-field domain in n-type GaAs. *Electron. Lett.* **7** (1971), p. 239.

8.13a Thim, H., Experimental verification of bistable switching with Gunn diodes. *Electron. Lett.* **7** (1971), p. 246.

8.13b Thim, H., Bosch, R. and Gramann, W., Bistable Gunn-effect switch for the sub-nanosecond range (in German). *Verhandl. Deutsche Physikal. Gesellschaft.* **7/VI** (1972), p. 521.

8.13c Thim, H. W., DiLorenzo, J. V., Gramann, W., Haydl, W. and Bosch, R., Bistable switching in high quality GaAs Gunn diodes with overcritical nL product. *Fourth Int. Symp. on Gallium Arsenide, Boulder, Col.*, 1972.

8.14   Hartnagel, H. L., Gunn-effect pulse code modulation. *Arch. Elektron. Übertrag.* **22** (1968), p. 225.

8.15   Szekely, V. and Tarnay, K., Buildup of Gunn domains. *Electron. Lett.* **4** (1968), p. 492.

8.16   Sugeta, T., Yanai, H. and Ikoma, T., Simplified analysis of transient characteristics of high-field dipole domains using piece-wise linear approximation. *Electron. Commun. Japan* **53–C** (1970), p. 118. Also (abstract): *Trans. IECE (Japan)* **53–C** (1970), p. 245.

8.17   Sugeta, T., Yanai, H. and Ikoma, T., Switching properties of digital devices using the Gunn effect. *Electron. Commun. Japan* **53–C** (1970), p. 127 and p. 253, respectively.

8.18   Izadpanah, S. H. and Hartnagel, H. L., Pulse gain and analogue-to-pulse conversion by Gunn diodes. *Electron. Lett.* **4** (1968), p. 26.

8.19   Elschner, H., Information storage on the basis of the Gunn effect (in German). *Nachrichtentechnik* **20** (1970), p. 265.

8.20   Guha, S., On the rate of growth and decay of high-field domains in n-type gallium arsenide. *Proc. IEEE* **59** (1971), p. 718.

8.21a Sugeta, T. and Yanai, H., Effect of parasitic elements on the output waveform of Gunn-effect digital devices (in Japanese). Jap. IECE National Convention Rec. No. 722 (Aug. 1970), p. 781.

8.21b Sugeta, T., Suzuki, N., Tanimoto, M. and Yanai, H., Gunn-effect functional device (in Japanese). Rept No. 45–11, Meeting Jap. IEE Prof. Group on Transistors, Sept. 1970.

8.22   Nakamura, M., Kurono, H., Hirao, M., Toyabe, T. and Kodera, H., High-speed pulse response of planar-type Gunn diodes. *Proc. IEEE* **59** (1971), p. 1039.

8.23a Yanai, H., Sugeta, T. and Sekido, K., Schottky-gate Gunn effect digital device. Paper at Int. Electron Devices Meeting, Washington, D.C., Oct. 11–13, 1971.

8.23b Sugeta, T., Yanai, H. and Sekido, K., Schottky-gate bulk-effect digital device. *Proc. IEEE* **59** (1971), p. 1629.

8.23c Sugeta, T. and Yanai, H., Gunn-effect digital functional devices and their performance evaluation (abstract, in English). *Trans. IECE (Japan)* **55–C** (1972), p. 437.

8.24   Mause, K., Fernmeldetechnisches Zentralamt der Deutschen Bundespost, private communcation, 1971; and: NTG Discussion Meeting, Bad Krozingen, Oct. 8–9, 1971.

8.25   Heinle, W., AEG-Telefunken Research Institute, private communication, 1972.

8.26a Sugeta, T. and Yanai, H., Signal-processing of a Gunn-effect digital functional device (in Japanese). Jap. IECE Techn. Group on Circuits and System Theory, Rept No. CT 71–71 (1972–02), Feb. 24, 1972. Also (abstract in English): *Trans. IECE (Japan)* **55–C** (1972), p. 445.

8.26b Sugeta, T., Tanimoto, M. and Yanai, H., Gunn effect digital functional device. *J. Fac. Engng Univ. Tokyo (B)* **31** (1972), p. 773. See also: *IEEE Trans. Electron Dev.* **ED–21** (1974), p. 504.

8.27   U.K. Patent No. 1,109,250 (inventors: B. G. Bosch, H. Pollmann and G. Schickle; priority: Nov. 12, 1965).

8.28   Hayashi, T., Three-terminal GaAs switches. *IEEE Trans. Electron Dev.* **ED–15** (1968), p. 105.

8.29   Anonymous, Epitaxial Gunn-effect device. *Wireless World* (Sept. 1965), p. 425. (Standard Telecommun. Labs.).

8.30   U.K. Patent No. 1,131,187 (inventors: B. G. Bosch and G. Schickle; priority: Dec. 15. 1965).

8.31a Yanai, H., Ikoma, T. and Sugeta, T., Switching device using the Gunn effect (in Japanese). Rept Meeting Jap. IEE Prof. Group on Transistors, June 1967.

8.31b Yanai, H., Ikoma, T. and Sugeta, T., Digital device using the Gunn effect (in Japanese). *Proc. Fall Meeting Jap. IECE*, No. 682 (Oct. 1967).

8.32   German Federal Patent Disclosure Bulletin No. 1,516,893 (inventor: F. Dörbeck; priority: June 22,1966).

8.33a Haydl, W. H., A wide-range variable-frequency Gunn oscillator. *Appl. Phys. Lett.* **12** (1968), p. 357.

8.33b Adams, R. F. and Schulte, H. J., Optically triggerable domains in GaAs Gunn diodes. *Appl. Phys. Lett.* **15** (1969), p. 265.

8.34   Mause, K., Simple integrated circuit with Gunn devices. *Electron. Lett.* **8** (1972), p. 62.

8.35   Sugeta, T., Yanai, H. and Ikoma, T., Switching properties of bulk-effect digital devices. *IEEE Trans. Electron Dev.* **ED–17** (1970), p. 940.

8.36   Hartnagel, H. L., Planar Gunn-effect devices for high microwave powers. *Proc. 1971 Eur. Microwave Conf., Stockholm, Aug. 23–28, 1971*, vol. 1, p. A1/1:1. (Stockholm, The Royal Swedish Academy of Engineering Sciences, 1971).

8.37   Yanai, H., Ikoma, T. and Sugeta, T., Neuron digital semiconductors using the Gunn effect (in Japanese). Meeting Jap. IECE Techn. Group on Semiconductors and Semiconductor Devices, Rept No. SSD 67–36, Jan. 1968.

8.38   Hashizume, N., Kawashima, M. and Kataoka, S., Nucleation and control of departure of a high-field domain by a gate electrode. *Electron. Lett.* **7** (1971), p. 195.

8.39   Mause, K., Salow, H., Schlachetzki, A., Bachem, K. H. and Heime, K., Circuit integration with gate controlled Gunn devices. *Proc. 4th Int. Symp. on GaAs and Related Comp., Boulder, Colorado, 1972*, p. 275 (London, Inst. Physics and Phys. Soc., Conf. Ser. no. 17, 1972).

8.40   Zuleeg, R., A GaAs pn-junction FET and gate-controlled Gunn effect device. *Proc. Int. Symp. Gallium Arsenide, Dallas, Oct. 1968*, p. 181 (Institute of Physics and Physical Society, London, 1969).

8.41   U.K. Patent No. 1,116,169 (inventors: B. G. Bosch, H. Pollmann and G. Schickle; priority: Nov. 11, 1965 to Feb. 18, 1966).

8.42a Hartnagel, H. L., Ultrafast pulse-signal processing using Gunn-effect diodes. Paper presented at South West. IEEE Conf. and Exhib. 1970.

8.42b Hartnagel, H. L., Pulse communication at microwave bit rates using Gunn domains. *Acta Electronica*, **15** (1972), p. 217.

8.43   Riesz, R. P., Optical interaction with high-field domain nucleation in GaAs. *IEEE Trans. Electron Dev.* **ED–17** (1970), p. 81.

8.44   Hartnagel, H. L., Theory of Gunn-effect logic. *Solid-State Electronics* **12** (1969), p. 19.

8.45   Hobson, G. S. and Izadpanah, S. H., Random domain triggering in Gunn effect pulse regenerators. *Solid-State Electronics* **13** (1970), p. 937.

8.46   Bell, D. A., *Electrical Noise* (New York, Van Nostrand, 1960).

8.47 Bittel, H. and Storm, L., *Noise* (in German. Berlin/Heidelberg/New York, Springer, 1971).

8.48 Yanai, H., Sugeta, T., Tanimoto, M., Ishibashi, R. and Suzuki, N., Investigation ot performance of Gunn-effect functional devices (in Japanese). *Ann. Rept Engng Res. Inst., Univ. of Tokyo* **30** (1971), p. 229.

8.49 Fallmann, W., Mathur, P. C. and Hartnagel, H. L., Minimum signal pulse width for Gunn domain nucleation. *Phys. Lett.* **34A** (1971), p. 445.

8.50 Myers, F. A., McStay, J. and Taylor, B. C., Variable length Gunn oscillator. *Electron. Lett.* **4** (1968), p. 386.

8.51 Böer, K. W. and Wilhelm, W. E., Artificial initiation of layer-like field inhomogeneities in CdS single crystals. *Phys. Stat. Sol.* **4** (1964), p. 237.

8.52 Yanai, H. and Sugeta, T., Some features and characteristics of the Gunn effect digital device (in Japanese). *Jap. IECE Nat. Conv. Rec.* No. 717 (Sept. 1969), p. 808.

8.53 Heime, K., Planar Schottky-gate Gunn devices. *Electron. Lett.* **7** (1971), p. 610, and: Planar Gunn elements with Schottky trigger electrode (in German). Techn. Rept No. A651 TBr 2, Fernmeldetechnisches Zentralamt der Deutschen Bundespost, Darmstadt, Sept. 1971.

8.54 Cobbold, R. S. C., *Field-effect Transistors* (New York and London, John Wiley & Sons: Wiley-Interscience, 1970).

8.55 Boccon-Gibod, D. and Teszner, J. L., Experimental evidence of bistable switching in a Gunn epitaxial coplanar diode by anode-surface loading. *Electron. Lett.* **7** (1971), p. 468.

8.56 Boccon-Gibod, D. and Teszner, J. L., Lateral capacitive probing of an anode-loaded epitaxial coplanar gallium-arsenide diode. *Electron. Lett.* **7** (1971), p. 469.

8.57 Shoji, M., Functional bulk semiconductor oscillators. *IEEE Trans. Electron Dev.* **ED–14** (1967), p. 535.

8.58 Teszner, J. L., *J. Appl. Phys.*, to be published.

8.59 Newton, C. O., Theoretical analysis of domain propagation in Gunn diodes with annular geometry. *J. Phys.: Appl. Phys.* **2** (1969), p. 341.

8.60 Robrock II, R. B., Analysis and simulation of domain propagation in nonuniformly doped bulk GaAs. *IEEE Trans. Electron Dev.* **ED–16** (1969), p. 647.

8.61 Bhattacharya, T. K., A simple analysis of tapered Gunn oscillators. *Phys. Stat. Sol. (A)* **1** (1970), p. 757.

8.62 Shah, P. L. and Rabson, T. A., Combined doping and geometry effects on transferred-electron bulk instabilities. *IEEE Trans. Electron Dev.* **ED–18** (1971), p. 170.

8.63 Shoji, M. and Dorman, P. W., Capacitively coupled GaAs current waveform generator. *Proc. IEEE* **56** (1968), p. 1613.

8.64 Hofmann, K. R., Gunn oscillations in thin samples with capacitive surface loading. *Electron. Lett.* **5** (1969), p. 289.

8.65 Engelbrecht, R. S., Solid-state bulk phenomena and their application to integrated electronics. *Int. Solid-State Circuits Conference, Philadelphia, Pa., Feb. 1968,* Digest of Technical Papers, p. 33.

8.66 Robertson, G. I. and Sandbank, C. P., Integrated solid state display needs no complex circuitry. *Electronics* **42** (1969), Feb. 3, p. 100.

8.67 Shoji, M., Coexistence of two stable high-field domains in single multi-terminal Gunn oscillators. *IEEE Trans. Electron Dev.* **ED–17** (1970), p. 80.

8.68 Shoji, M., Two-dimensional Gunn-domain dynamics. *IEEE Trans. Electron Dev.* **ED–16** (1969), p. 748.

8.69 Izadpanah, S. H. and Hartnagel, H. L., Gunn-effect pulse and logic devices. *Rad. Electron. Engng* **39** (1970), p. 329.

8.70 U.K. Patent No. 1,120,567 (inventor: B. G. Bosch, priority: Dec. 15, 1965).

8.71 Hartnagel, H. L., Microwave logic with Gunn diodes. *Proc. 1968 MOGA Conference, Nachrichtentechn. Fachberichte* **35** (1968), p. 461.

8.72 Hartnagel, H. L., Pulse communication using Gunn diodes and hetero-junction lasers. *Arch. Elektron. Übertrangungstechn.* **25** (1971), p. 51.

8.73 Lanza, C. and Esposito, R. M., Bulk negative resistance device operated in a relaxation mode. *Solid-State Electronics* **12** (1969), p. 463.

8.74 Petrov, V. A. and Prokhorov, E. D., Some characteristic features of Gunn diodes operating in inductive circuits. *Sov. Radio Engng Electron. Physics* **14** (1969), p. 1493.

8.75 Boccon-Gibod, D., Rectangular-pulse generator and memory element with a Gunn diode. *Electron. Lett.* **5** (1969), p. 91.

8.76 U.K. Patent No. 1,111,409 (inventor: G. Schickle; priority: March 1, 1966).

8.77 Lewis II, P. M. and Coates, C. L., *Threshold Logic* (New York/London/Sydney, J. Wiley & Sons, 1967).

8.78 Thim, M. W., Dawson, L. R., DiLorenzo, J. V., Dyment, J. C., Hwang, C. J. and Rode, D. L., Subnanosecond PCM of GaAs lasers by Gunn-effect switches. *Int. Solid-State Circuits Conference, Philadelphia, Pa., Feb. 1973*, Digest of Techn. Papers, p. 92.

8.79 Newton, C. O. and Bew, G., Frequency measurements on Gunn effect devices with concentric electrodes. *J. Phys. D: Appl. Phys.* **3** (1970), p. 1189.

8.80 Shoji, M., Temperature-gradient controlled voltage tunable bulk semiconductor oscillator. *Proc. IEEE* **55** (1967), p. 1646.

8.81 Sugeta, T. and Yanai, H., Logic and memory applications of the Schottky-gate Gunn-effect digital device. *Proc. IEEE* **60** (1972), p. 238.

8.82 Becker, R., Bosch, B. G. and Engelmann, R. W. H., Domains and guided electromagnetic waves in GaAs stripline. *Electron. Lett.* **6** (1970), p. 604.

8.83 Hampel, D. and Winder, R. O., Threshold logic. *IEEE Spectrum* **8** (1971), May, p. 32.

8.84 Tanimoto, M., Yanai, H. and Sugeta, T., Thermally induced FM noise in Gunn oscillators and jitter in Gunn-effect digital devices. *IEEE Trans. Electron Dev.* **ED-21** (1974), p. 258.

8.85 Hartnagel, H. L., Some basic logic circuits employing Gunn effect devices. *Solid-State Electronics* **11** (1968), p. 568.

8.86 White, G. and Adams, R. F., A 2-GHz multiple Gunn device logic circuit. *Proc. IEEE* **57** (1969), p. 1684.

8.87 Boccon-Gibod, D. and Veilex, R., Application possibilities of the Gunn effect in digital electronics (in French). *L'Onde Électrique* **49** (1969), p. 671.

8.88 Hartnagel, H. L., Three-level Gunn-effect logic. *Solid-State Electronics* **14** (1971), p. 439.

8.89a Kataoka, S., Komamiya, K. and Morisue, M., A high-speed adder using Gunn diodes. *Proc. IEEE* **59** (1971), p. 1526.

8.89b Kataoka, S., Tateno, H. and Kawashima, M., A proposal for a full adder using Gunn diodes. *Proc. IEEE* **59** (1971), p. 1527.

8.90 Izadpanah, S. H., Ph.D. Thesis, University of Sheffield, 1969.

8.91 Guétin, P., Gunn effect with two samples in parallel. *Electronics Lett.* **4** (1968), p. 63.

8.92 Izadpanah, S. H. and Hartnagel, H. L., Memory loop with Gunn-effect pulse diodes. *Electronics Lett.* **5** (1969), p. 53.

8.93 Fallmann, W. F., Hartnagel, H. L. and Srivastava, G. P., Microwave pulse processing using Gunn diodes. 1970 Symp. on GaAs and Related Compounds, Conf. Series No. 9, p. 148 (London and Bristol, The Institute of Physics, 1971). Also: A Gunn-effect shift register. *Arch. Elektron. Übertragung* **24** (1970), p. 473.

8.94 Engelmann, R. W. H., 1967, patent application.

List of Symbols

General Notations

$\vec{E}$	vector (of electric field strength E)
E_x	x-component (of vector $\vec{E}$)
$\overset{\leftrightarrow}{\sigma}'$	tensor (of differential conductivity σ')
σ'_x	xx-component (of diagonal tensor $\overset{\leftrightarrow}{\sigma}'$)
E_0	d.c. component (of E)
ΔE	a.c. component (of E)
$\hat{E}$	amplitude (of ΔE)
$\tilde{E}$	$= \hat{E}\exp(-jk_z z)$, space-variant factor (of ΔE)
$\bar{n}$	average value (of electron density n)
$\langle W \rangle$	average value (of electron kinetic energy); notion used in Ch. 2 only
$E_{1\infty}$	steady-state value (of E_1)
f_{max}	maximum value (of frequency f)
λ_{min}	minimum value (of wavelength λ)
$(\delta f)_{rms}$	root-mean-square value (of δf)

Following a common practice in electron physics, the electric field strength E is defined to be positive in the electron drift direction and the magnetic field strength H in direction from north to south pole, in order to avoid numerous confusing minus signs in the equations. Units are generally given in the V, A, cm, s system.

Specific Symbols

a	half-layer thickness of TE semiconductor
a_0	lattice constant
A	current carrying area of TE device
$A^\star$	spreading area of heat flow
A_b	area of Schottky barrier
b	dipole-domain width
$b^\star$	value of b during domain extinction
B_0	external d.c. magnetic induction field
B_L	load admittance
c	phase velocity of electromagnetic waves
c_0	$= 2 \cdot 998 \times 10^{10}$ cm s^{-1}, free-space velocity of light
c_v	specific heat capacity
C	capacitance, in particular equivalent parallel capacitance, $\mathrm{Im}(Y)/\omega$, of TE device
C'	capacitance per unit area
C_b	Schottky-barrier capacitance
C_c	equivalent case (package) capacitance; coupling capacitance
C_D	dipole-domain capacitance
$C_{D,N}$	noise value of C_D
C_{eff}	effective parallel capacitance of TE device due to electron scattering
C_g	mounting-gap capacitance
C_G	capacitance in gate lead

C_h	homogeneous (geometrical) capacitance of TE device
C_j	varactor junction capacitance
C_L	parallel load capacitance
C_p	phase-shift capacitance of electron scattering causing C_{eff}; waveguide post capacitance
C_r	capacitance of low-field region in TE device
C_{sh}	parasitic shunt capacitance
d	thickness or diameter (transverse dimension) of TE semiconductor
$d^\star$	transverse segment of TE semiconductor
d_c	thickness of contact
d_d	depletion layer width
d_h	thermally effective thickness of heat sink
d_p	thickness of passive semiconductor
d_s	thickness of semi-insulating semiconductor
d_{tol}	tolerable value of $d = w$ establishing T_{tol} in active TE semiconductor
D	electron diffusion constant
D_h	$= D(E_h)$
D_1	D of central-valley electrons
D_2	D of satellite-valley electrons
e	$= 1 \cdot 60 \times 10^{-19}$ C, elementary charge
E	electric field strength defined to be positive in *electron* drift direction
E_a	electric-field parameter in analytical $v(E)$ expression
$\hat{E}_a$	amplitude of homogeneous-field oscillation
E_A	$= V_A/L$
$\hat{E}_b$	electric-field amplitude of space-charge wave
E_{bd}	dielectric breakdown field
E_B	$= V_B/L$
E_c	value of $\bar{E}$ in cathode ragion
E_G	value of E below gate electrode
E_h	$= V/L$, homogeneous (space-averaged) component of E
E_H	value of E in homogeneous region of TE device with gate
E_1	$= V_1/L$, lower turning point of alternating E_h
E_p	value of E at drift-velocity peak
E_S	$= V_S/L$, dipole-domain sustaining value of E_h
$E_S^\star$	$= V_S^\star/L$, pre-triggering value of E_S if $R_L \neq 0$
E_T	$= V_T/L$, dipole-domain formation threshold value of E_h
E_u	$= V_u/L$, upper turning point of alternating E_h
E_v	value of E at drift velocity valley
E_1	value of E outside of dipole domain (low-field value)
E_2	domain peak value of E (high-field) value
$E_{2,bd}$	value of E_2 at breakdown $\simeq E_{bd}$
E_3	value of E at anode boundary when dipole domain dissolves
f	frequency, specifically operating frequency of TE device
$f_{A/P}$	analogue-to-pulse conversion frequency
$f_{c.i.}$	upper frequency limit of controlled-injection device
f_{ex}	external signal frequency

405

$f(E_1, E_2)$	characteristic area in $v(E)$ plane
$f_{hyb.}$	upper frequency limit of hybrid mode
f_i	idler frequency
f_{LSA}	LSA oscillation frequency, specifically its upper limit
f_m	modulation frequency
$f_{n.c.}$	upper frequency limit of negative conductance
f_p	$= \omega_p/2\pi$, precession frequency of electron spins
$f_{q.m.d.}$	upper frequency limit of quenched-multiple-dipole mode
$f_{q.s.d.}$	upper frequency limit of quenched-single-dipole mode
f_r	pulse repetition frequency; circuit resonant frequency
f_s	resonant frequency of stabilizing cavity
$f_{s.c.w.}$	upper frequency limit of space-charge-wave growth
f_{TE}	upper frequency limit of TE effect
f_τ	$= 1/\tau$, transit-time frequency
f_1	primary oscillation frequency of TE device
f_2	secondary operating frequency of TE device
$f_{2n.c.}$	upper frequency limit of secondary negative conductance
$f^{(1)} \ldots f^{(5)}$	frequency constants of μ'_m
F	noise figure
F_ν	electron distribution function in νth valley
g	power gain; geometrical factor in expression for heat resistance
g_f	value of g in flat-gain region Δf
g_t	gate trigger factor
g_v	voltage gain
G	equivalent r.f. parallel conductance, $Re(Y)$, of TE device
G_D	r.f. domain conductance
G_{eff}	$= en_0\mu_{eff}L/A$, effective G_t
G_{h0}	$= en_0\mu'_{h0}L/A$, homogeneous r.f. conductance of TE device at d.c. bias point
G_L	$= 1/R_L$, load conductance
G_r	conductance of low-field region of TE device
G_t	total conductance of TE device
G_Ω	(ohmic) conductance
G_1	$= en_0\mu_1L/A$, low-field (homogeneous) conductance of TE device
h	$= 6\cdot62 \times 10^{-34}$ Js, Planck's constant; waveguide height
$\hbar$	$= h/2\pi$
h_N	$= \varepsilon/(-e\mu_N)$, space-charge growth parameter
h_P	$= \varepsilon/(e\mu_P)$, space-charge decay parameter
H	magnetic field strength
i	integer number; general index; noise current
i_A	noise current of active device
I	drift current
I_0	d.c. component of current
I_A	$\simeq I_T$, drift-current level for domain triggering
$I_{ch}^\star$	charging current of metal shield
$I_{ch}^{\star\star}$	charging current of metal shield through load resistance R_{Lm}

I_d	discharging current of dipole domain during extinction
I_D	dipole-domain drift current
I_1	value of I at E_1
I_s	value of I_o in saturation region characterized by $E_{h0} > E_T$
I_t	total current
I_u	value of I at E_u
j	$= \sqrt{-1}$, imaginary unit
J	$= env$, drift-current density
J_c	carrier-current density
J_{ch}	displacement-(charging) current density
J_G	value of J below gate region of TE device
J_H	value of J in homogeneous region of TE device with gate
J_p	$= en_0 v_p$, peak value of J
J_t	total current density
J_v	$= en_0 v_v$, valley value of J
k	electron wave propagation number, $2\pi/\lambda$; complex propagation number, $\beta + j\alpha$, of space-charge waves
k_B	$= 1 \cdot 38 \times 10^{-23}\ \mathrm{JK^{-1}}$, Boltzmann's constant
k_φ	(reduced) wave propagation number of phonons
l	length of unit cell; length of resonator cavity
l_e	electrical length of TE travelling-wave device
l_m	length of metal shield
l_+	length of positively charged part of metal shield
L	drift length of TE semiconductor; inductance
L_a	length of anode part of TE semiconductor
L_c	length of cathode part of TE semiconductor, specifically of 'control section'
L_{Db}	Debye length
L_{eff}	effective value of drift length L if $R_L \neq 0$
L_f	equivalent series inductance of TE device due to electron scattering
L_g	centre-conductor inductance at mounting gap
L_I	length of inactive cathode section of TE semiconductor
L_1	series load inductance at position of cavity coupling
L_L	parallel load inductance
L_n	length of doping notch
L_p	waveguide post inductance
m	lateral mode order of space-charge waves; number of dipole domains; transverse resonant order of cavity; fan-out of TE device; threshold level factor
$m^\star$	effective electron mass
n_0	equilibrium value of electron density n
$\dot{m}_1, m_2$	components of relative-permeability tensor
$m_1^\star$	effective mass of central-valley electrons
$m_2^{\star(N)}$	density-of-states effective mass of satellite-valley electrons

M	noise measure
M_i	internal M of space-charge-wave amplifier
n	electron density;
	main resonant order of cavity;
	characteristic oscillation order of TE device;
	number of logic-gate inputs
n_0	equilibrium value of electron-density n
n_1	density of central-valley electrons
n_2	density of satellite-valley electrons
N	number of quanta involved in digitalizing process
$N^\star$	effective density of states of conduction-band valley
N_A	ionized acceptor density
N_d	net doping density
N_D	ionized donor density
p	instantaneous electron momentum;
	complex frequency parameter $j\omega$
p_φ	phonon momentum
p_0	d.c. power density
$p_1 \ldots p_n$	zero values of $Z(p)$
P	r.f. output power of TE device
P_{ac}	a.c. power absorbed by drifting electron
P_{dc}	d.c. power absorbed by drifting electron
P_{dis}	power dissipated in active semiconductor volume
P_{ex}	external power injected into oscillator cavity
P_N	noise power
P_{NA}	P_N from active device
$P_{N,AM}$	P_N amplitude-modulated on carrier in single sideband
$P_{N,FM}$	P_N frequency-modulated on carrier in single sideband
P_{NS}	P_N from source impedance
P_t	total r.f. output power
P_0	d.c. input power
$P_1 \ldots P_n$	r.f. output power in 1st . . . nth harmonic
Q	quality factor of a circuit
Q_e	effective Q_L of oscillator resonant circuit
Q_L	loaded-circuit Q
Q_m	charge on metal shield
Q_p	Q of parallel circuit
Q_s	Q of series circuit;
	Q of stabilizing cavity
r	complex reflection coefficient;
	radius
r_L	$= R_L/R_r$, normalized load resistance
R	resistance, specifically equivalent r.f. parallel resistance, $1/\text{Re}(Y)$, of TE device
R_{ac}	a.c. resistance of TE-device region oscillating with homogeneous field

R_c	parasitic contact resistance
R_d	varactor-junction series resistance
R_D	r.f. dipole-domain resistance
R_{em}	differential emitter-junction resistance
R_G	resistance in gate lead
R_{h0}	$= 1/G_{h0}$, homogeneous r.f. resistance of TE device at d.c. bias point
R_H	Hall constant
R_I	resistance of inactive cathode section of TE device
R_1	series load resistance at cavity coupling position
R_L	parallel load resistance, $\mathrm{Re}(1/Y_L)$, of TE device
$R_L^\star$	parallel combination of R_L and R_1
R_{La}	load resistance in anode lead
R_{Lc}	load resistance in cathode lead
R_{Lm}	load resistance of metal shield
$R_{L,m}$	R_L at microwave limit $f \rightarrow f_{TE}$
$R_{L1} \ldots R_{Ln}$	R_L at 1st ... nth harmonic
R_M	magnetoresistance
R_p	phase-shift loss resistance of electron scattering
R_{ph}	photoresistance
R_s	parasitic series resistance
R_{sr}	equivalent r.f. series resistance, $\mathrm{Re}(Z)$, of TE device
R_{sh}	shunt resistance across anode part of TE device
R_{th}	thermal resistance
R_{tot}	$= R_1 + R_c$
R_W	equivalent noise resistance of space-charge waves
R_τ	$= \tau/C_h$, normalization resistance
R_Ω	(ohmic) resistance
R_1	$= 1/G_1$, low-field (homogeneous) resistance of TE device
R_{1a}	R_1 of part (a) of device
R_{1b}	R_1 of part (b) of device
R_{1c}	R_1 of part (c) of device
R_{1G}	low-field resistance of TE device with gate
s	saturation parameter associated with amplitude modulation of an oscillator
S_i	spectral density of mean-square noise current
S_{iA}	S_i of active device
S_u	spectral density of mean-square noise voltage
S_{uA}	S_u of active device
S_{uR}	S_u of R_{sr}
S_{uS}	S_u of source impedance
S_{uW}	S_u of space-charge waves
t	time
t_{crit}	critical t_w for domain triggering
t_d	delay time of output pulse
$t_d^\star$	delay time in domain formation
t_f	fall time of output-current pulse

t_r	rise time of output-current pulse
t_s	space-charge growth time during LSA cycle
t_w	(trigger) pulse width
T	duration of oscillation period;
	absolute temperature
T_e	electron temperature
T_{e1}	T_e in central valley
T_{e2}	T_e in satellite valley
T_L	lattice temperature of TE semiconductor
T_N	noise temperature
T_{ND}	T_N of dipole domain
T_{Neq}	equivalent T_N
T_{Nh}	T_N of homogeneous part of TE device
T_{tol}	tolerable T_{max}
T_w	width of output pulse of digital TE device
T_0	standard temperature, generally 290 K;
	ambient temperature
u	instantaneous carrier velocity;
	noise voltage
u_A	noise voltage of active device
u_S	noise voltage of source impedance
u_W	noise voltage of space-charge waves
v	electron drift velocity
v_a	accumulation-layer velocity
v_D	steady-state (drift) velocity of dipole domain
v_{gr}	$= d\omega/d\beta$, group velocity of space-charge waves
v_h	$= v(E_h)$
v_1	$= v(E_1)$
v_p	$= v(E_p)$, drift-velocity peak value
v_{ph}	$= \omega/\beta$, phase velocity of space-charge waves
v_u	$= v(E_u)$
v_v	$= v(E_v)$, drift-velocity valley value
v_1	$= v(E_1)$
v_2	$= v(E_2)$
V	terminal voltage of TE device
V_A	value of V required to trigger a dipole domain
V_b	terminal voltage of Schottky barrier
V_{bd}	value of V at dielectric breakdown in dipole domain
V_B	battery (source) voltage
V_{BA}	value of V_B required to trigger a dipole domain
V_{df}	diffusion voltage of Schottky barrier
V_D	voltage across dipole domain
V_G	gate voltage
V_{GT}	threshold value of V_G
V_1	lower turning point of alternating voltage V
V_{po}	channel pinch-off value of $(-V_G + V_{df})$

V_r	voltage across low-field region of TE device
V_S	dipole-domain sustaining value of V
$V_S^\star$	pre-triggering value of V_S if $R_L \neq 0$
V_T	threshold value of V for dipole-domain formation
V_u	upper turning point of alternating voltage V
w	width of TE semiconductor
w_m	width of metal shield
W	electron energy
W_a	thermal activation energy
W_G	semiconductor energy band gap
W_i	impact ionization value of W
W_φ	phonon energy
x	Cartesian co-ordinate in w direction of TE device; composition coefficient; fraction of output pulse height; section of cavity length
y	Cartesian co-ordinate in d direction of TE device
Y	r.f. admittance of TE device
Y_D	r.f. admittance of dipole domain
Y_L	load admittance at TE device terminals
Y_m	microwave value of Y in the limit $f \to f_{TE}$
Y_r	r.f. admittance of low-field region of TE device
z	Cartesian co-ordinate in L direction of TE device or in crystal growth direction
$z^\star$	$= z - v_D t$
z_a	accumulation-layer position in TE device
z_G	gate position in TE device
Z	$= 1/Y$, r.f. impedance of TE device
Z_d	overall r.f. impedance of TE device
Z_1	load impedance at cavity-coupling position
Z_L	$= 1/Y_L$, load impedance at TE-device terminals
Z_N	input impedance of equalizing network
Z_S	source impedance of amplifier
Z_0	characteristic impedance of space-charge waves or of transmission line
α	amplitude constant or growth coefficient of wave; coefficient of thermal expansion; current-gain factor of pulse amplifier
α_R	$= -\omega_R/v_{h0}$, growth coefficient of negative dieletric relaxation
β	$= 2\pi/\lambda$, phase constant of wave
β_e	$= \omega/v_{h0}$, electronic phase constant
β_g	$= 2\pi/\lambda_g$, β of guided wave
γ	current-tube parameter; $= j(\beta_e - k_z)$

γ_A	bias trigger capability
γ_G	gate trigger capability
Γ	space-charge growth factor of total LSA cycle
Γ_c	critical limit for Γ_N
Γ_d	critical limit for Γ_P
Γ_N	>1, space-charge growth factor
Γ_P	<1, space-charge decay factor
δ	$= D/(vL)$, diffusion parameter; skin depth; indicating small deviation of quantity denoted by immediately following symbol
δ'	$= dD/d(vL)$, differential diffusion parameter
δE_N	deviation of noise electric field
δf	FM frequency deviation
δn_0	statistical net-doping deviation
$\delta\Phi_{D,N}$	noise fluctuation of Φ_D
Δ	indicating difference or small-signal values of quantity denoted by immediately following symbol
ΔE_A	$= E_A - E_{h0}$
ΔE_-	$= E_2 - E_1$
Δf	noise or amplifier bandwidth; $= f_{ex} - f$; $= f_s - f$
Δf_m	modulation bandwidth
ΔI	$= I_0 - I_\infty$
ΔI_A	$= I_A - I_0 \simeq I_T - I_0$
ΔT	temperature rise
ΔW	energy separation of conduction-band valleys
Δz_G	gate length
$\Delta\varphi$	phase delay
ε	permittivity
ε_0	$= 8.85 \times 10^{-14}$ F cm^{-1}, permittivity of free space
ζ_G	$= z_G/L$
η	d.c.-to-r.f. power conversion efficiency, specifically for fundamental-frequency oscillation
η_m	η in microwave limit $f \to f_{TE}$
η_t	d.c.-to-total-r.f.-power conversion efficiency
$\eta_1 \ldots \eta_n$	imaginary part of $\vartheta_1 \ldots \vartheta_n$
ϑ	scaling factor; $= jkL$ or jk_zL
$\vartheta_1 \ldots \vartheta_n$	zero positions of $Z(\vartheta)$
θ	$= \beta_e L$, transit angle
κ	thermal conductivity
$\kappa^\star$	$= \kappa T$
λ	wavelength, specifically (longitudinal) wavelength of space-charge waves, and TEM resonator wavelength

List of Symbols

λ_g	wavelength in guiding structure
μ	electron drift (total) mobility, v/E;
	permeability
μ'	differential electron mobility, dv/dE
μ'_A	$= \mu'(E_A)$
μ'_D	$= dv_D/dE_{2\infty}$, differential electron mobility in dipole domain
μ_{eff}	effective value of μ
μ'_h	$= \mu'(E_h)$
μ_{LSA}	equivalent r.f. electron mobility for semiconductor chip oscillating in LSA mode
μ'_m	microwave value of μ' in the limit $f \rightarrow f_{TE}$
μ_n	maximum value of $-\mu'$
μ_N	time-averaged μ' in space-charge build-up interval t_s
μ_P	time-averaged μ' in space-charge decay interval $T - t_s$
μ'_r	$= \mu'_z/\mu'_y$
μ_T	time-averaged μ' of total LSA cycle T
μ_0	$= 4\pi \times 10^{-9}$ H cm^{-1}, permeability of free space
μ_1	μ of central-valley electrons;
	low-field value of μ for TE semiconductor
μ_2	μ of satellite-valley electrons
$\mu^{(1)} \dots \mu^{(5)}$	(complex) mobility parameters for μ'_m
ν	subscript referring to type of conduction band valley;
	collision frequency $1/\tau_p$
$\xi_1 \dots \xi_n$	real part of $\vartheta_1 \dots \vartheta_n$
Ξ	deformation potential
ρ	space-charge density
ρ_A	area-charge density
ρ_s	surface-charge density
σ	total conductivity, J/E
σ'	differential conductivity, dJ/dE
σ_{ph}	photoconductivity
σ_1	low-field value of σ (ohmic range)
τ	electron transit time
τ_D	dipole-domain charging (formation) or discharging (dissolution) time
τ_e	energy relaxation time of electrons
τ_{eff}	effective relaxation time of electrons
τ_f	inductive time constant of electron scattering
τ_n	density relaxation time of electrons
τ_p	momentum relaxation time of electrons
τ_R	$= \varepsilon/(en_0\mu)$, dielectric relaxation time of electrons
τ_{RA}	τ_R for μ'_A
τ_{RD}	τ_R for μ'_D
τ_{RN}	τ_R for μ_N
τ_{RP}	τ_R for μ_P
τ_{R1}	τ_R for μ_1

413

τ_s	$= R_s C_h$, series-resistance time constant of TE device
τ_{sc}	scattering relaxation time of electrons
$\tau^{(1)} \ldots \tau^{(5)}$	relaxation time constants for μ'_m
φ	phase difference between two signals
Φ	electric potential
$\hat{\Phi}_A$	potential amplitude of antisymmetric space-charge wave
Φ_D	dipole-domain excess potential
$\Phi_{D,crit}$	critical value of Φ_D leading to mature dipole domain
$\Phi_{D,n}$	Φ_D value of nucleation domain
$\hat{\Phi}_s$	potential amplitude of symmetric space-charge wave
ω	$= 2\pi f$, angular frequency
ω_p	precession angular frequency of electron spins
ω_{pl}	plasma frequency of electrons
ω_R	$= 1/\tau_R$, dielectric relaxation frequency of electrons
ω_τ	$= 2\pi/\tau$, angular transit-time frequency
$\omega^{(1)} \ldots \omega^{(5)}$	(complex) relaxation frequency constants of μ'_m

Microwave Frequency-band symbols *

Band symbol	Frequency range (GHz)
L	1·00– 2·35
S	2·35– 5·20
C	3·6 – 8·65
J	4·9 – 8·65
X	7·5 –13·2
Ku	11·0 –23·5
K	16 –35
V	24 –42·25
Q	33 –60

*Other notations in use deviate somewhat.

Subject Index

Author Index

425

DATE DUE

MA